Universitext

Universitext

Universitext is a series of textbooks that presents material from a wide variety of mathematical disciplines at master's level and beyond. The books, often well class-tested by their author, may have an informal, personal even experimental approach to their subject matter. Some of the most successful and established books in the series have evolved through several editions, always following the evolution of teaching curricula, into very polished texts.

Thus as research topics trickle down into graduate-level teaching, first textbooks written for new, cutting-edge courses may make their way into *Universitext*.

More information about this series at http://www.springer.com/series/223

Caroline Gruson · Vera Serganova

A Journey Through Representation Theory

From Finite Groups to Quivers via Algebras

 Springer

Caroline Gruson
Institut Elie Cartan, UMR 7502 du CNRS
Université de Lorraine, CNRS, IESL
Nancy, France

Vera Serganova
Department of Mathematics
University of California, Berkeley
Berkeley, CA, USA

ISSN 0172-5939 ISSN 2191-6675 (electronic)
Universitext
ISBN 978-3-319-98269-4 ISBN 978-3-319-98271-7 (eBook)
https://doi.org/10.1007/978-3-319-98271-7

Library of Congress Control Number: 2018950805

Mathematics Subject Classification (2010): 20-XX

This Springer imprint is published by the registered company Springer Nature Switzerland AG
The registered company address is: Gewerbestrasse 11, 6330 Cham, Switzerland

To Andrei Zelevinsky
who had such a great influence on our journey

Preface

Representation theory is a very active research topic in mathematics.

There are representations associated to several algebraic structures, representations of algebras and groups (finite or infinite). Roughly speaking, a representation is a vector space equipped with a linear action of the algebraic structure. For example, $\mathbb{C}^n$ is naturally a representation of the algebra of $n \times n$ matrices. A slightly more complicated example is the action of the group $GL(n, \mathbb{C})$ by conjugation on the space of $n \times n$-matrices.

When representations were first introduced, there was no tendency to classify all the representations of a given object. The first result in this direction is due to Frobenius, who was interested in the general theory of finite groups. If G is a finite group, a representation V of G is a complex vector space V together with a morphism of groups $\rho : G \to GL(V)$. One says V is irreducible if (1) there exists no non-zero proper subspace $W \subset V$ such that W is stable under all $\rho(g), g \in G$ and (2) $V \neq \{0\}$. Frobenius showed there are finitely many irreducible representations of G and that they are completely determined by their characters: the character of V is the complex-valued function $g \in G \mapsto \mathrm{Tr}(\rho(g))$ where Tr is the trace of the endomorphism. These characters form a basis of the complex-valued functions on G that are invariant under conjugation. Then Frobenius proceeded to compute the characters of symmetric groups in general. His results inspired Schur, who was able to relate them to the theory of complex finite dimensional representations of $GL(n, \mathbb{C})$ through the Schur–Weyl duality. In both cases, every finite dimensional representation of the group is a direct sum of irreducible representations (we say that the representations are completely reducible).

The representation theory of symmetric groups and the related combinatorics turn out to be very useful in a lot of questions. We decided to follow Zelevinsky and his book [38] and employ a Hopf algebra approach. This is an early example of categorification, which was born before the fashionable term categorification was invented.

Most of the results about representations of finite groups can be generalized to compact groups. In particular, once more, the complex finite dimensional representations of a compact group are completely reducible. Moreover, the regular representation in the space of continuous functions on the compact group contains every irreducible finite dimensional representation of the group. This theory was developed by H. Weyl and the original motivation came from quantum mechanics. The first examples of continuous compact groups are the group $SO(2)$ of rotations of the plane (the circle) and the group $SO(3)$ of rotations of the 3-dimensional space. In the former case, the problem of computing the Fourier series for a function on the circle is equivalent to the decomposition of the regular representation. More generally, the study of complex representations of compact groups helps to understand Fourier analysis on such groups.

If a topological group is not compact, for example, the group of real numbers under addition, the representation theory of such a group involves more complicated analysis (Fourier transform instead of Fourier series). The representation theory of real non-compact groups was initiated by Harish-Chandra and by the Russian school led by Gelfand. Here emphasis is on the classification of unitary representations due to applications from physics. It is also worth mentioning that this theory is closely related to harmonic analysis, and many special functions (such as Legendre polynomials) naturally appear in the context of representation theory.

In the theory of finite groups one can drop the assumption that the characteristic of the ground field is zero. This leads immediately to the loss of complete reducibility. This representation theory was initiated by Brauer, and it is more algebraic. If one turns to algebras, a representation of an algebra is, by definition, the same as a module over this algebra. Let k be a field. Let A be a k-algebra which is finite dimensional as a vector space. It is a well-known fact that A-modules are not, in general, completely reducible: for instance, if $A = k[X]/X^2$ and $M = A$, the module M contains kX as a submodule which has no A-stable complement. An indecomposable A-module is a non-zero module which has no non-trivial decomposition as a direct sum. It is also interesting to attempt a classification of A-modules. This is a very difficult task in general. Nevertheless, the irreducible A-modules are finite in number. The radical R of A is defined as the ideal of A which annihilates each of those irreducible modules. It is a nilpotent ideal. Assuming k is algebraically closed, the quotient ring A/R is a product of matrix algebras over k, $A/R = \Pi_i End_k(S_i)$ where S_i runs along the irreducible A-modules.

If G is a finite group, the algebra $k(G)$ of k-valued functions on G, the composition law being the convolution, is a finite dimensional k-algebra, with a zero radical as long as the characteristic of the field k does not divide the order of G. The irreducible modules of $k(G)$ are exactly the finite dimensional representations of the group G, and the action of G extends linearly to $k(G)$. This shows that all $k(G)$-modules are completely reducible (Maschke's theorem).

In order to study representations of finite dimensional k-algebras more generally, it is useful to introduce quivers. Let A be a finite dimensional k-algebra, denote $S_1,\ldots, S_n$ its irreducible representations, and draw the following graph, called the quiver associated to A: the vertices are labelled by the S_is and we put l arrows between S_i and S_j, pointing at S_j, if $Ext^1(S_i, S_j)$ is of dimension l (the explicit definition of Ext^1 requires some homological algebra which is difficult to summarize in such a short introduction).

More generally, a quiver is an oriented graph with any number of vertices. Let Q be a quiver. A representation of Q is a set of vector spaces indexed by the vertices of Q together with linear maps associated to the arrows of Q. Those objects were first systematically used by Gabriel in the early 70s and studied by a lot of people ever since. The aim is to characterize the finitely represented algebras, or in other terms the algebras with a finite number of indecomposable modules (up to isomorphism).

When we get to representations of quivers (Chapters 7, 8 and 9), we will sometimes need some notions associated to algebraic groups. We do not provide a course in algebraic groups in this book; hence we refer the reader to the books of Humphreys and Springer cited in the bibliography.

Today, representation theory has many flavours. In addition to the above mentioned, one should add representations over non-Archimedean local fields with its applications to number theory, representations of infinite-dimensional Lie algebras with applications to number theory and physics, and representations of quantum groups. However, in all these theories certain main ideas appear again and again, very often in disguise. Due to technical details it may be difficult for a neophyte to recognize them. The goal of this book is to present some of these ideas in their most elementary incarnation.

We will assume that the reader is familiar with linear algebra (including the theory of Jordan forms and tensor products of vector spaces) and the basic theory of groups and rings.

The book is organized as follows. In the first two chapters we deal with the basic representation theory of finite groups over fields of characteristic zero. Some of these results extend to compact groups, see Chapter 3. Our aim in Chapter 4 is to provide examples where Fourier analysis plays a key role in unitary representations of locally compact groups. Since we need a lot of algebra later on, Chapter 5 is a collection of algebraic tools. Chapter 6 deepens the study of representations of symmetric groups and links them with representations of $GL_n(\mathbb{F}_q)$. Chapters 7 and 8 are an introduction to quivers and their representation theory. Finally, Chapter 9 gives some applications of quivers. Chapters 3 and 4 are not used in the rest of the book and can be omitted. We did not try to give a complete bibliography on the subject and cited only those books and papers which were directly used in the text.

Acknowledgements: First of all, our warmest thanks to Laurent Gruson, who helped us a lot with Chapter 6, which we wrote together. Throughout the writing of this book, he frequently encouraged us and showed a real interest in our efforts. We are also thankful to Alex Sherman for his help in preparing the final version of the manuscript.

We are very grateful to the referees of this book who pointed out numerous errors and misprints.

Vera gave these notes to her UC Berkeley graduate students while they were studying for her course on Representation Theory in 2016–17 and they were kind enough to share with us their lists of typos and questions: we thank them heartfully.

Vera Serganova was supported by NSF grant 1701532.

Nancy, France   Caroline Gruson
Berkeley, USA   Vera Serganova

Contents

CHAPTER 1

Introduction to representation theory of finite groups

Beauty is the first test: there is no permanent place in the world for ugly mathematics. (G.H. Hardy)

In which we have a first encounter with representations of finite groups, discover that they are (if the base field is compliant enough) completely reducible and get acquainted with irreducible representations and their characters. Not to mention Schur's lemma. In the end of the chapter, there is a mysterious appearance of the quaternions.

1. Definitions and examples

Let k be a field, V a vector space over k. The group of all invertible linear operators in V, under composition, is denoted by $\mathrm{GL}(V)$. If $\dim V = n$, then $\mathrm{GL}(V)$ is isomorphic to the group of invertible $n \times n$ matrices with entries in k.

DEFINITION 1.1. A *(linear) representation* of a group G on V is a group homomorphism

$$\rho : G \to \mathrm{GL}(V).$$

The number $\dim V$ is called the *degree* or the *dimension* of the representation ρ (it may be infinite). For any $g \in G$ we denote by ρ_g the image of g in $\mathrm{GL}(V)$ and for any $v \in V$ we denote by $\rho_g v$ the image of v under the action of ρ_g.

The following properties are direct consequences of the definition

- $\rho_g \rho_h = \rho_{gh}$;
- $\rho_1 = \mathrm{Id}$;
- $\rho_g^{-1} = \rho_{g^{-1}}$;
- $\rho_g(av + bw) = a\rho_g v + b\rho_g w$.

DEFINITION 1.2. Let X be a set (not necessarily finite). Let G be a group. A *left action* of G on X is a homomorphism of G to the permutation group of X.

The associated *permutation representation* of G is the vector space with basis $\{e_x, x \in X\}$ [1], together with the action $\rho_g(\sum_{x \in X} a_x e_x) := \sum_{x \in X} a_x e_{gx}$.

---

[1]Bourbaki usually denotes this vector space $k^{(X)}$ in his *Éléments de Mathématiques*.

© Springer Nature Switzerland AG 2018
C. Gruson and V. Serganova, *A Journey Through Representation Theory*,
Universitext, https://doi.org/10.1007/978-3-319-98271-7_1

EXAMPLE 1.3.   (1) Let us consider the abelian group of integers $\mathbb{Z}$ with operation of addition. Let V be the plane $\mathbb{R}^2$ and for every $n \in \mathbb{Z}$, we set $\rho_n = \begin{pmatrix} 1 & n \\ 0 & 1 \end{pmatrix}$. The reader can check that this defines a representation of degree 2 of $\mathbb{Z}$.

(2) For any group G (finite or infinite) the *trivial representation* is the homomorphism $\rho : G \to GL(1, k) = k^*$ such that $\rho_s = 1$ for all $s \in G$.

(3) Let G be the symmetric group S_n, $V = k^n$. For every $s \in S_n$ and $(x_1, \ldots, x_n) \in k^n$ set

$$\rho_s(x_1, \ldots, x_n) = \left(x_{s^{-1}(1)}, \ldots, x_{s^{-1}(n)} \right).$$

In this way we obtain a representation of the symmetric group S_n which is called the *natural permutation representation*.

(4) The *group algebra* $k(G)$ is, by definition, the vector space of all finite linear combinations $\sum_{g \in G} c_g g$, $c_g \in k$ together with the natural multiplication. We define the *regular representation* as the permutation representation associated to the left action of G on itself, $R : G \to GL(k(G))$, namely

$$R_g \left(\sum_{h \in G} c_h h \right) = \sum_{h \in G} c_h g h.$$

(5) A right action of G on X is a map $\cdot : X \times G \to X$ satisfying $x \cdot (gh) = (x \cdot g) \cdot h$ and $x \cdot 1 = x$. Let X be a set with a right action of G. Consider the space $\mathcal{F}(X)$, of k-valued functions on X. Then the formula

$$\rho_g \varphi(x) := \varphi(x \cdot g)$$

defines a representation of G in $\mathcal{F}(X)$.

(6) In particular, if $X = G$,

$$\mathcal{F}(G) = \{ \varphi : G \to k \}.$$

For any $g, h \in G$ and $\varphi \in \mathcal{F}(G)$, let

$$\rho_g \varphi(h) = \varphi(hg).$$

Then $\rho : G \to GL(\mathcal{F}(G))$ is a linear representation.

DEFINITION 1.4. Two representations of a group G, $\rho : G \to GL(V)$ and $\sigma : G \to GL(W)$ are called *equivalent* or *isomorphic* if there exists an invertible linear operator $T : V \to W$ such that $T \circ \rho_g = \sigma_g \circ T$ for any $g \in G$.

EXAMPLE 1.5. If G is a finite group, then the representations numbered (4) and (6) in Example 1.3 are equivalent. Indeed, define $T : \mathcal{F}(G) \to k(G)$ by the formula

$$T(\varphi) = \sum_{h \in G} \varphi(h) h^{-1}.$$

Then for any $\varphi \in \mathcal{F}(G)$ and $g \in G$ we have

$$T\left(\rho_g \varphi\right) = \sum_{h \in G} \rho_g \varphi\left(h\right) h^{-1} = \sum_{h \in G} \varphi\left(hg\right) h^{-1} = \sum_{l \in G} \varphi\left(l\right) gl^{-1} = R_g\left(T(\varphi)\right).$$

EXERCISE 1.6. Let G be a group and X a set. Consider a *left* action $l : G \times X \to X$ of G on X. For every $\varphi \in \mathcal{F}(X)$, $g \in G$ and $x \in X$ set

$$\sigma_g \varphi(x) = \varphi(g^{-1} \cdot x).$$

(a) Prove that σ is a representation of G in $\mathcal{F}(X)$.
(b) Define a right action $r : X \times G \to X$ by

$$x \cdot g := g^{-1} \cdot x,$$

and consider the representation ρ of G in $\mathcal{F}(X)$ associated with this action. Check that ρ and σ are equivalent representations.

REMARK 1.7. In other words, the previous exercise shows that (1) if we are given a left action of G on X, there is a canonical way to produce a right action of G on X and (2) that the associated representations in $\mathcal{F}(X)$ are equivalent.

2. Ways to produce new representations

Let G be a group.

Restriction. If H is a subgroup of G and $\rho : G \to \mathrm{GL}(V)$ is a representation of G, the restriction of the homomorphism ρ to H gives a representation of H which we call the *restriction* of ρ to H. We denote by $\mathrm{Res}_H \rho$ the restriction of ρ on H.

Lift. Let $p : G \to H$ be a homomorphism of groups. Then for every representation $\rho : H \to \mathrm{GL}(V)$, the composite homomorphism $\rho \circ p : G \to \mathrm{GL}(V)$ gives a representation of G on V. This construction is frequently used in the following case: let N be a normal subgroup of G, H denote the quotient group G/N and p be the natural projection. In this case p is obviously surjective. Note that in the general case we do not require p to be surjective.

Direct sum. If we have two representations $\rho : G \to \mathrm{GL}(V)$ and $\sigma : G \to \mathrm{GL}(W)$, then we can define $\rho \oplus \sigma : G \to \mathrm{GL}(V \oplus W)$ by the formula

$$(\rho \oplus \sigma)_g (v, w) = (\rho_g v, \sigma_g w).$$

Tensor product. The tensor product of two representations $\rho : G \to \mathrm{GL}(V)$ and $\sigma : G \to \mathrm{GL}(W)$ is defined by

$$(\rho \otimes \sigma)_g (v \otimes w) = \rho_g v \otimes \sigma_g w.$$

Exterior tensor product. Let G and H be two groups. Consider representations $\rho : G \to \mathrm{GL}(V)$ and $\sigma : H \to \mathrm{GL}(W)$ of G and H, respectively. One defines their *exterior tensor product* $\rho \boxtimes \sigma : G \times H \to \mathrm{GL}(V \otimes W)$ by the formula

$$(\rho \boxtimes \sigma)_{(g,h)} v \otimes w = \rho_g v \otimes \sigma_h w.$$

EXERCISE 2.1. If $\delta : G \to G \times G$ is the diagonal embedding, show that for any representations ρ and σ of G

$$\rho \otimes \sigma = (\rho \boxtimes \sigma) \circ \delta.$$

Dual representation. Let V^* denote the dual space of V and $\langle \cdot, \cdot \rangle$ denote the natural pairing between V and V^*. For any representation $\rho : G \to \mathrm{GL}(V)$ one can define the dual representation $\rho^* : G \to \mathrm{GL}(V^*)$ by the formula

$$\langle \rho_g^* \varphi, v \rangle = \langle \varphi, \rho_g^{-1} v \rangle$$

for every $v \in V, \varphi \in V^*$.

Let V be a finite-dimensional representation of G with a fixed basis. Let A_g for $g \in G$ be the matrix of ρ_g in this basis. Then the matrix of ρ_g^* in the dual basis of V^* is equal to $(A_g^t)^{-1}$.

EXERCISE 2.2. Show that if G is finite, then its regular representation is self-dual (isomorphic to its dual).

More generally, if $\rho : G \to \mathrm{GL}(V)$ and $\sigma : G \to \mathrm{GL}(W)$ are two representations, then one can naturally define a representation τ of G on $\mathrm{Hom}_k(V, W)$ by the formula

$$\tau_g \varphi = \sigma_g \circ \varphi \circ \rho_g^{-1}, \ g \in G, \ \varphi \in \mathrm{Hom}_k(V, W).$$

EXERCISE 2.3. Show that if V and W are finite dimensional, then the representation τ of G on $\mathrm{Hom}_k(V, W)$ is isomorphic to $\rho^* \otimes \tau$.

Intertwining operators. A linear operator $T : V \to W$ is called an *intertwining* operator if $T \circ \rho_g = \sigma_g \circ T$ for any $g \in G$. The set of all intertwining operators will be denoted by $\mathrm{Hom}_G(V, W)$. It is clearly a vector space. Moreover, if $\rho = \sigma$, then $\mathrm{End}_G(V) := \mathrm{Hom}_G(V, V)$ has a natural structure of associative k-algebra with multiplication given by composition.

EXERCISE 2.4. Consider the regular representation of G in $k(G)$. Prove that the algebra of intertwiners $\mathrm{End}_G(k(G))$ is isomorphic to $k(G)$. (Hint: $\varphi \in \mathrm{End}_G(k(G))$ is completely determined by $\varphi(1)$.)

3. Invariant subspaces and irreducibility

3.1. Invariant subspaces and subrepresentations. Consider a representation $\rho : G \to \mathrm{GL}(V)$. A subspace $W \subset V$ is called G-*invariant* if $\rho_g(W) \subset W$ for any $g \in G$.

If W is a G-invariant subspace, then there are two representations of G naturally associated with it: the representation in W which is called a *subrepresentation* and the representation in the quotient space V/W which is called a *quotient* representation.

EXERCISE 3.1. Let $\rho : S_n \to \mathrm{GL}(k^n)$ be the permutation representation, then

$$W = \{x(1, \ldots, 1) \mid x \in k\}$$

and

$$W' = \{(x_1, \ldots, x_n) \mid x_1 + x_2 + \cdots + x_n = 0\}$$

are invariant subspaces.

EXERCISE 3.2. Let G be a finite group of order $|G|$. Prove that any representation of G contains an invariant subspace of dimension less than or equal to $|G|$.

3.2. Maschke's theorem.

THEOREM 3.3. (Maschke) Let G be a finite group such that char k does not divide $|G|$. Let $\rho : G \to \mathrm{GL}(V)$ be a representation and $W \subset V$ be a G-invariant subspace. Then there exists a complementary G-invariant subspace, i.e. a G-invariant subspace $W' \subset V$ such that $V = W \oplus W'$.

PROOF. Let W'' be a subspace (not necessarily G-invariant) such that $W \oplus W'' = V$. Consider the projector $P : V \to V$ onto W with kernel W'': $P^2 = P$. Now we construct a new operator

$$\bar{P} := \frac{1}{|G|} \sum_{g \in G} \rho_g \circ P \circ \rho_g^{-1}.$$

An easy calculation shows that $\rho_g \circ \bar{P} \circ \rho_g^{-1} = \bar{P}$ for all $g \in G$, and therefore $\rho_g \circ \bar{P} = \bar{P} \circ \rho_g$. In other words, $\bar{P} \in \mathrm{End}_G(V)$.

On the other hand, $\bar{P}|_W = \mathrm{Id}$ and $\mathrm{Im}\,\bar{P} = W$. Hence $\bar{P}^2 = \bar{P}$.

Let $W' = \mathrm{Ker}\,\bar{P}$. First, we claim that W' is G-invariant. Indeed, let $w \in W'$, then $\bar{P}(\rho_g w) = \rho_g(\bar{P}w) = 0$ for all $g \in G$, hence $\rho_g w \in \mathrm{Ker}\,\bar{P} = W'$.

Now we prove that $V = W \oplus W'$. Indeed, $W \cap W' = 0$, since $\bar{P}|_W = Id$. On the other hand, for any $v \in V$, we have $w = \bar{P}v \in W$ and $w' = v - \bar{P}v \in W'$. Thus, $v = w + w'$, and therefore $V = W + W'$. □

Remarks. If char k divides $|G|$ or G is infinite, the conclusion of Maschke's theorem does not hold anymore. Indeed, in the example of Exercise 3.1 W and W' are complementary if and only if char k does not divide n. Otherwise, $W \subset W' \subset V$, and one can show that neither W nor W' have a G-invariant complement.

In the case of an infinite group, consider the representation of $\mathbb{Z}$ in $\mathbb{R}^2$ as in the first example of Section 1. The span of $(1,0)$ is the only G-invariant line. Therefore it can not have a G-invariant complement in $\mathbb{R}^2$.

3.3. Irreducible representations and Schur's lemma.

DEFINITION 3.4. A non-zero representation is called *irreducible* if it does not contain any proper non-zero G-invariant subspace.

EXERCISE 3.5. Show that the dimension of any irreducible representation of a finite group G is not bigger than its order $|G|$.

The following elementary statement plays a key role in representation theory.

LEMMA 3.6. *(Schur) Let $\rho : G \to \mathrm{GL}(V)$ and $\sigma : G \to \mathrm{GL}(W)$ be two irreducible representations. If $T \in \mathrm{Hom}_G(V, W)$, then either $T = 0$ or T is an isomorphism.*

PROOF. Note that $\mathrm{Ker}\, T$ and $\mathrm{Im}\, T$ are G-invariant subspaces of V and W, respectively. Then by irreducibility of ρ, either $\mathrm{Ker}\, T = V$ or $\mathrm{Ker}\, T = 0$, and by irreducibility of σ, either $\mathrm{Im}\, T = W$ or $\mathrm{Im}\, T = 0$. Hence the statement.   □

COROLLARY 3.7. *(a) Let $\rho : G \to \mathrm{GL}(V)$ be an irreducible representation. Then $\mathrm{End}_G(V)$ is a division ring.*
(b) If the characteristic of k does not divide $|G|$, $\mathrm{End}_G(V)$ is a division ring if and only if ρ is irreducible.
(c) If k is algebraically closed and ρ is irreducible, then $\mathrm{End}_G(V) = k$.

PROOF. (a) is an immediate consequence of Schur's Lemma.

To prove (b) we use Maschke's theorem. Indeed, if V is reducible, then $V = V_1 \oplus V_2$ for some proper subspaces V_1 and V_2. Let p_1 be the projector on V_1 with kernel V_2 and p_2 be the projector onto V_2 with kernel V_1. Then $p_1, p_2 \in \mathrm{End}_G(V)$ and $p_1 \circ p_2 = 0$. Hence $\mathrm{End}_G(V)$ has zero divisors.

Let us prove (c). Consider $T \in \mathrm{End}_G(V)$. Then T has an eigenvalue $\lambda \in k$ and $T - \lambda \, \mathrm{Id} \in \mathrm{End}_G(V)$. Since $T - \lambda \, \mathrm{Id}$ is not invertible, it must be zero by (a). Therefore $T = \lambda \, \mathrm{Id}$.   □

3.4. Complete reducibility.

DEFINITION 3.8. A representation is called *completely reducible* if it splits into a direct sum of irreducible subrepresentations. (This direct sum might be infinite.)

THEOREM 3.9. *Let $\rho : G \to GL(V)$ be a representation of a group G. The following conditions are equivalent.*
(a) ρ is completely reducible;
(b) For any G-invariant subspace $W \subset V$ there exists a complementary G-invariant subspace W'.

PROOF. This theorem is easier in the case of finite-dimensional V. To prove it for arbitrary V and G we need Zorn's lemma. First, note that if V is non-zero and finite dimensional, then it always contains an irreducible subrepresentation. Indeed, we can take a subrepresentation of minimal positive dimension. If V is infinite dimensional then this is not true in general. Let us show two Lemmas before finishing the proof.

LEMMA 3.10. *If ρ satisfies (b), any subrepresentation and any quotient of ρ also satisfy (b).*

PROOF. To prove that any subrepresentation satisfies (b) consider a flag of G-invariant subspaces $U \subset W \subset V$. Let $U' \subset V$ and $W' \subset V$ be G-invariant subspaces such that $U \oplus U' = V$ and $W \oplus W' = V$. Then $W = U \oplus (U' \cap W)$.

The statement about quotients is dual and we leave it to the reader as an exercise.   □

LEMMA 3.11. *Let ρ satisfy (b). Then it contains an irreducible subrepresentation.*

PROOF. Pick up a non-zero vector $v \in V$ and let V' be the span of $\rho_g v$ for all $g \in G$. Consider the set of G-invariant subspaces of V' which do not contain v, with partial order given by inclusion. For any linearly ordered subset $\{X_i\}_{i \in I}$ there exists a maximal element, given by the union $\bigcup_{i \in I} X_i$. Hence there exists a proper maximal G-invariant subspace $W \subset V'$, which does not contain v. By the previous lemma one can find a G-invariant subspace $U \subset V'$ such that $V' = W \oplus U$. Then U is isomorphic to the quotient representation V'/W, which is irreducible by the maximality of W in V'. $\qquad\square$

Now we will prove that (a) implies (b). We write

$$V = \bigoplus_{i \in I} V_i$$

for a family of irreducible G-invariant subspaces V_i. Let $W \subset V$ be some G-invariant subspace. By Zorn's lemma there exists a maximal subset $J \subset I$ such that

$$W \cap \bigoplus_{j \in J} V_j = 0.$$

We claim that $W' := \bigoplus_{j \in J} V_j$ is complementary to W. Indeed, it suffices to prove that $V = W + W'$. For any $i \notin J$ we have $(V_i \oplus W') \cap W \neq 0$. Therefore there exists a non-zero vector $v \in V_i$ equal to $w + w'$ for some $w \in W$ and $w' \in W'$. Hence $V_i \cap (W' + W) \neq 0$ and by irreducibility of V_i, we have $V_i \subset W + W'$. Therefore $V = W + W'$.

To prove that (b) implies (a) consider the family of all irreducible subrepresentations $\{W_k\}_{k \in K}$ of V. Note that $\sum_{k \in K} W_k = V$ because otherwise $\sum_{k \in K} W_k$ has a G-invariant complement which contains an irreducible subrepresentation. Again due to Zorn's lemma one can find a maximal $J \subset K$ such that $\sum_{j \in J} W_j = \bigoplus_{j \in J} W_j$. Then again using existence of a complementary subspace we have $V = \bigoplus_{j \in J} W_j$. $\quad\square$

The next statement follows from Maschke's theorem and Theorem 3.9.

PROPOSITION 3.12. *Let G be a finite group and k be a field such that char k does not divide $|G|$. Then every representation of G is completely reducible.*

4. Characters

4.1. Definition and main properties. For a linear operator T in a finite-dimensional vector space V we denote by $\mathrm{Tr}\, T$ the trace of T.

For any finite-dimensional representation $\rho : G \to \mathrm{GL}(V)$ the function $\chi_\rho : G \to k$ defined by

$$\chi_\rho(g) = \mathrm{Tr}\ \rho_g.$$

is called the *character* of the representation ρ.

EXERCISE 4.1. Check the following properties of characters.

(1) $\chi_\rho(1) = \dim \rho$;
(2) if $\rho \cong \sigma$, then $\chi_\rho = \chi_\sigma$;
(3) $\chi_{\rho \oplus \sigma} = \chi_\rho + \chi_\sigma$;
(4) $\chi_{\rho \otimes \sigma} = \chi_\rho \chi_\sigma$;
(5) $\chi_{\rho^*}(g) = \chi_\rho(g^{-1})$;
(6) $\chi_\rho(ghg^{-1}) = \chi_\rho(h)$.

EXERCISE 4.2. Calculate the character of the permutation representation of S_n (see the first example of Section 1).

EXAMPLE 4.3. If R is the regular representation of a finite group, then $\chi_R(g) = 0$ for any $g \neq 1$ and $\chi_R(1) = |G|$.

EXAMPLE 4.4. Let $\rho : G \to \mathrm{GL}(V)$ be a representation of dimension n and assume char $k \neq 2$. Consider the representation $\rho \otimes \rho$ in $V \otimes V$ and the decomposition

$$V \otimes V = S^2 V \oplus \Lambda^2 V.$$

The subspaces $S^2 V$ and $\Lambda^2 V$ are G-invariant. Denote by sym and alt the subrepresentations of G in $S^2 V$ and $\Lambda^2 V$, respectively. Let us compute the characters χ_{sym} and χ_{alt}.

Let $g \in G$ and denote by $\lambda_1, \ldots, \lambda_n$ the eigenvalues of ρ_g (taken with multiplicities). Then the eigenvalues of alt_g are the products $\lambda_i \lambda_j$ for all $i < j$ while the eigenvalues of sym_g are $\lambda_i \lambda_j$ for $i \leq j$. This leads to

$$\chi_{\mathrm{sym}}(g) = \sum_{i \leq j} \lambda_i \lambda_j,$$

$$\chi_{\mathrm{alt}}(g) = \sum_{i < j} \lambda_i \lambda_j.$$

Hence

$$\chi_{\mathrm{sym}}(g) - \chi_{\mathrm{alt}}(g) = \sum_i \lambda_i^2 = \mathrm{Tr}\ \rho_{g^2} = \chi_\rho(g^2).$$

On the other hand by properties (3) and (4)

$$\chi_{\mathrm{sym}}(g) + \chi_{\mathrm{alt}}(g) = \chi_{\rho \otimes \rho}(g) = \chi_\rho^2(g).$$

Thus, we get

(1.1) $$\chi_{\mathrm{sym}}(g) = \frac{\chi_\rho^2(g) + \chi_\rho(g^2)}{2}, \quad \chi_{\mathrm{alt}}(g) = \frac{\chi_\rho^2(g) - \chi_\rho(g^2)}{2}.$$

LEMMA 4.5. *If $k = \mathbb{C}$ and G is finite, then for any finite-dimensional represen-tation ρ and any $g \in G$ we have*

$$\chi_\rho(g) = \overline{\chi_\rho(g^{-1})}.$$

PROOF. Indeed, $\chi_\rho(g)$ is the sum of all the eigenvalues of ρ_g. Since g has finite order, every eigenvalue of ρ_g is a root of 1. Therefore the eigenvalues of $\rho_{g^{-1}}$ are the complex conjugates of the eigenvalues of ρ_g. □

4.2. Orthogonality relations. In this subsection we assume that G is finite and the characteristic of the ground field k is zero. Introduce a non-degenerate symmetric bilinear form on the space of functions $\mathcal{F}(G)$ by the formula

$$(1.2) \qquad (\varphi, \psi) = \frac{1}{|G|} \sum_{s \in G} \varphi\left(s^{-1}\right) \psi\left(s\right).$$

If $\rho : G \to \mathrm{GL}(V)$ is a representation, then we denote by V^G the subspace of G-invariant vectors, i.e.

$$V^G = \{v \in V | \rho_g(v) = v, \forall g \in G\}.$$

LEMMA 4.6. *If $\rho : G \to \mathrm{GL}(V)$ is a representation, then*

$$\dim V^G = (\chi_\rho, \chi_{triv}),$$

where χ_{triv} denotes the character of the trivial representation, i.e. $\chi_{triv}(g) = 1$ for all $g \in G$.

PROOF. Consider the linear operator $P \in \mathrm{End}_G(V)$ defined by the formula

$$P = \frac{1}{|G|} \sum_{g \in G} \rho_g.$$

Note that $P^2 = P$ and $\mathrm{Im}\, P = V^G$. Thus, P is a projector on V^G. Since char $k = 0$ we have

$$\mathrm{Tr}\, P = \dim \mathrm{Im}\, P = \dim V^G.$$

On the other hand, by direct calculation we get $\mathrm{Tr}\, P = (\chi_\rho, \chi_{triv})$, and the lemma follows. □

Note that for two representations $\rho : G \to \mathrm{GL}(V)$ and $\sigma : G \to \mathrm{GL}(W)$ we have

$$(1.3) \qquad \mathrm{Hom}_k(V, W)^G = \mathrm{Hom}_G(V, W) = (V^* \otimes W)^G.$$

Therefore we have the following

COROLLARY 4.7. *One has*

$$\dim \mathrm{Hom}_G(V, W) = (\chi_\rho, \chi_\sigma).$$

PROOF. The statement is a consequence of the following computation:

$$(\chi_\rho, \chi_\sigma) = \frac{1}{|G|} \sum_{g \in G} \chi_\rho(g^{-1}) \chi_\sigma(g) = \frac{1}{|G|} \sum_{g \in G} \chi_{\rho^* \otimes \sigma}(g) = (\chi_{\rho^* \otimes \sigma}, \chi_{triv}).$$

$\square$

The following theorem is usually called the *orthogonality relations* for characters.

THEOREM 4.8. *Let ρ, σ be irreducible representations over a field of characteristic zero.*
   (a) If $\rho : G \to GL(V)$ and $\sigma : G \to GL(W)$ are not isomorphic, then $(\chi_\rho, \chi_\sigma) = 0$.
   (b) Assume that the ground field is algebraically closed. If ρ and σ are equivalent, then $(\chi_\rho, \chi_\sigma) = 1$.

PROOF. By Schur's lemma

$$\mathrm{Hom}_G(V, W) = 0.$$

Therefore Corollary 4.7 implies (a).
   Assertion (b) follows from Corollary 3.7 (c) and Corollary 4.7. $\square$

This theorem has several important corollaries.

COROLLARY 4.9. *Let*

$$\rho = m_1 \rho_1 \oplus \cdots \oplus m_r \rho_r$$

be a decomposition into a sum of irreducible representations, where $m_i \rho_i$ is the direct sum of m_i copies of ρ_i. Then $m_i = \dfrac{(\chi_\rho, \chi_{\rho_i})}{(\chi_{\rho_i}, \chi_{\rho_i})}$.

The number m_i is called the *multiplicity* of an irreducible representation ρ_i in ρ.

COROLLARY 4.10. *Two finite-dimensional representations ρ and σ are equivalent if and only if their characters coincide.*

In the rest of this section we assume that the ground field is algebraically closed.

COROLLARY 4.11. *A representation ρ is irreducible if and only if $(\chi_\rho, \chi_\rho) = 1$.*

EXERCISE 4.12. Let ρ and σ be irreducible representations of finite groups G and H, respectively.
   (a) If the ground field is algebraically closed, then the exterior product $\rho \boxtimes \sigma$ is an irreducible representation of $G \times H$.
   (b) Give a counterexample to (a) in the case when the ground field is not algebraically closed.

THEOREM 4.13. *Every irreducible representation ρ appears in the regular representation with multiplicity $\dim \rho$.*

PROOF. The statement is a direct consequence of the following computation

$$(\chi_\rho, \chi_R) = \frac{1}{|G|}\chi_\rho(1)\chi_R(1) = \dim \rho.$$

$\square$

COROLLARY 4.14. *A finite group has finitely many non-isomorphic irreducible representations.*

COROLLARY 4.15. *Let $\rho_1, \ldots, \rho_r$ be all irreducible representations of G (up to isomorphism) and denote $n_i = \dim \rho_i$. Then*

$$n_1^2 + \cdots + n_r^2 = |G|.$$

PROOF. Indeed,

$$\dim R = |G| = \chi_R(1) = \sum_{i=1}^{r} n_i \chi_{\rho_i}(1) = \sum_{i=1}^{r} n_i^2.$$

$\square$

EXAMPLE 4.16. Let G act on a finite set X and

$$k(X) = \left\{ \sum_{x \in X} b_x x \mid b_x \in k \right\}.$$

Define $\rho : G \to \mathrm{GL}(k(X))$ by

$$\rho_g \left(\sum_{x \in X} b_x x \right) = \sum_{x \in X} b_x g \cdot x.$$

It is easy to check that ρ is a representation and that

$$\chi_\rho(g) = |\{x \in X \mid g \cdot x = x\}|.$$

Clearly, ρ contains the trivial representation as a subrepresentation. To find the multiplicity of the trivial representation in ρ, we have to calculate $(1, \chi_\rho)$:

$$(1, \chi_\rho) = \frac{1}{|G|} \sum_{g \in G} \chi_\rho(g) = \frac{1}{|G|} \sum_{g \in G} \sum_{g \cdot x = x} 1 = \frac{1}{|G|} \sum_{x \in X} \sum_{g \in G_x} 1 = \frac{1}{|G|} \sum_{x \in X} |G_x|,$$

where

$$G_x = \{g \in G \mid g \cdot x = x\}.$$

Let $X = X_1 \sqcup \cdots \sqcup X_m$ be the decomposition of X in disjoint G-orbits. Then $|G_x| = \frac{|G|}{|X_i|}$ for each $x \in X_i$ and therefore

$$(1, \chi_\rho) = \frac{1}{|G|} \sum_{i=1}^{m} \sum_{x \in X_i} \frac{|G|}{|X_i|} = m.$$

Now, let us evaluate (χ_ρ, χ_ρ):

$$(\chi_\rho, \chi_\rho) = \frac{1}{|G|} \sum_{g \in G} \left(\sum_{g \cdot x = x} 1 \right)^2 = \frac{1}{|G|} \sum_{g \in G} \sum_{g \cdot x = x, g \cdot y = y} 1 = \frac{1}{|G|} \sum_{(x,y) \in X \times X} |G_{(x,y)}|.$$

Let σ be the representation associated with the action of G on $X \times X$. Then the last formula implies

$$(\chi_\rho, \chi_\rho) = (1, \chi_\sigma).$$

Thus, ρ is irreducible if and only if $|X| = 1$, and ρ has two irreducible components if and only if the action of G on $(X \times X) \setminus \Delta$, where Δ is the diagonal, is transitive.

4.3. The number of irreducible representations of a finite group.

DEFINITION 4.17. Let

$$\mathcal{C}(G) = \left\{ \varphi \in \mathcal{F}(G) \mid \varphi \left(ghg^{-1} \right) = \varphi(h) \right\}.$$

The elements of $\mathcal{C}(G)$ are called *class functions*.

EXERCISE 4.18. Check that the restriction of $(\cdot, \cdot)$ on $\mathcal{C}(G)$ is non-degenerate.

THEOREM 4.19. *The collection of the characters of irreducible representations of G is an orthonormal basis of $\mathcal{C}(G)$.*

PROOF. We have to show that if $\varphi \in \mathcal{C}(G)$ and $(\varphi, \chi_\rho) = 0$ for any irreducible representation ρ, then $\varphi = 0$. The following lemma is straightforward.

LEMMA 4.20. *Let $\rho : G \to \mathrm{GL}(V)$ be a representation, $\varphi \in \mathcal{C}(G)$ and set*

$$T = \frac{1}{|G|} \sum_{g \in G} \varphi \left(g^{-1} \right) \rho_g.$$

Then $T \in \mathrm{End}_G V$ and $\mathrm{Tr}\, T = (\varphi, \chi_\rho)$.

Thus, for any irreducible representation ρ we have

$$(1.4) \qquad \frac{1}{|G|} \sum_{g \in G} \varphi \left(g^{-1} \right) \rho_g = 0.$$

But then the same is true for any representation ρ, since any representation is a direct sum of irreducible representations. Apply (1.4) to the case when $\rho = R$ is the regular representation. Then

$$\frac{1}{|G|} \sum_{g \in G} \varphi \left(g^{-1} \right) R_g(1) = \frac{1}{|G|} \sum_{g \in G} \varphi \left(g^{-1} \right) g = 0.$$

Hence $\varphi \left(g^{-1} \right) = 0$ for all $g \in G$, i.e. $\varphi = 0$. $\qquad \square$

COROLLARY 4.21. *The number of isomorphism classes of irreducible representations equals the number of conjugacy classes in the group G.*

COROLLARY 4.22. *If G is a finite abelian group, then every irreducible representation of G is one-dimensional and the number of irreducible representations is the order of the group G.*

For any group G (not necessarily finite) let G^* denote the set of all one-dimensional representations of G.

EXERCISE 4.23. (a) Show that G^* is a group with respect to the operation of tensor product.

(b) Show that the kernel of any $\rho \in G^*$ contains the commutator $[G, G]$. Hence we have $G^* \simeq (G/[G, G])^*$.

(c) Show that if G is a finite abelian group, then $G^* \simeq G$. (This isomorphism is not canonical.)

EXERCISE 4.24. Consider the symmetric group S_n for $n \geq 2$.

(a) Prove that the commutator $[S_n, S_n]$ coincides with the subgroup A_n of all even permutations.

(b) Show that S_n has two one-dimensional representations (up to isomorphism): the trivial one and the sign representation $\varepsilon : S_n \to \{1, -1\}$.

EXERCISE 4.25. Let ρ be a one-dimensional representation of a finite group G and σ another representation of G. Show that σ is irreducible if and only if $\rho \otimes \sigma$ is irreducible.

Moreover, check that there is a 1-1 correspondence between the subrepresentations of σ and those of $\rho \otimes \sigma$.

4.4. Isotypic components. Consider the decomposition of some representation $\rho : G \to GL(V)$ into a direct sum of irreducible representations

$$\rho = m_1 \rho_1 \oplus \cdots \oplus m_r \rho_r.$$

Denote by V_i the space of the representation ρ_i. Then the subspace W_i of V which corresponds to $m_i \rho_i$ is isomorphic to $V_i^{\oplus m_i}$. The following Lemma (Lemma 4.26) ensures that the decomposition $V = \bigoplus W_i$ is unique. The space W_i is called the *isotypic component* of type ρ_i of V.

LEMMA 4.26. *Let n_i denote the dimension of the irreducible representation ρ_i and set*

$$\pi_i := \frac{n_i}{|G|} \sum_{g \in G} \chi_i(g^{-1}) \rho_g.$$

Then π_i is the projector on the isotypic component W_i of type ρ_i.

PROOF. Define a linear operator on V_j by the formula

$$\pi_{ij} := \frac{n_i}{|G|} \sum_{g \in G} \chi_i(g^{-1})(\rho_j)_g.$$

By construction $\pi_{ij} \in \mathrm{End}_G(V_j)$. Corollary 3.7 (c) implies that $\pi_{ij} = \lambda \,\mathrm{Id}$. By Theorem 4.8

$$\mathrm{Tr}\ \pi_{ij} = n_i(\chi_i, \chi_j) = n_i \delta_{ij}.$$

Now we write

$$\pi_i = \sum_{j=1}^{r} \pi_{ij}.$$

Hence

$$\pi_i|_{W_j} = \delta_{ij} \,\mathrm{Id}\,.$$

The statement follows.   □

4.5. Faithful representations. A representation $\rho : G \to GL(V)$ is called *faithful* if ρ is injective.

THEOREM 4.27. *Let* $\rho : G \to GL(V)$ *be a faithful representation of a finite group* G. *Then every irreducible representation of* G *occurs in some tensor power of* ρ.

PROOF. We have to show that for any irreducible representation ρ_i there exists m such that $(\chi_\rho^m, \chi_i) \neq 0$. Assume that the statement is false for some i. Consider the generating functions

$$\sum_{m=0}^{\infty} \chi_\rho^m(g) t^m = \frac{1}{1 - t\chi_\rho(g)}.$$

By our assumption, for i, we have

$$\sum_{g \in G} \frac{\chi_i(g^{-1})}{1 - t\chi_\rho(g)} = 0.$$

Rewrite the above identity in the form

$$\frac{n_i}{1 - tn} = -\sum_{g \in G\backslash\{1\}} \frac{\chi_i(g^{-1})}{1 - t\chi_\rho(g)},$$

where n denotes the dimension of ρ.

Both sides of the above identity are rational functions. If they are equal then $\chi_\rho(g) = n$ for at least one $g \neq 1$. Let $\varepsilon_1, \dots, \varepsilon_n$ be the eigenvalues of ρ_g. Since g has finite order, all ε_i-s are roots of 1. The condition

$$\chi_\rho(g) = \varepsilon_1 + \cdots + \varepsilon_n = n$$

implies $\varepsilon_1 = \cdots = \varepsilon_n = 1$. Hence ρ_g is the identity operator, which contradicts the assumption that ρ is faithful.   □

5. Examples

In the examples below we assume that the ground field is $\mathbb{C}$.

EXAMPLE 5.1. Let $G = S_3$. There are three conjugacy classes in G, and each class is denoted by some element in this class: 1,(12),(123). Therefore there are three irreducible representations; denote their characters by χ_1, χ_2 and χ_3. It is not difficult to see that S_3 has the following table of characters

	1	(12)	(123)
χ_1	1	1	1
χ_2	1	−1	1
χ_3	2	0	−1

The characters of one-dimensional representations are given in the first and the second row (those are the trivial representation and the sign representation, see Exercise 4.24), the last character χ_3 can be obtained by using the identity

(1.5)
$$\chi_{\text{perm}} = \chi_1 + \chi_3,$$

where χ_{perm} stands for the character of the permutation representation, see Exercise 4.2.

EXAMPLE 5.2. Let $G = S_4$. In this case we have the following character table (in the first row we write the number of elements in each conjugacy class).

	1	6	8	3	6
	1	(12)	(123)	(12)(34)	(1234)
χ_1	1	1	1	1	1
χ_2	1	−1	1	1	−1
χ_3	3	1	0	−1	−1
χ_4	3	−1	0	−1	1
χ_5	2	0	−1	2	0

The first two rows are the characters of the one-dimensional representations. The third one can again be obtained from (1.5). When we take the tensor product $\rho_4 := \rho_2 \otimes \rho_3$ we get a new 3-dimensional irreducible representation, see Exercise 4.25 whose character χ_4 is equal to the product $\chi_2\chi_3$. The last character can be obtained through Theorem 4.8. An alternative way to describe ρ_5 is to consider S_4/K_4, where

$$K_4 = \{1, (12)(34), (13)(24), (14)(23)\}$$

is the Klein subgroup. Observe that $S_4/K_4 \cong S_3$, and therefore the two-dimensional representation σ of S_3 can be lifted to a representation of S_4 by

$$\rho_5 = \sigma \circ p, \cdot$$

where $p : S_4 \to S_3$ is the natural projection.

EXAMPLE 5.3. Now let $G = A_5$. There are 5 irreducible representations of G over $\mathbb{C}$. Here is the character table

	1	20	15	12	12
	1	(123)	$(12)(34)$	(12345)	(12354)
χ_1	1	1	1	1	1
χ_2	4	1	0	-1	-1
χ_3	5	-1	1	0	0
χ_4	3	0	-1	$\frac{1+\sqrt{5}}{2}$	$\frac{1-\sqrt{5}}{2}$
χ_5	3	0	-1	$\frac{1-\sqrt{5}}{2}$	$\frac{1+\sqrt{5}}{2}$

To obtain χ_2 we use the permutation representation and (1.5) once more. In order to construct new irreducible representations we consider the characters χ_{sym} and χ_{alt} of the second symmetric and the second exterior powers of ρ_2, respectively. Using (1.1) we compute

	1	(123)	$(12)(34)$	(12345)	(12354)
χ_{sym}	10	1	2	0	0
χ_{alt}	6	0	-2	1	1

It is easy to check that

$$(\chi_{sym}, \chi_{sym}) = 3, \ (\chi_{sym}, \chi_1) = (\chi_{sym}, \chi_2) = 1.$$

Therefore

$$\chi_3 = \chi_{sym} - \chi_1 - \chi_2$$

is the character of another irreducible representation of dimension 5. We still miss two.

To find them we use χ_{alt}. We have

$$(\chi_{alt}, \chi_{alt}) = 2, \ (\chi_{alt}, \chi_1) = (\chi_{alt}, \chi_2) = (\chi_{alt}, \chi_3) = 0.$$

Therefore $\chi_{alt} = \chi_4 + \chi_5$ is the sum of two irreducible characters. First we compute the dimensions of ρ_4 and ρ_5 using

$$1^2 + 4^2 + 5^2 + n_4^2 + n_5^2 = 60.$$

We obtain $n_4 = n_5 = 3$.

Next, we use Theorem 4.8 to compute some other values of χ_4 and χ_5. The equations

$$(\chi_4, \chi_1 + \chi_2) = 0, \quad (\chi_4, \chi_3) = 0$$

imply

$$\chi_4\left((123)\right) = 0, \quad \chi_4\left((12)(34)\right) = -1.$$

The same argument applied to χ_5 gives

$$\chi_5\left((123)\right) = 0, \quad \chi_5\left((12)(34)\right) = -1.$$

Finally let us denote

$$x = \chi_4\left((12345)\right), \quad y = \chi_4\left((12354)\right)$$

and write down the equation arising from $(\chi_4, \chi_4) = 1$:

$$\frac{1}{60}\left(9 + 15 + 12x^2 + 12y^2\right) = 1,$$

or more simply

(1.6) $$x^2 + y^2 = 3.$$

On the other hand, $(\chi_4, \chi_1) = 0$, which gives

$$3 - 15 + 12\left(x + y\right) = 0,$$

or simply

(1.7) $$x + y = 1.$$

The system (1.6), (1.7) has two solutions

$$x_1 = \frac{1 + \sqrt{5}}{2}, y_1 = \frac{1 - \sqrt{5}}{2}, \quad x_2 = \frac{1 - \sqrt{5}}{2}, y_2 = \frac{1 + \sqrt{5}}{2}.$$

They give the characters χ_4 and χ_5.

Now that we have the character table of A_5, we would like to explain a geometric construction related to it. First, we observe that the previous constructions work over the ground field $\mathbb{R}$ of real numbers. In particular, the representations ρ_4 and ρ_5 are defined over $\mathbb{R}$. Indeed, they are subrepresentations of the second exterior power of ρ_2 and by Lemma 4.26 the corresponding projectors are defined over $\mathbb{R}$. Therefore we have an action of A_5 in $\mathbb{R}^3$. This action preserves an Euclidean scalar product. This is a consequence of the more general result:

LEMMA 5.4. *Let V be a finite-dimensional vector space over $\mathbb{R}$ and ρ be a representation of some finite group G in V. There exists a positive definite scalar product $B : V \times V \to \mathbb{R}$ such that $B(\rho_g u, \rho_g v) = B(u, v)$ for any $u, v \in V$ and $g \in G$.*

Remark. Such a scalar product is called *G-invariant*.

PROOF. Let $C : V \times V \to \mathbb{R}$ be some positive definite scalar product. Set

$$B(u,v) := \sum_{g \in G} C(\rho_g u, \rho_g v).$$

Then B satisfies the conditions of the lemma. $\qquad\square$

Dodecahedron. We have constructed two 3-dimensional irreducible representations of A_5. We can use either of them to construct a dodecahedron, i.e. a regular polyhedron with 12 pentagonal faces and 20 vertices. For instance, let us take $\rho = \rho_4$. By Lemma 5.4 we may assume that for all $g \in G$, ρ_g acts on $\mathbb{R}^3$ by an orthogonal matrix. We claim that ρ_g preserves the orientation in $\mathbb{R}^3$, in other words the determinant $\det \rho_g$ is 1 for all $g \in G$. We already know that $\det \rho_g = \pm 1$. Therefore if g is of odd order the determinant is necessarily 1. If g is of even order, it belongs to the conjugacy class of $(12)(34)$. Hence it is an involution with trace -1, thus a rotation by $180°$. Recall that any isometry in $\mathbb{R}^3$ preserving the orientation is a rotation.

Let $g = (123)$, it is of order 3. Therefore, ρ_g is a rotation by $120°$. Pick up a non-zero x fixed by ρ_g. Consider its orbit $S = \{\rho_g(x) | g \in A_5\}$. Since the stabilizer of x in A_5 is the cyclic group generated by ρ_g, we know that S has 20 points. Moreover, all points of S lie on a sphere, and hence the convex hull Δ of S is a polytope with vertices in S. We will show that Δ is a regular polytope whose faces are regular pentagons.

Let $h = (12345)$. Consider the subgroup $H \subset A_5$ generated by h. Then ρ_h is a rotation by $72°$. Without loss of generality we may assume the axis of ρ_h is vertical. Hence the different orbits of H in S lie on 4 horizontal planes. The top and the bottom plane sections are faces of Δ. Thus, we can conclude that at least some faces of Δ are regular pentagons.

Next, we claim that any vertex of Δ belongs to exactly three pentagonal faces. Indeed, this follows from the fact that the stabilizer of any $s \in S$ has order three and it acts on the set of pentagonal faces containing s. The transitivity of A_5-action on S implies the transitivity of A_5-action on the set of all faces. Hence all the faces of Δ are regular pentagons and the total number of faces is 12.

Note that we have also proven that the group of rotations of a dodecahedron is isomorphic to A_5.

EXERCISE 5.5. Let D_4 denote the dihedral group of order 8 and H_8 denote the multiplicative subgroup of quaternions consisting of $\pm 1, \pm i, \pm j, \pm k$. Compute the character tables of both groups and verify that those tables coincide.

6. Invariant forms

We assume here that char $k = 0$. Recall that a bilinear form on a vector space V is a map $B : V \times V \to k$ satisfying
  (1) $B(cv, dw) = cdB(v, w)$;
  (2) $B(v_1 + v_2, w) = B(v_1, w) + B(v_2, w)$;

(3) $B(v, w_1 + w_2) = B(v, w_1) + B(v, w_2)$.

If V is finite-dimensional, one can also think about a bilinear form as a vector in $V^* \otimes V^*$ or as the homomorphism $B : V \to V^*$ given by the formula $B_v(w) = B(v, w)$. A bilinear form is symmetric if $B(v, w) = B(w, v)$ and skew-symmetric if $B(v, w) = -B(w, v)$. Every bilinear form can be uniquely written as a sum $B = B^+ + B^-$ where B^+ is symmetric and B^- skew-symmetric form,

$$B^{\pm}(v, w) = \frac{B(v, w) \pm B(w, v)}{2}.$$

Such a decomposition corresponds to the decomposition

(1.8) $$V^* \otimes V^* = S^2 V^* \oplus \Lambda^2 V^*.$$

A bilinear form is *non-degenerate* if $B : V \to V^*$ is an isomorphism, in other words if $B(v, V) = 0$ implies $v = 0$.

Let $\rho : G \to \mathrm{GL}(V)$ be a representation. We say that a bilinear form B on V is *G-invariant* if

$$B(\rho_g v, \rho_g w) = B(v, w)$$

for any $v, w \in V$, $g \in G$. If there is no possible confusion we use the word invariant instead of G-invariant.

EXERCISE 6.1. Check the following

(1) If $W \subset V$ is an invariant subspace, then $W^{\perp} = \{v \in V \mid B(v, W) = 0\}$ is invariant. In particular, Ker B is invariant.
(2) $B : V \to V^*$ is invariant if and only if $B \in \mathrm{Hom}_G(V, V^*)$.
(3) If B is invariant, then B^+ and B^- are invariant.

LEMMA 6.2. *Let $\rho : G \to GL(V)$ be an irreducible representation of G, then any non-zero invariant bilinear form on V is non-degenerate. If k is algebraically closed, then such a bilinear form is unique up to scalar multiplication.*

Remark. Lemma 6.2 holds for a field of arbitrary characteristic.

PROOF. Follows from Exercise 6.1 (2) and Schur's lemma. □

LEMMA 6.3. *Let $\rho : G \to GL(V)$ be an irreducible representation of G. Then it admits a non-zero invariant form if and only if $\chi_\rho(g) = \chi_\rho(g^{-1})$ for any $g \in G$.*

PROOF. Since every non-trivial invariant bilinear form establishes an isomorphism between ρ and ρ^*, the statement follows from Corollary 4.10. □

LEMMA 6.4. *(a) If k is algebraically closed, then every non-zero invariant bilinear form on an irreducible representation ρ is either symmetric or skew-symmetric.*
(b) Define

$$m_\rho = \frac{1}{|G|} \sum_{g \in G} \chi_\rho(g^2).$$

Then $m_\rho = 1, 0$ or -1.

(c) If $m_\rho = 0$, then ρ does not admit an invariant form. If $m_\rho = 1$ (resp. $m_\rho = -1$), then ρ admits a symmetric (resp. skew-symmetric) invariant form.

PROOF. First, (a) is a consequence of Lemma 6.2 and Exercise 6.1.
Let us prove (b) and (c). Recall that $\rho \otimes \rho = \rho_{\text{alt}} \oplus \rho_{\text{sym}}$. Using 1.1 we obtain

$$(\chi_{\text{sym}}, \chi_{\text{triv}}) = \frac{1}{|G|} \sum_{g \in G} \frac{\chi_\rho^2(g) + \chi_\rho(g^2)}{2},$$

$$(\chi_{\text{alt}}, \chi_{\text{triv}}) = \frac{1}{|G|} \sum_{g \in G} \frac{\chi_\rho^2(g) - \chi_\rho(g^2)}{2}.$$

Note that

$$\frac{1}{|G|} \sum_{g \in G} \chi_\rho^2(g) = (\chi_\rho, \chi_{\rho^*}).$$

Therefore

$$(\chi_{\text{sym}}, \chi_{\text{triv}}) = \frac{(\chi_\rho, \chi_{\rho^*}) + m_\rho}{2},$$

$$(\chi_{\text{alt}}, \chi_{\text{triv}}) = \frac{(\chi_\rho, \chi_{\rho^*}) - m_\rho}{2}.$$

We have the following trichotomy

- ρ does not have an invariant form, if and only if ρ is not isomorphic to ρ^*. In this case $(\chi_\rho, \chi_{\rho^*}) = 0$ and $(\chi_{\text{sym}}, \chi_{\text{triv}}) = (\chi_{\text{alt}}, \chi_{\text{triv}}) = 0$. Therefore $m_\rho = 0$.
- ρ has a symmetric invariant form if and only if $(\chi_\rho, \chi_{\rho^*}) = 1$ and $(\chi_{\text{sym}}, \chi_{\text{triv}}) = 1$. This implies $m_\rho = 1$.
- ρ admits a skew-symmetric invariant form if and only if $(\chi_\rho, \chi_{\rho^*}) = 1$ and $(\chi_{\text{alt}}, \chi_{\text{triv}}) = 1$. This implies $m_\rho = -1$.

$\square$

Let $k = \mathbb{C}$. An irreducible representation of a finite group G is called *real* if $m_\rho = 1$, *complex* if $m_\rho = 0$ and *quaternionic* if $m_\rho = -1$.

Remark. Since $\chi_\rho(s^{-1}) = \bar{\chi}_\rho(s)$, χ_ρ takes only real values for real and quaternionic representations. If ρ is complex there is at least one $g \in G$ such that $\chi_\rho(g) \notin \mathbb{R}$. This terminology will become clear in Section 8.

EXERCISE 6.5. Show that
(a) All irreducible representation of S_4 are real.
(b) All non-trivial irreducible representations of $\mathbb{Z}_3$ are complex.
(c) The two-dimensional representation of the quaternionic group H_8 is quaternionic (see Exercise 5.5).

EXERCISE 6.6. Assume that the order of G is odd. Show that all non-trivial irreducible representation of G are complex. (Hint: prove that $m_\rho = (\chi_\rho, \chi_{\text{triv}})$.)

7. Representations over ℝ

Let us recall that by Lemma 5.4 every representation of a finite group over ℝ admits an invariant positive definite scalar product. Assume the representation $\rho: G \to GL(V)$ is irreducible. Denote by $B(\cdot, \cdot)$ an invariant scalar product and let $Q(\cdot, \cdot)$ denote another invariant symmetric form on V. These two forms can be simultaneously diagonalized. Therefore there exists $\lambda \in \mathbb{R}$, such that $\mathrm{Ker}\,(Q - \lambda B) \neq 0$. Since $\mathrm{Ker}\,(Q - \lambda B) \neq 0$ is G-invariant and ρ is irreducible, this implies $Q = \lambda B$. Therefore we have

LEMMA 7.1. *Let $\rho: G \to GL(V)$ be an irreducible representation of G over ℝ. There is exactly one invariant symmetric form on V up to scalar multiplication.*

THEOREM 7.2. *Let K such that $\mathbb{R} \subset K$ be a division ring which is also a finite-dimensional algebra over ℝ. Then $\mathbb{K}$ is isomorphic to $\mathbb{R}, \mathbb{C}$ or $\mathbb{H}$.*

PROOF. If K is a field, then $K \cong \mathbb{R}$ or $\mathbb{C}$, because $\mathbb{C} = \bar{\mathbb{R}}$ and $[\mathbb{C} : \mathbb{R}] = 2$.

Assume that K is not commutative. Then it contains a subfield isomorphic to $\mathbb{C}$, obtained by taking $x \in K \backslash \mathbb{R}$ and considering $\mathbb{R}[x]$. Therefore, without loss of generality, we may assume that $\mathbb{R} \subset \mathbb{C} \subset K$.

Consider the involutive $\mathbb{C}$-linear automoprhism of K defined by the formula:

$$f(x) = ixi^{-1}.$$

Look at the eigenspace decomposition of K with respect to f:

$$K = K_1 \oplus K_{-1},$$

where

$$K_{\pm 1} = \{x \in K \mid f(x) = \pm x\}.$$

One can easily check the following inclusions

$$K_1 K_1 \subset K_1, \quad K_{-1} K_{-1} \subset K_1, \quad K_1 K_{-1} \subset K_{-1}, \quad K_{-1} K_1 \subset K_{-1}.$$

The eigenspace K_1 coincides with the centralizer of $\mathbb{C}$ in K. Therefore $K_1 = \mathbb{C}$.

Choose a non-zero $y \in K^-$. The left multiplication by y defines an isomorphism of ℝ vector spaces between K_1 and K_{-1}. Hence $\dim_{\mathbb{R}} K_1 = \dim_{\mathbb{R}} K_{-1} = 2$ and $\dim K = 4$.

For any $z = a + bi \in K_1$ and any $w \in K_{-1}$, we have

$$w\bar{z} = wa - wbi = aw + biw = zw.$$

Since w^2 belongs to $\mathbb{C}$ and commutes with w, we have $w^2 \in \mathbb{R}$. We claim that w^2 is negative since otherwise $w^2 = c^2$ for some real c and $(w - c)(w + c) = 0$, which is impossible, since K is a division ring. Set $j := \frac{w}{\sqrt{-w^2}}$. Then $j^2 = -1$ and $ij = -ji$. So if we set $k := ij$, then $1, i, j, k$ form the standard basis of $\mathbb{H}$. □

LEMMA 7.3. *Let $\rho: G \to GL(V)$ be an irreducible representation over ℝ, then there are three possibilities:*

(1) $End_G(V) = \mathbb{R}$ and $(\chi_\rho, \chi_\rho) = 1$;
(2) $End_G(V) \cong \mathbb{C}$ and $(\chi_\rho, \chi_\rho) = 2$;
(3) $End_G(V) \cong \mathbb{H}$ and $(\chi_\rho, \chi_\rho) = 4$.

PROOF. Corollary 3.7 and Theorem 7.2 imply that $End_G(V)$ is isomorphic to $\mathbb{R}, \mathbb{C}$ or $\mathbb{H}$, $(\chi_\rho, \chi_\rho) = 1, 2$ or 4 as follows from Corollary 4.7.

□

8. Relationship between representations over $\mathbb{R}$ and over $\mathbb{C}$

Hermitian invariant form. Recall that a Hermitian form is a binary additive form on a complex vector space, satisfying the conditions:

$$H(av, bw) = \bar{a}bH(v, w), \, H(w, v) = \bar{H}(v, w).$$

The following Lemma can be proven exactly as Lemma 7.1.

LEMMA 8.1. *Every representation of a finite group over $\mathbb{C}$ admits a positive definite invariant Hermitian form. If the representation is irreducible, then any two invariant Hermitian forms on it are proportional.*

Let $\rho : G \to GL(V)$ be a representation of dimension n over $\mathbb{C}$. Denote by $V^{\mathbb{R}}$ the space V considered as a vector space over $\mathbb{R}$ of dimension $2n$. Denote by $\rho^{\mathbb{R}}$ the representation of G in $V^{\mathbb{R}}$.

EXERCISE 8.2. Show that

$$\chi_{\rho^{\mathbb{R}}} = \chi_\rho + \bar{\chi}_\rho.$$

The exercise implies that $(\chi_{\rho^{\mathbb{R}}}, \chi_{\rho^{\mathbb{R}}})$ is either 2 or 4. Hence, $\dim End_G(V^{\mathbb{R}})$ is either 2 or 4. Moreover, $\mathbb{C}$ is a self-centralizing subalgebra in $End_G(V^{\mathbb{R}})$. Therefore $End_G(V^{\mathbb{R}})$ is isomorphic to $\mathbb{C}, \mathbb{H}$ or to the ring $M_2(\mathbb{R})$ of real matrices of size 2×2.

PROPOSITION 8.3. *Let $\rho : G \to GL(V)$ be an irreducible representation over $\mathbb{C}$. Then one of the following three cases occur.*

(1) $End_G(V^{\mathbb{R}}) \simeq M_2(\mathbb{R})$. *Then there exists a basis of V such that the matrices ρ_g for all $g \in G$ have real entries. In this case V admits an invariant scalar product.*
(2) $End_G(V^{\mathbb{R}}) \simeq \mathbb{C}$. *Then ρ is complex, i.e. ρ does not admit any invariant bilinear form.*
(3) $End_G(V^{\mathbb{R}}) \simeq \mathbb{H}$. *Then ρ admits an invariant skew-symmetric form.*

PROOF. (1) If $End_G(V^{\mathbb{R}}) \simeq M_2(\mathbb{R})$, then $V^{\mathbb{R}} = W \oplus W = \mathbb{C}W$ for some irreducible representation W of G over $\mathbb{R}$. The existence of an invariant scalar product follows from Lemma 7.1. For (2) use Exercise 8.2. Since $(\chi_{\rho^{\mathbb{R}}}, \chi_{\rho^{\mathbb{R}}}) = 2$ by Lemma 7.3, then $\chi_\rho \neq \bar{\chi}_\rho$, and therefore ρ is complex.

Finally let us prove (3). Let $j \in \mathrm{End}_G\left(V^{\mathbb{R}}\right) = \mathbb{H}$, then $j\left(bv\right) = \bar{b}jv$ for all $b \in \mathbb{C}$. Let H be a positive definite invariant Hermitian form on V. Then

$$Q\left(v, w\right) = H\left(jw, jv\right)$$

is another invariant positive definite Hermitian form. By Lemma 8.1, $Q = \lambda H$ and λ must be positive because Q is also positive definite. Since $j^2 = -1$, then $\lambda^2 = 1$ and therefore $\lambda = 1$. Thus,

$$H\left(v, w\right) = H\left(jw, jv\right).$$

Set

$$B\left(v, w\right) = H\left(jv, w\right).$$

Then B is a bilinear invariant form, and

$$B\left(w, v\right) = H\left(jw, v\right) = H\left(jv, j^2 w\right) = -H\left(jv, w\right) = -B\left(v, w\right),$$

hence B is skew-symmetric. $\qquad\square$

COROLLARY 8.4. *Let σ be an irreducible representation of G over $\mathbb{R}$. There are three possibilities for σ:*

(1) *$\chi_\sigma = \chi_\rho$ for some real representation ρ of G over $\mathbb{C}$;*
(2) *$\chi_\sigma = \chi_\rho + \bar{\chi}_\rho$ for some complex representation ρ of G over $\mathbb{C}$;*
(3) *$\chi_\sigma = 2\chi_\rho$ for some quaternionic representation ρ of G over $\mathbb{C}$.*

THEOREM 8.5. *Let G be a finite group, r denote the number of conjugacy classes and s denote the number of classes which are stable under inversion. Then $\frac{r+s}{2}$ is the number of irreducible representations of G over $\mathbb{R}$.*

PROOF. Recall that $\mathcal{C}(G)$ is the space of complex-valued class functions on G. Consider the involution $\theta : \mathcal{C}(G) \to \mathcal{C}(G)$ given by:

$$\theta\varphi(g) = \varphi(g^{-1}).$$

An easy calculation shows that $\dim \mathcal{C}(G)^\theta = s + \frac{r-s}{2} = \frac{r+s}{2}$.

Denote by $\chi_1, \ldots, \chi_r$ the irreducible characters of G over $\mathbb{C}$. Recall that $\chi_1, \ldots, \chi_r$ form a basis of $\mathcal{C}(G)$. Observe that for any character χ_ρ

$$\theta(\chi_\rho) = \chi_{\rho^*}.$$

Therefore, θ permutes the irreducible characters $\chi_1, \ldots, \chi_r$. Corollary 8.4 implies that the number of irreducible representations of G over $\mathbb{R}$ equals the number of self-dual irreducible representations over $\mathbb{C}$ plus half the number of those which are not self-dual. Therefore this number is equal to $\dim \mathcal{C}(G)^\theta$. $\qquad\square$

CHAPTER 2

Modules with applications to finite groups

In which modules over associative rings come into play, we discover that they are actually represen-tations in disguise and uncover properties of the group algebra visible through this lens. Not to mention our powerful new tool, induction.

1. Modules over associative rings

1.1. The notion of module.

DEFINITION 1.1. Let R be an associative ring with identity element $1 \in R$. An abelian group M is called a *(left) R-module* if there is a map $R \times M \to M$, $(a, m) \mapsto am$ such that for all $a, b \in R$ and $m, n \in M$ we have

(1) $(ab) m = a (bm)$;
(2) $1m = m$;
(3) $(a + b) m = am + bm$;
(4) $a (m + n) = am + an$.

One can define in a similar way a right R-module. Unless otherwise stated we only consider left modules and we say module for left module.

EXAMPLE 1.2. If R is a field then the R-modules are the vector spaces over R.

EXAMPLE 1.3. Let G be a group and $k(G)$ be its group algebra over k. Then every $k(G)$-module V is a vector space over k equipped with a G-action. Set

$$\rho_g v := gv$$

for all $g \in G \subset k(G)$, $v \in V$. This defines a representation $\rho : G \to GL(V)$.

Conversely, if V is a vector space over k and $\rho \colon G \to \mathrm{GL}(V)$ is a representation, the formula

$$\left(\sum_{g \in G} a_g g \right) v := \sum_{g \in G} a_g \rho_g v.$$

defines a $k(G)$-module structure on V.

In other words, to study representations of G over k is exactly the same as to study $k(G)$-modules. Hence from now on we will talk indifferently of $k(G)$-modules, representations of G over k or just simply G-modules over k.

© Springer Nature Switzerland AG 2018
C. Gruson and V. Serganova, *A Journey Through Representation Theory*,
Universitext, https://doi.org/10.1007/978-3-319-98271-7_2

DEFINITION 1.4. Let M be an R-module. A *submodule* $N \subset M$ is a subgroup which is invariant under the R-action. If $N \subset M$ is a submodule then the quotient M/N has a natural R-module structure. A non-zero module M is *simple* or *irreducible* if all submodules are either zero or M.

REMARK 1.5. Sums and intersections of submodules are submodules.

EXAMPLE 1.6. If R is an arbitrary ring, then R is a left R-module for the action given by the left multiplication. Its submodules are the left ideals.

Let $\{M_j\}_{j \in J}$ be a family of R-modules. We define the direct sum $\bigoplus_{j \in J} M_j$ and the direct product $\prod_{j \in J} M_j$ in the obvious way. An R-module is *free* if it is isomophic to a direct sum of I copies of R (I can be infinite).

EXERCISE 1.7. If R is a division ring, then every non-zero R-module is free.

EXERCISE 1.8. Let $R = \mathbb{Z}$ be the ring of integers.
(a) Show that any simple $\mathbb{Z}$-module is isomorphic to $\mathbb{Z}/p\mathbb{Z}$ for some prime p.
(b) Let M be a $\mathbb{Z}$-module. We call $m \in M$ a *torsion* element if $rm = 0$ for some non-zero $r \in \mathbb{Z}$. Prove that the subset M^{tor} of all torsion elements is a submodule.
(c) We say M is *torsion-free* if $M^{\text{tor}} = 0$. Prove that M/M^{tor} is torsion-free.
(d) Give an example of a non-zero torsion-free $\mathbb{Z}$-module which is not free.

Let M and N be R-modules. In the same way as in the group case, we define the abelian group $\text{Hom}_R(M, N)$ of R-invariant homomorphisms from M to N and the ring $\text{End}_R(M)$ of R-invariant endomorphisms of M. In particular if k is a field and V is an n-dimensional vector space, then $\text{End}_k(V)$ is the matrix ring $M_n(k)$.

In this context we have the following formulation of Schur's Lemma. Its proof is the same as in the group case.

LEMMA 1.9. *Let M and N be simple R-modules. If $\varphi \in \text{Hom}_R(M, N)$ is not zero then it is an isomorphism.*
If M is a simple module, then $\text{End}_R(M)$ is a division ring.

1.2. A group algebra is a product of matrix rings. Recall that for every ring R one defines R^{op} as the ring with the same abelian group structure together with the new multiplication $*$ given by

$$a * b = ba.$$

LEMMA 1.10. *The ring $\text{End}_R(R)$ is isomorphic to R^{op}.*

PROOF. For all $a \in R$, define $\varphi_a \in \text{End}(R)$ by the formula:

$$\varphi_a(x) = xa.$$

It is easy to check that

- $\varphi_a \in \mathrm{End}_R(R)$,
- $\varphi_{ba} = \varphi_a \circ \varphi_b$.

In this way we have constructed a homomorphism

$$\varphi : R^{\mathrm{op}} \to \mathrm{End}_R(R).$$

All we have to show that φ is an isomorphism.

Injectivity: Assume $\varphi_a = \varphi_b$. Then $\varphi_a(1) = \varphi_b(1)$, hence $a = b$.

Surjectivity: let $\gamma \in \mathrm{End}_R(R)$. If $x \in R$, then

$$\gamma(x) = \gamma(x1) = x\gamma(1).$$

Therefore $\gamma = \varphi_{\gamma(1)}$. $\qquad\square$

LEMMA 1.11. *Let $\rho_i : G \to \mathrm{GL}(V_i)$, $i = 1, \ldots, l$, be a finite set of pairwise non-isomorphic irreducible representations of a finite group G over an algebraically closed field k, and set*

$$V = V_1^{\oplus m_1} \oplus \cdots \oplus V_l^{\oplus m_l}.$$

Then

$$\mathrm{End}_G(V) \cong M_{m_1}(k) \times \cdots \times M_{m_l}(k).$$

PROOF. If φ is an element of $\mathrm{End}_G(V)$, then Schur's Lemma implies that φ preserves isotypic components. Therefore, we have an isomorphism

$$\mathrm{End}_G(V) \cong \mathrm{End}_G\left(V_1^{\oplus m_1}\right) \times \cdots \times \mathrm{End}_G\left(V_l^{\oplus m_l}\right).$$

Thus it suffices to prove the following

LEMMA 1.12. *Let G be a finite group, k be an algebraically closed field and W be a simple $k(G)$-module. Then $\mathrm{End}_G(W^{\oplus m})$ is isomorphic to the matrix ring $M_m(k)$.*

PROOF. For all $i, j = 1, \ldots, m$, denote by p_j the canonical projection of $W^{\oplus m}$ onto its j-th factor and by q_i the embedding of W as the i-th factor into $W^{\oplus m}$. Take $\varphi \in \mathrm{End}_G(W^{\oplus m})$. For all $i, j = 1, \ldots, m$, denote by φ_{ij} the composition map

$$W \xrightarrow{q_j} W^{\oplus m} \xrightarrow{\varphi} W^{\oplus m} \xrightarrow{p_i} W.$$

Since $\varphi_{ij} \in \mathrm{End}_G(W)$, Schur's Lemma implies

$$\varphi_{ij} = c_{ij}\,\mathrm{Id}_W$$

for some $c_{ij} \in k$. Thus we obtain a map

$$\Phi : \mathrm{End}\left(W^{\oplus m}\right) \to M_m(k).$$

Moreover, φ can be written uniquely as

$$\varphi = \sum_{i,j=1}^{m} c_{ij} q_i \circ p_j.$$

If ψ is another element in End $(W^{\oplus m})$, we write

$$\psi = \sum_{i,j=1}^{m} d_{ij} q_i \circ p_j.$$

Then the composition is:

$$\varphi \circ \psi = \sum_{i,j,k=1}^{m} c_{ik} d_{kj} q_i \circ p_j.$$

This shows that Φ is a homomorphism of rings. Injectivity and surjectivity of Φ are direct consequences of the definition. □

□

THEOREM 1.13. *Let G be a finite group. Assume k is algebraically closed and char $k = 0$. Then*

$$k(G) \cong M_{n_1}(k) \times \cdots \times M_{n_r}(k),$$

where $n_1, \ldots, n_r$ are the dimensions of all irreducible representations, up to isomorphism.

PROOF. By Lemma 1.10

$$\mathrm{End}_{k(G)}(k(G)) \cong k(G)^{\mathrm{op}}.$$

Moreover, the map $g \mapsto g^{-1}$ gives an isomorphism

$$k(G)^{\mathrm{op}} \cong k(G).$$

On the other hand, by Theorem 4.13 Chapter 1, the following equality holds:

$$k(G) = V_1^{\oplus n_1} \oplus \cdots \oplus V_r^{\oplus n_r},$$

where $V_1, \ldots, V_r$ are simple G-modules. Applying Lemma 1.11, we get the theorem. □

2. Finitely generated modules and Noetherian rings

DEFINITION 2.1. An R-module M is *finitely generated* if there exist finitely many elements $x_1, \ldots, x_n \in M$ such that $M = Rx_1 + \cdots + Rx_n$.

LEMMA 2.2. *Let*

$$0 \to N \xrightarrow{q} M \xrightarrow{p} L \to 0$$

be an exact sequence of R-modules.
   (a) If M is finitely generated, then L is finitely generated.
   (b) If N and L are finitely generated, then M is finitely generated.

PROOF. The first assertion is obvious. For the second let

$$L = Rx_1 + \cdots + Rx_n, \quad N = Ry_1 + \cdots + Ry_m,$$

then $M = Rp^{-1}(x_1) + \cdots + Rp^{-1}(x_n) + Rq(y_1) + \cdots + Rq(y_m)$. □

LEMMA 2.3. *Let R be a ring. The following conditions are equivalent*

(1) *Every increasing chain of left ideals in R is finite, in other words for any sequence $I_1 \subset I_2 \subset \ldots$ of left ideals, there exists n_0 such that for all $n > n_0$, $I_n = I_{n_0}$.*

(2) *Every left ideal is a finitely generated R-module.*

PROOF. $(1) \Rightarrow (2)$. Assume that some left ideal I is not finitely generated. Then there exists an infinite sequence of $x_n \in I$ such that

$$x_{n+1} \notin Rx_1 + \cdots + Rx_n.$$

But then $I_n = Rx_1 + \cdots + Rx_n$ form an infinite increasing chain of ideals which does not stabilize.

$(2) \Rightarrow (1)$. Let $I_1 \subset I_2 \subset \ldots$ be an increasing chain of ideals. Consider

$$I := \bigcup_n I_n.$$

Then by (2) I is finitely generated. Therefore $I = Rx_1 + \cdots + Rx_s$ for some $x_1, \ldots x_s \in I$. Then there exists n_0 such that $x_1, \ldots, x_s \in I_{n_0}$. Hence $I = I_{n_0}$ and the chain stabilizes. $\qquad \square$

DEFINITION 2.4. *A ring satisfying the conditions of Lemma 2.3 is called (left) Noetherian.*

LEMMA 2.5. *Let R be a left Noetherian ring and M be a finitely generated R-module. Then every submodule of M is finitely generated.*

PROOF. First, we prove the statement when M is free. Then M is isomorphic to R^n for some n and we use induction on n. For $n = 1$ the statement follows from the definition. Consider the exact sequence

$$0 \to R^{n-1} \to R^n \to R \to 0.$$

Let N be a submodule of R^n. Consider the exact sequence obtained by restriction to N

$$0 \to N \cap R^{n-1} \to N \to N' \to 0.$$

By induction assumption $N \cap R^{n-1}$ is finitely generated and $N' \subset R$ is finitely generated. Therefore by Lemma 2.2 (b), N is finitely generated.

In the general case, M is a quotient of a free module of finite rank. We use the exact sequence

$$0 \to K \to R^n \xrightarrow{p} M \to 0.$$

If N is a submodule of M, then $p^{-1}(N) \subset R^n$ is finitely generated. Therefore by Lemma 2.2 (a), N is also finitely generated. $\qquad \square$

EXERCISE 2.6. (a) Show that a principal ideal domain is a Noetherian ring. In particular, $\mathbb{Z}$ and the polynomial ring $k[X]$ are Noetherian.

(b) Show that the polynomial ring $k[X_1, \ldots, X_n, \ldots]$ on infinitely many variables is not Noetherian.

(c) A subring of a Noetherian ring is not automatically Noetherian. For example, let R be the subring of $\mathbb{C}[X, Y]$ consisting of polynomial functions which are constant on the cross $X^2 - Y^2 = 0$. Show that R is not Noetherian.

Let R be a commutative ring. An element $r \in R$ is called *integral over* $\mathbb{Z}$ if there exists a monic polynomial $p(X) \in \mathbb{Z}[X]$ such that $p(r) = 0$.

EXERCISE 2.7. Check that r is integral over $\mathbb{Z}$ if and only if $\mathbb{Z}[r] \subset R$ is a finitely generated $\mathbb{Z}$-module.

Remark. The complex numbers which are integral over $\mathbb{Z}$ are usually called algebraic integers. All the rational numbers which are integral over $\mathbb{Z}$ belong to $\mathbb{Z}$.

LEMMA 2.8. *Let R be a commutative ring and S be the set of elements which are integral over $\mathbb{Z}$. Then S is a subring of R.*

PROOF. Let $x, y \in S$. By assumption, $\mathbb{Z}[x]$ and $\mathbb{Z}[y]$ are finitely generated $\mathbb{Z}$-modules. Then $\mathbb{Z}[x, y]$ is also finitely generated. Since $\mathbb{Z}$ is Noetherian ring, Lemma 2.5 implies that, for every $s \in \mathbb{Z}[x, y]$, the $\mathbb{Z}$-submodule $\mathbb{Z}[s]$ is finitely generated. $\square$

3. The centre of the group algebra $k(G)$

Throughout this section, we assume that k is algebraically closed of characteristic 0 and G is a finite group. In this section we obtain some results about the centre $Z(G)$ of the group ring $k(G)$. It is clear that $Z(G)$ can be identified with the subspace of class functions:

$$Z(G) = \left\{ \sum_{s \in G} f(s) s \mid f \in \mathcal{C}(G) \right\}.$$

Recall that if $n_1, \ldots, n_r$ are the dimensions of isomorphism classes of simple G-modules, then by Theorem 1.13 we have an isomorphism

$$k(G) \simeq M_{n_1}(k) \times \cdots \times M_{n_r}(k).$$

If $e_i \in k(G)$ denotes the element corresponding to the identity matrix in $M_{n_i}(K)$, the elements $e_1, \ldots, e_r$ form a basis of $Z(G)$ and

$$e_i e_j = \delta_{ij} e_i$$

$$1_G = e_1 + \cdots + e_r.$$

If $\rho_j : G \to \mathrm{GL}(V_j)$ is an irreducible representation, then e_j acts on V_j as the identity element and we have

(2.1) $\rho_j(e_i) = \delta_{ij} \mathrm{Id}_{V_j}.$

LEMMA 3.1. *If χ_i is the character of the irreducible representation ρ_i of dimension n_i, then*

$$(2.2) \qquad e_i = \frac{n_i}{|G|} \sum_{g \in G} \chi_i \left(g^{-1}\right) g.$$

PROOF. We have to check (2.1). Since $\rho_j(e_i)$ belongs to $\mathrm{End}_G(V_j)$, Schur's Lemma implies $\rho_j(e_i) = \lambda \,\mathrm{Id}$ for some λ. Now we use orthogonality relations, Theorem 4.8 in Chapter 1:

$$\mathrm{Tr}\, \rho_j(e_i) = \frac{n_i}{|G|} \sum \chi_i\left(g^{-1}\right) \chi_j(g) = n_i\left(\chi_i, \chi_j\right) = \delta_{ij} n_i.$$

Therefore, we have $n_j \lambda = \delta_{ij} n_i$, which implies $\lambda = \delta_{ij}$. $\qquad\square$

EXERCISE 3.2. Define $\omega_i : Z(G) \to k$ by the formula

$$\omega_i\left(\sum a_s s\right) = \frac{1}{n_i} \sum a_s \chi_i(s)$$

and $\omega : Z(G) \to k^r$ by

$$\omega = (\omega_1, \ldots, \omega_r).$$

Check that ω is an isomorphism of rings. Hint: check that $\omega_i(e_j) = \delta_{ij}$ using again the orthogonality relations.

For any conjugacy class C in G let

$$\eta_C := \sum_{g \in C} g.$$

Clearly, the set η_C for C running over the set of conjugacy classes is a basis in $Z(G)$.

LEMMA 3.3. *For any irreducible character χ_i and for any g in a conjugacy class C, the number $\frac{|C|}{n_i}\chi_i(g)$ is algebraic integer.*

PROOF. Observe that

$$\frac{|C|}{n_i}\chi_i(g) = \omega_i(\eta_C).$$

Since any homomorphism maps an algebraic integer to an algebraic integer, the statement follows. $\qquad\square$

LEMMA 3.4. *For any conjugacy class $C \subset G$ we have*

$$\eta_C = |C| \sum_{i=1}^{r} \frac{\chi_i(g)}{n_i} e_i,$$

where g is any element of C.

PROOF. If we extend by linearity $\chi_1, \ldots, \chi_r$ to linear functionals on $k(G)$, then (2.1) implies $\chi_j(e_i) = n_i \delta_{i,j}$. Thus, $\chi_1, \ldots, \chi_r$ form a basis in the dual space $Z(G)^*$. Therefore it suffices to check that

$$\chi_j(\eta_C) = |C| \sum_{i=1}^{r} \frac{\chi_i(g)}{n_i} \chi_j(e_i) = |C| \chi_j(g). \qquad \square$$

The next statement is sometimes called the second orthogonality relation.

LEMMA 3.5. If $g, h \in G$ lie in the same conjugacy class C, we have

$$\sum_{i=1}^{r} \chi_i(g) \chi_i(h^{-1}) = \frac{|G|}{|C|}.$$

If g and h are not conjugate we have

$$\sum_{i=1}^{r} \chi_i(g) \chi_i(h^{-1}) = 0.$$

PROOF. The statement follows from Lemma 3.1 and Lemma 3.4. Indeed, if g is in the conjugacy class C, we have

$$\eta_C = |C| \sum_{i=1}^{r} \frac{\chi_i(g)}{n_i} e_i = \frac{|C|}{|G|} \sum_{i=1}^{r} \sum_{h \in G} \chi_i(g) \chi_i(h^{-1}) h.$$

The coefficient of h in the last expression is 1 if $h \in C$ and zero otherwise. This implies the lemma. $\qquad \square$

LEMMA 3.6. Let $u = \sum_{g \in G} a_g g \in Z(G)$. If all a_g are algebraic integers, then u is integral over $\mathbb{Z}$.

PROOF. Consider the basis η_C of $Z(G)$. Every η_C is integral over $\mathbb{Z}$ since the subring generated by all η_C is a finitely generated $\mathbb{Z}$-module. Now the statement follows from Lemma 2.8. $\qquad \square$

THEOREM 3.7. For all $i = 1, \ldots, r$ the dimension n_i divides $|G|$.

PROOF. For every $g \in G$, all eigenvalues of $\rho(g)$ are roots of 1. Therefore $\chi_\rho(s)$ is an algebraic integer. By Lemma 3.6 $u = \sum_{s \in G} \chi_i(s) s^{-1}$ is integral over $\mathbb{Z}$. Recall the homomorphism ω_i from Exercise 3.2. Since $\omega_i(u)$ is an algebraic integer we have

$$\omega_i(u) = \frac{1}{n_i} \sum_{s \in G} \chi_i(s) \chi_i(s^{-1}) = \frac{|G|}{n_i} (\chi_i, \chi_i) = \frac{|G|}{n_i}.$$

Therefore $\frac{|G|}{n_i} \in \mathbb{Z}$. $\qquad \square$

THEOREM 3.8. Let Z be the centre of G and ρ be an irreducible n-dimensional representation of G. Then n divides $\frac{|G|}{|Z|}$.

Proof. Let G^m be the direct product of m copies of G and ρ^m be the exterior product of m copies of ρ. The dimension of ρ^m is n^m. Furthermore, ρ^m is irreducible by Exercise 4.12. Consider the normal subgroup N of G^m defined by

$$N = \{(z_1, \ldots, z_m) \in Z^m \mid z_1 z_2 \ldots z_m = 1\}.$$

We have $|N| = |Z|^{m-1}$. Furthermore, N lies in the kernel of ρ^m. Therefore ρ^m is a representation of the quotient group $H = G/N$. Hence, by Theorem 3.7, n^m divides $\frac{|G|^m}{|Z|^{m-1}}$ for every $m > 0$. It follows from prime factorization that n divides $\frac{|G|}{|Z|}$. □

4. One application

We will now show one application of the results of the previous section. It is well known that any finite group whose order is a power of a prime number is solvable. The following statement is a generalization of this result. The proof essentially uses representation theory.

THEOREM 4.1. (Burnside). *Let G be a finite group of order $p^s q^t$ for some prime numbers p, q and some $s, t \in \mathbb{N}$. Then G is solvable.*

Proof. For the proof of the theorem we need the following lemma: □

LEMMA 4.2. *Let ρ be a complex representation of a finite group G of dimension n. Let $g \in G$ be such that $\frac{\chi_\rho(g)}{n}$ is an algebraic integer. Then ρ_g is a scalar operator.*

Proof. Let m be the order of g and $\varepsilon_1, \ldots, \varepsilon_n$ be the eigenvalues of ρ_g. Let ζ be a primitive m-th root of 1 and denote by Γ the Galois group of $\mathbb{Q}(\zeta)$ over $\mathbb{Q}$. Each ε_i is a power of ζ. Assume that $\varepsilon_i \neq \varepsilon_j$ for some $i \neq j$. By the triangle inequality, we have

$$\left|\frac{\chi_\rho(g)}{n}\right| = \left|\frac{\varepsilon_1 + \cdots + \varepsilon_n}{n}\right| < 1,$$

and for any $\gamma \in \Gamma$,

$$\left|\frac{\gamma(\chi_\rho(g))}{n}\right| = \left|\frac{\gamma(\varepsilon_1) + \cdots + \gamma(\varepsilon_n)}{n}\right| \leq 1.$$

Set

$$d := \prod_{\gamma \in \Gamma} \frac{\gamma(\chi_\rho(g))}{n}.$$

Then d belongs to $\mathbb{Z}$, but $|d| < 1$: we obtain a contradiction. □

To prove the theorem, it suffices to show that G has a non-trivial proper normal subgroup. Indeed, if N is such subgroup, then, using induction on the order of the group, we may assume that G/N and N are solvable and hence G is also solvable.

Consider a Sylow subgroup P of G of order p^s. Then P has a non-trivial centre, let $g \neq 1$ be an element in this centre. Denote by C the conjugacy class of g. If $|C| = 1$, then G has a non-trivial centre and we are done. Assume that $|C| > 1$. Since the centralizer of g in G contains P, we have $|C| = q^m$ for some $m > 0$.

Let $\chi_1, \ldots, \chi_r$ be the irreducible characters of G, denote by $n_1, \ldots, n_r$ the dimensions of the corresponding irreducible representations $\rho_1, \ldots, \rho_r$. Assume that ρ_1 is the trivial representation. The relation

$$\sum_{i=1}^{r} n_i^2 = 1 + \sum_{i=2}^{r} n_i^2 = |G|$$

implies that there exists $i > 1$ such that $(n_i, q) = 1$. We claim that i can be chosen in such a way that $\chi_i(g) \neq 0$. Indeed, by Lemma 3.5 we have

$$0 = \sum_{i=1}^{r} \chi_i(g)\chi_i(1) = \sum_{i=1}^{r} n_i\chi_i(g) = 1 + \sum_{i=2}^{r} n_i\chi_i(g).$$

If we assume that $\chi_i(g) = 0$ for all $i > 1$ such that $(n_i, q) = 1$, then we have $1 + qa = 0$ for some algebraic integer a and this is impossible.

By Lemma 3.3 the number $\frac{|C|\chi_i(g)}{n_i}$ is an algebraic integer. Recall that $(n_i, |C|) = 1$. Hence, there exist integers a, b such that $a|C| + bn_i = 1$. Therefore $\frac{\chi_i(g)}{n_i} = a\frac{|C|\chi_i(g)}{n_i} + b\chi_i(g)$ is an algebraic integer. By Lemma 4.2 we obtain that $\rho_i(g)$ is a scalar operator.

Let N be the subset of all elements $h \in G$ such that $\rho_i(h)$ is a scalar operator. It is easy to check that N is a normal subgroup of G. Since $g \in N$, we know that $N \neq \{1\}$. If $N = G$, then $n_i = 1$, and therefore the commutator group $[G, G]$ is nontrivial, otherwise, N is a proper normal subgroup of G. The proof of the theorem is complete. $\qquad\square$

5. General facts on induced modules

Let A be a ring, B be a subring of A and M be a B-module. Consider the abelian group $A \otimes_B M$ defined by generators and relations in the following way. The generators are all elements of the Cartesian product $A \times M$ and the relations:

(2.3) $\qquad (a_1 + a_2) \times m - a_1 \times m - a_2 \times m, \qquad a_1, a_2 \in A, m \in M,$

(2.4) $\qquad a \times (m_1 + m_2) - a \times m_1 - a \times m_2, \qquad a \in A, m_1, m_2 \in M,$

(2.5) $\qquad ab \times m - a \times bm, \qquad a \in A, b \in B, m \in M.$

This group is an A-module, the action of A being by left multiplication. For every $a \in A$ and $m \in M$, we denote by $a \otimes m$ the corresponding element in $A \otimes_B M$.

DEFINITION 5.1. The A-module $A \otimes_B M$ is called the induced module (from B to A).

EXERCISE 5.2. (a) Show that $A \otimes_B B$ is isomorphic to A.

(b) Show that if M_1 and M_2 are two B-modules, then there exists a canonical isomorphism of A-modules

$$A \otimes_B (M_1 \oplus M_2) \simeq A \otimes_B M_1 \oplus A \otimes_B M_2.$$

(c) Check that for all $n \in \mathbb{Z}$,

$$\mathbb{Q} \otimes_{\mathbb{Z}} (\mathbb{Z}/n\mathbb{Z}) = 0.$$

THEOREM 5.3. *(Frobenius reciprocity.) For every B-module M and for every A-module N, there is an isomorphism of abelian groups*

$$\operatorname{Hom}_B (M, N) \cong \operatorname{Hom}_A (A \otimes_B M, N).$$

PROOF. Let M be a B-module and N be an A-module. Consider $j : M \to A \otimes_B M$ defined by

$$j(m) := 1 \otimes m,$$

which is a homomorphism of B-modules. □

LEMMA 5.4. *For every $\varphi \in \operatorname{Hom}_B (M, N)$ there exists a unique $\psi \in \operatorname{Hom}_A (A \otimes_B M, N)$ such that $\psi \circ j = \varphi$. In other words, the following diagram is commutative*

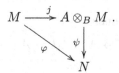

PROOF. We define ψ by the formula

$$\psi(a \otimes m) := a\varphi(m),$$

for all $a \in A$ and $m \in M$. The reader can check that ψ is well defined, i.e. that the relations defining $A \otimes_B M$ are preserved by ψ. That proves the existence of ψ.

To check the uniqueness, we just note that for all $a \in A$ and $m \in M$, ψ must satisfy the relation

$$\psi(a \otimes m) = a\psi(1 \otimes m) = a\varphi(m).$$

□

To prove the theorem we observe that by the above lemma the map $\psi \mapsto \varphi := \psi \circ j$ gives an isomorphism between $\operatorname{Hom}_A (A \otimes_B M, N)$ and $\operatorname{Hom}_B (M, N)$. □

REMARK 5.5. For the readers who are familiar with category theory, the former theorem can be reformulated as follows: since any A-module M is automatically a B-module, we have a natural functor Res from the category of A-modules to the category of B-modules. This functor is usually called the *restriction* functor. The *induction* functor Ind from the category of B-modules to the category of A-modules which sends M to $A \otimes_B M$ is the left-adjoint of Res.

EXAMPLE 5.6. Let $k \subset F$ be a field extension. For any vector space M over k, $F \otimes_k M$ is a vector space of the same dimension over F. If we have an exact sequence of vector spaces

$$0 \to N \to M \to L \to 0,$$

then the sequence

$$0 \to F \otimes_k N \to F \otimes_k M \to F \otimes_k L \to 0$$

is also exact. In other words the induction in this situation is an exact functor.

EXERCISE 5.7. Let A be a ring and B be a subring of A.
(a) Show that if a sequence of B-modules

$$N \to M \to L \to 0$$

is exact, then the sequence

$$A \otimes_B N \to A \otimes_B M \to A \otimes_B L \to 0$$

of induced modules is also exact. In other words the induction functor is right-exact.

(b) Assume that A is a free B-module, then the induction functor is exact. In other words, if a sequence

$$0 \to N \to M \to L \to 0$$

of B-modules is exact, then the sequence

$$0 \to A \otimes_B N \to A \otimes_B M \to A \otimes_B L \to 0$$

is also exact.

(c) Let $A = \mathbb{Z}[X]/(X^2, 2X)$ and $B = \mathbb{Z}$. Consider the exact sequence

$$0 \to \mathbb{Z} \xrightarrow{\varphi} \mathbb{Z} \to \mathbb{Z}/2\mathbb{Z} \to 0,$$

where φ is the multiplication by 2. Check that after applying the induction functor, we obtain a sequence of A-modules

$$0 \to A \to A \to A/2A \to 0,$$

which is not exact.

Later, we will discuss general properties of induction, but now we are going to study induction in the case of groups.

6. Induced representations for groups

Let G be a finite group. Let H be a subgroup of G and $\rho : H \to \mathrm{GL}\,(V)$ be a representation of H with character χ. Then the induced representation $\mathrm{Ind}_H^G \rho$ is by definition the $k\,(G)$-module

$$k\,(G) \otimes_{k(H)} V.$$

The following lemma has a straightforward proof.

LEMMA 6.1. *The dimension of* $\mathrm{Ind}_H^G \rho$ *equals the product of* $\dim \rho$ *and the index* $[G : H]$ *of* H. *More precisely, let* S *be a set of representatives of left cosets in* G/H, *i.e.*

$$G = \coprod_{s \in S} sH,$$

then

(2.6)
$$k\left(G\right) \otimes_{k(H)} V = \bigoplus_{s \in S} s \otimes V.$$

Moreover, for any $g \in G$, $s \in S$, there exists a unique $s' \in S$ such that $(s')^{-1}gs \in H$. Then the action of g on $s \otimes v$ for all $v \in V$ is given by

(2.7)
$$g\left(s \otimes v\right) = s' \otimes \rho_{(s')^{-1}gs}v.$$

EXAMPLE 6.2. Let ρ be the trivial representation of H. Then $\mathrm{Ind}_H^G \rho$ is the permutation representation of G obtained from the natural left action of G on the set of left cosets G/H; see Definition 1.2 Chapter 1.

LEMMA 6.3. We keep the notations of the previous lemma. Denote by $\mathrm{Ind}_H^G \chi$ the character of the induced representation. Then, for $g \in G$

(2.8)
$$\mathrm{Ind}_H^G \chi\left(g\right) = \sum_{s \in S, s^{-1}gs \in H} \chi\left(s^{-1}gs\right).$$

PROOF. (2.6) and (2.7) imply

$$\mathrm{Ind}_H^G \chi\left(g\right) = \sum_{s \in S} \delta_{s,s'} \, \mathrm{Tr}\, \rho_{(s')^{-1}gs}.$$

$\square$

COROLLARY 6.4. In the notations of Lemma 6.3 we have

$$\mathrm{Ind}_H^G \chi\left(g\right) = \frac{1}{|H|} \sum_{u \in G, u^{-1}gu \in H} \chi\left(u^{-1}gu\right).$$

PROOF. If $s^{-1}gs \in H$, then for all $u \in sH$ we have $\chi(u^{-1}gu) = \chi(s^{-1}gs)$. Therefore

$$\chi(s^{-1}gs) = \frac{1}{|H|} \sum_{u \in sH} \chi\left(u^{-1}gu\right).$$

Hence the statement follows from (2.8).

$\square$

COROLLARY 6.5. Let H be a normal subgroup in G. Then $\mathrm{Ind}_H^G \chi\left(g\right) = 0$ for any $g \notin H$.

EXERCISE 6.6. (a) Let $G = S_3$ and $H = A_3$ be its normal cyclic subgroup. Consider a one-dimensional representation of H such that $\rho(123) = \varepsilon$, where ε is a primitive 3-rd root of 1. Show that then

$$\mathrm{Ind}_H^G \chi_\rho\left(1\right) = 2,$$

$$\mathrm{Ind}_H^G \chi_\rho\left(12\right) = 0,$$

$$\mathrm{Ind}_H^G \chi_\rho\left(123\right) = -1.$$

Therefore $\mathrm{Ind}_H^G \rho$ is the irreducible two-dimensional representation of S_3.

(b) Next, consider the 2-element subgroup K of $G = S_3$ generated by the transposition (12), and let σ be the (unique) non-trivial one-dimensional representation of K. Show that

$$\operatorname{Ind}_K^G \chi_\sigma (1) = 3,$$

$$\operatorname{Ind}_K^G \chi_\sigma (12) = -1,$$

$$\operatorname{Ind}_K^G \chi_\rho (123) = 0.$$

Therefore $\operatorname{Ind}_K^G \sigma$ is the direct sum of the sign representation and the 2-dimensional irreducible representation.

Now we assume that k has characteristic zero. Let us recall that, in Section 4.2 Chapter 1, we defined a scalar product on the space $\mathcal{C}(G)$ of class functions by (1.2). When we consider several groups at the same time we specify the group by the a lower index.

THEOREM 6.7. *Consider two representations* $\rho : G \to \operatorname{GL}(V)$ *and* $\sigma \colon H \to \operatorname{GL}(W)$. *Then we have the identity*

(2.9) $$\left(\operatorname{Ind}_H^G \chi_\sigma, \chi_\rho \right)_G = \left(\chi_\sigma, \operatorname{Res}_H \chi_\rho \right)_H .$$

PROOF. The statement follows from Frobenius reciprocity (Theorem 5.3) and Corollary 4.7 in Chapter 1, since

$$\dim \operatorname{Hom}_G \left(\operatorname{Ind}_H^G W, V \right) = \dim \operatorname{Hom}_H (W, V) .$$

$\square$

EXERCISE 6.8. Prove Theorem 6.7 directly from Corollary 6.4: define two maps

$$\operatorname{Res}_H : \mathcal{C}(G) \to \mathcal{C}(H) , \operatorname{Ind}_H^G : \mathcal{C}(H) \to \mathcal{C}(G) ,$$

the former being the restriction on a subgroup, the latter being defined by (2.8). Show that for any $\varphi \in \mathcal{C}(G) , \psi \in \mathcal{C}(H)$

$$\left(\operatorname{Ind}_H^G \varphi, \psi \right)_G = (\varphi, \operatorname{Res}_H \psi)_H .$$

7. Double cosets and restriction to a subgroup

If K and H are subgroups of G one can define the equivalence relation on G: $s \sim t$ if and only if $s \in KtH$. The equivalence classes are called *double cosets*. We can choose a set of representative $T \subset G$ such that

$$G = \coprod_{t \in T} KtH.$$

We denote the set of double cosets by $K \backslash G / H$. One can identify $K \backslash G / H$ with K-orbits on $S = G/H$ in the obvious way, and with G-orbits on $G/K \times G/H$ by the formula

$$KtH \mapsto G(K, tH) .$$

EXAMPLE 7.1. Let $\mathbb{F}$ be a field. Let $G = \mathrm{GL}_2(\mathbb{F})$ be the group of all invertible 2×2 matrices with coefficients in $\mathbb{F}$. Consider the natural action of G on $\mathbb{F}^2$. Consider the subgroup B of upper triangular matrices in G. We denote by $\mathbb{P}^1$ the projective line which is the set of all one-dimensional linear subspaces of $\mathbb{F}^2$. Clearly, G acts on $\mathbb{P}^1$.

EXERCISE 7.2. Prove that G acts transitively on $\mathbb{P}^1$ and that the stabilizer of any point in $\mathbb{P}^1$ is isomorphic to B.

By the above exercise one can identify G/B with the set of lines $\mathbb{P}^1$. The set of double cosets $B \backslash G / B$ can be identified with the set of G-orbits in $\mathbb{P}^1 \times \mathbb{P}^1$ or with the set of B-orbits in $\mathbb{P}^1$.

EXERCISE 7.3. Check that G has only two orbits on $\mathbb{P}^1 \times \mathbb{P}^1$: the diagonal and its complement. Thus, $|B \backslash G / B| = 2$ and

$$G = B \cup BsB,$$

where

$$s = \begin{pmatrix} 0 & 1 \\ 1 & 0 \end{pmatrix}.$$

THEOREM 7.4. Let $T \subset G$ such that $G = \coprod_{s \in T} KsH$. Then

$$\mathrm{Res}_K \mathrm{Ind}_H^G \rho = \oplus_{s \in T} \mathrm{Ind}_{K \cap sHs^{-1}}^K \rho^s,$$

where

$$\rho_h^s \overset{def}{=} \rho_{s^{-1}hs},$$

for any $h \in sHs^{-1}$.

PROOF. Let $s \in T$ and $W^s = k(K)(s \otimes V)$. Then by construction, W^s is K-invariant and

$$k(G) \otimes_{k(H)} V = \oplus_{s \in T} W^s.$$

Thus, we need to check that the representation of K in W^s is isomorphic to $\mathrm{Ind}_{K \cap sHs^{-1}}^K \rho^s$. We define a homomorphism

$$\alpha : \mathrm{Ind}_{K \cap sHs^{-1}}^K V \to W^s$$

by $\alpha(t \otimes v) = ts \otimes v$ for any $t \in K, v \in V$. It is well defined

$$\alpha(th \otimes v - t \otimes \rho_h^s v) = ths \otimes v - ts \otimes \rho_{s^{-1}hs}v = ts(s^{-1}hs) \otimes v - ts \otimes \rho_{s^{-1}hs}v = 0$$

and obviously surjective. Injectivity can be proven by counting the dimensions. □

EXAMPLE 7.5. Let us go back to our example $B \subset \mathrm{SL}_2(\mathbb{F})$ (see Exercise 7.3). We now assume that $\mathbb{F} = \mathbb{F}_q$ is the finite field with q elements. Theorem 7.4 tells us that for any representation ρ of B

$$\mathrm{Res}_B \mathrm{Ind}_B^G \rho = \rho \oplus \mathrm{Ind}_H^B \rho',$$

where $H = B \cap sBs^{-1}$ is the subgroup of diagonal matrices and

$$\rho' \begin{pmatrix} a & 0 \\ 0 & b \end{pmatrix} = \rho \begin{pmatrix} b & 0 \\ 0 & a \end{pmatrix}.$$

COROLLARY 7.6. *If H is a normal subgroup of G, then*

$$\operatorname{Res}_H \operatorname{Ind}_H^G \rho = \oplus_{s \in G/H} \rho^s.$$

8. Mackey's criterion

In order to compute $\left(\operatorname{Ind}_H^G \chi, \operatorname{Ind}_H^G \chi \right)$, we use Frobenius reciprocity and Theorem 7.4. The following equation holds:

$$\left(\operatorname{Ind}_H^G \chi, \operatorname{Ind}_H^G \chi \right)_G = \left(\operatorname{Res}_H \operatorname{Ind}_H^G \chi, \chi \right)_H = \sum_{s \in T} \left(\operatorname{Ind}_{H \cap sHs^{-1}}^H \chi^s, \chi \right)_H =$$

$$= \sum_{s \in T} \left(\chi^s, \operatorname{Res}_{H \cap sHs^{-1}} \chi \right)_{H \cap sHs^{-1}} = (\chi, \chi)_H + \sum_{s \in T \setminus \{1\}} \left(\chi^s, \operatorname{Res}_{H \cap sHs^{-1}} \chi \right)_{H \cap sHs^{-1}}.$$

We call two representation *disjoint* if they do not have any irreducible component in common, or in other words if their characters are orthogonal.

THEOREM 8.1. *(Mackey's criterion) The representation $\operatorname{Ind}_H^G \rho$ is irreducible if and only if ρ is irreducible and ρ^s and ρ are disjoint representations of $H \cap sHs^{-1}$ for all $s \in T \setminus \{1\}$ (see Theorem 7.4 for the definition of T).*

PROOF. Write the condition

$$\left(\operatorname{Ind}_H^G \chi, \operatorname{Ind}_H^G \chi \right)_G = 1$$

and use the above formula. □

COROLLARY 8.2. *Let H be a normal subgroup of G and ρ be an irreducible representation of H. Then $\operatorname{Ind}_H^G \rho$ is irreducible if and only if ρ^s is not isomorphic to ρ for any $s \in G/H$, $s \notin H$.*

REMARK 8.3. Note that if H is normal, then G/H acts on the set of representations of H. In fact, this is a part of the action of the group $\operatorname{Aut} H$ of automorphisms of H on the set of representation of H. Indeed, if $\varphi \in \operatorname{Aut} H$ and $\rho : H \to \operatorname{GL}(V)$ is a representation, then $\rho^\varphi : H \to \operatorname{GL}(V)$ defined by

$$\rho_t^\varphi = \rho_{\varphi(t)},$$

is a new representation of H.

9. Hecke algebras, a first glimpse

DEFINITION 9.1. Let G be a group, $H \subset G$ a subgroup, consider $\mathcal{H}(G, H) \subset k(G)$ defined by:

$$\mathcal{H}(G, H) := \operatorname{End}_G(\operatorname{Ind}_H^G \operatorname{triv}).$$

This is the *Hecke algebra* associated to the pair (G, H).

Define the projector

$$\Pi_H := \frac{1}{|H|} \sum_{h \in H} h \in k(G).$$

EXERCISE 9.2. Show that

$$\operatorname{Ind}_H^G \operatorname{triv} = k(G) \Pi_H.$$

Applying Frobenius reciprocity, one gets:

$$\operatorname{End}_G \operatorname{Ind}_H^G \operatorname{triv} = \operatorname{Hom}_H(\operatorname{triv}, \operatorname{Ind}_H^G \operatorname{triv}).$$

We can identify the Hecke algebra with $\Pi_H k(G) \Pi_H$. Therefore a basis of the Hecke algebra can be enumerated by the double cosets, i.e. elements of $H \backslash G / H$.

Set, for $g \in G$,

$$\eta_g := \Pi_H g \Pi_H.$$

it is clear that those functions are constant on double cosets. If we fix a set S of representatives of double cosets $S = H \backslash G / H$, then $\{\eta_s\}_{s \in S}$ is a basis of the Hecke algebra. Then, the multiplication is given by the formula

$$(2.10) \qquad \eta_g \eta_{g'} = \sum_{g'' \in S} \frac{1}{|H|} |gHg' \cap Hg''H| \eta_{g''}.$$

EXERCISE 9.3. Consider the pair $G = GL_2(\mathbb{F}_q)$, $H = B$ the subgroup of upper triangular matrices. Then by Exercise 7.3 we know that the Hecke algebra $\mathcal{H}(G, B)$ is two-dimensional. The identity element η_e corresponds to the double coset B. The second element of the basis is η_s. Let us compute η_s^2 using (2.10). We have

$$\eta_s^2 = a\eta_e + b\eta_s,$$

where

$$a = \frac{|sBs \cap B|}{|B|}, \qquad b = \frac{sBs \cap BsB}{|B|}.$$

Since sBs is the subgroup of the lower triangular matrices in G, the intersection subgroup $sBs \cap B$ is the subgroup of diagonal matrices. Therefore we have

$$|B| = (q - 1)^2 q, \quad |sBs \cap B| = (q - 1)^2, \quad a = \frac{1}{q}, \quad b = 1 - a = \frac{q - 1}{q}.$$

DEFINITION 9.4. We say that a G-module V is *multiplicity free* if any simple G-module appears in V with multiplicity either 0 or 1.

PROPOSITION 9.5. *Assume that k is algebraically closed. The following conditions on the pair $H \subset G$ are equivalent*

(1) *The G-module Ind_H^G triv is multiplicity free;*
(2) *For any simple G-module M the dimension of subspace M^H of H-invariants is at most one;*
(3) *The Hecke algebra $\mathcal{H}(G, H)$ is commutative.*

PROOF. (1) is equivalent to (2) by Frobenius reciprocity. Equivalence of (1) and (3) follows from Lemma 1.11. $\qquad\square$

LEMMA 9.6. *Let G be a finite group and $H \subset G$ be a subgroup. Let $\varphi : G \to G$ be antiautomorphism of G such that for any $g \in G$ we have $\varphi(g) \in HgH$. Then $\mathcal{H}(G, H)$ is commutative.*

PROOF. Extend φ to the whole group algebra $k(G)$ by linearity. Then φ is an antiautomorphism of $k(G)$ and for all $g \in G$ we have $\varphi(\eta_g) = \eta_g$. Therefore for any $g, h \in H \backslash G / H$ we have

$$\eta_g \eta_h = \sum c_{g,h}^u \eta_u = \sum_{u \in H \backslash G / H} c_{g,h}^u \varphi(\eta_u) = \varphi(\eta_g \eta_h) = \varphi(\eta_h) \varphi(\eta_g) = \eta_h \eta_g.$$

$\qquad\square$

EXERCISE 9.7. Let G be the symmetric group S_n and $H = S_p \times S_{n-p}$. Prove that $\mathcal{H}(G, H)$ is abelian. Hint: consider $\varphi(g) = g^{-1}$ and apply Lemma 9.6.

10. Some examples

Let H be a subgroup of G of index 2. Then H is normal and $G = H \cup sH$ for some $s \in G \backslash H$. Suppose that ρ is an irreducible representation of H. There are two possibilities

(1) ρ^s is isomorphic to ρ;
(2) ρ^s is not isomorphic to ρ.

Hence there are two possibilities for $\mathrm{Ind}_H^G \rho$:

(1) $\mathrm{Ind}_H^G \rho = \sigma \oplus \sigma'$, where σ and σ' are two non-isomorphic irreducible representations of G;
(2) $\mathrm{Ind}_H^G \rho$ is irreducible.

For instance, let $G = S_5$, $H = A_5$ and $\rho_1, \ldots, \rho_5$ be the irreducible representations of H introduced in Example 5.3 in Chapter 1. Then for $i = 1, 2, 3$

$$\mathrm{Ind}_H^G \rho_i = \sigma_i \oplus (\sigma_i \otimes \mathrm{sgn}),$$

where sgn denotes the sign representation. Furthermore, the induced modules $\mathrm{Ind}_H^G \rho_4$ and $\mathrm{Ind}_H^G \rho_5$ are isomorphic and irreducible. Thus in dimensions 1, 4 and 5, S_5 has two non-isomorphic irreducible representations and only one in dimension 6.

Now let G be the subgroup of $\mathrm{GL}_2\left(\mathbb{F}_q\right)$ consisting of matrices of shape

$$\left(\begin{array}{cc} a & b \\ 0 & 1 \end{array}\right),$$

where $a \in \mathbb{F}_q^*$ and $b \in \mathbb{F}_q$. Let us classify complex irreducible representations of G. The order of G is equal to $q^2 - q$. Furthermore G has q conjugacy classes with the following representatives

$$\left(\begin{array}{cc} 1 & 0 \\ 0 & 1 \end{array}\right), \left(\begin{array}{cc} 1 & 1 \\ 0 & 1 \end{array}\right), \left(\begin{array}{cc} a & 0 \\ 0 & 1 \end{array}\right),$$

(in the last case $a \neq 1$). Note that

$$H = \left\{\left(\begin{array}{cc} 1 & b \\ 0 & 1 \end{array}\right), b \in \mathbb{F}_q\right\}$$

is a normal subgroup of G and the quotient G/H is isomorphic to $\mathbb{F}_q^*$ which is cyclic of order $q - 1$.

Therefore G has $q - 1$ one-dimensional representations which can be lifted from G/H. That leaves one more representation, its dimension must be $q - 1$. Let us try to obtain it using induction from H. Let σ be a non-trivial irreducible representation of H, its dimension is automatically 1. Then the dimension of the induced representation $\mathrm{Ind}_H^G \sigma$ is equal to $q - 1$ as required. We claim that it is irreducible. Indeed, if ρ is a one-dimensional representation of G, then by the Frobenius reciprocity, Theorem 6.7, we have

$$\left(\mathrm{Ind}_H^G \sigma, \rho\right)_G = (\sigma, \mathrm{Res}_H \rho)_H = 0,$$

since $\mathrm{Res}_H \rho$ is trivial. Therefore $\mathrm{Ind}_H^G \sigma$ is irreducible.

EXERCISE 10.1. Compute the character of this representation.

11. Some general facts about field extension

LEMMA 11.1. *If char $k = 0$ and G is finite, then a representation $\rho : G \to \mathrm{GL}\left(V\right)$ is irreducible if and only if $\mathrm{End}_G\left(V\right)$ is a division ring.*

PROOF. In one direction it is Schur's Lemma. In the opposite direction if V is not irreducible, then $V = V_1 \oplus V_2$ and the projectors p_1 and p_2 are intertwiners such that $p_1 \circ p_2 = 0$. □

For any extension F of k and any representation $\rho : G \to \mathrm{GL}\left(V\right)$ over k we denote by ρ_F the representation $G \to \mathrm{GL}\left(F \otimes_k V\right)$.

For any representation $\rho : G \to \mathrm{GL}\,(V)$ we denote by V^G the subspace of G-invariants in V, i.e.

$$V^G = \{v \in V \mid \rho_s v = v, \forall s \in G\}.$$

LEMMA 11.2. *One has* $(F \otimes_k V)^G = F \otimes_k V^G$.

PROOF. The embedding $F \otimes_k V^G \subset (F \otimes_k V)^G$ is trivial. On the other hand, V^G is the image of the operator

$$p = \frac{1}{|G|} \sum_{s \in G} \rho_s,$$

in particular $\dim V^G$ equals the rank of p. Since rank p does not depend on the base field, we have

$$\dim F \otimes_k V^G = \dim (F \otimes_k V)^G.$$

$\square$

COROLLARY 11.3. *Let* $\rho : G \to \mathrm{GL}\,(V)$ *and* $\sigma : G \to \mathrm{GL}\,(W)$ *be two representations over* k. *Then*

$$\mathrm{Hom}_G\,(F \otimes_k V, F \otimes_k W) = F \otimes \mathrm{Hom}_G\,(V, W).$$

In particular,

$$\dim_k \mathrm{Hom}_G\,(V, W) = \dim_F \mathrm{Hom}_G\,(F \otimes_k V, F \otimes_k W).$$

PROOF.

$$\mathrm{Hom}_G\,(V, W) = (V^* \otimes W)^G.$$

$\square$

A representation $\rho : G \to \mathrm{GL}\,(V)$ over k is called *absolutely irreducible* if it remains irreducible after any extension of k. This property is equivalent to the equality $(\chi_\rho, \chi_\rho) = 1$.

A field K is called *splitting* for a group G if every irreducible representation of G over K is absolutely irreducible. It is not difficult to see that for a finite group G, there exists a finite extension of $\mathbb{Q}$ which is a splitting field for G.

12. Artin's theorem and representations over ℚ

Let G be a finite group and ρ be a representation of G over ℚ. Then the character χ_ρ is called a *rational character*. Note that $\chi_\rho(g) \in \mathbb{Z}$ since $\chi_\rho(g)$ is algebraic integer and rational. For a cyclic subgroup $H \subset G$ and a one-dimensional character $\theta : H \to \mathbb{C}^*$ we denote by F_H^θ the the character of the induced representation $\mathrm{Ind}_H^G \theta$. For an element $g \in G$ we denote by $\langle g \rangle$ the cyclic subgroup generated by g.

LEMMA 12.1. *If χ is a rational character and $\langle g \rangle = \langle h \rangle$, then $\chi(g) = \chi(h)$.*

PROOF. Let $\chi = \chi_\rho$. Let $\varepsilon_1, \ldots, \varepsilon_n$ be the eigenvalues of ρ_g counted with multiplicities. Then $\chi(g) = \varepsilon_1 + \cdots + \varepsilon_n$. Assume that the order of $\langle g \rangle$ is equal to m. Then $\varepsilon_1, \ldots, \varepsilon_n$ are m-th roots of 1 and therefore they are elements of the cyclotomic extension $\mathbb{Q}(\zeta)$, where ζ is a primitive m-th root of 1. Furthermore, $h = g^p$, where p is relatively prime to m, and $\chi(h) = \varepsilon_1^p + \cdots + \varepsilon_n^p$. Let σ be the element of the Galois group of $\mathbb{Q}(\zeta)$ such that $\sigma(\zeta) = \zeta^p$. Then $\chi(h) = \sigma(\chi(g))$. Since $\chi(g) \in \mathbb{Z}$, we obtain $\chi(g) = \chi(h)$. □

THEOREM 12.2. *(Artin) Let χ be a rational character of G. There exist algebraic integers $a_1, \ldots, a_m$, cyclic subgroups $H_1, \ldots, H_m$ and 1-dimensional characters $\theta_i : H_i \to \mathbb{C}^*$ such that*

$$\chi = \sum_{i=1}^m \frac{a_i}{|G|} F_{H_i}^{\theta_i}.$$

PROOF. Let H be a cyclic subgroup of G. Define a function $T_H : H \to \mathbb{Z}$ by the formula

$$T_H(x) = \begin{cases} |H| & \text{if } \langle x \rangle = H \\ 0 & \text{otherwise} \end{cases}.$$

EXERCISE 12.3. Let $N(H)$ denote the normalizer of H in G. Show that

$$\mathrm{Ind}_H^G T_H(x) = \begin{cases} |N(H)| & \text{if } \langle x \rangle \text{ is conjugate to } H \\ 0 & \text{otherwise} \end{cases}.$$

LEMMA 12.4. *For a cyclic group H, the function T_H is a linear combination of induced characters with algebraic integral coefficients.*

PROOF. We prove the statement by induction on the order of H. Assume that the statement is true for all proper subgroups of H. It follows easily from Exercise 12.3 that, for any $x \in H$, we have

$$\sum_{P \subset H} \mathrm{Ind}_P^H T_P(x) = |H|$$

where the summation is taken over all subgroups P of H. Hence we have

$$T_H(x) = |H| - \sum_{P \neq H} \mathrm{Ind}_P^H T_P(x).$$

By the induction hypothesis, $\operatorname{Ind}_P^H T_P(x)$ is a linear combination of induced characters for all proper subgroups P of H. Hence the same is true for T_H. $\quad\square$

Let us choose a set $H_1, \ldots, H_m$ of representatives of conjugacy classes of all cyclic subgroups of G and let $H_i = \langle g_i \rangle$. Define $\tau : G \to \mathbb{Z}$ by

$$\tau := |G| \sum_{j=1}^{m} \frac{\chi(g_i)}{|N(H_i)|} \operatorname{Ind}_{H_i}^G T_{H_i}.$$

For any element $x \in G$, there exists exactly one index i such that $T_{H_i}(x) = |N(H_i)|$. For all other j, we have $T_{H_j}(x) = 0$. This implies $\tau(x) = |G| \chi(g_i)$. By Lemma 12.1, $\chi(x) = \chi(g_i)$. Hence $\chi = \frac{\tau}{|G|}$ and the proof of Theorem is complete by Lemma 12.4. $\quad\square$

PROPOSITION 12.5. *Let G be a finite group. The number of irreducible representations of G over $\mathbb{Q}$ is equal to the number of conjugacy classes of all cyclic subgroups.*

PROOF. Let m be the number of conjugacy classes of cyclic subgroups and q be the number of irreducible rational representations. By Lemma 12.1, $q \leq m$. On the other hand, let $H_1, \ldots, H_m$ be all cyclic subgroups of G, up to conjugacy. Then by Exercise 12.3 and Lemma 12.4, $\{\operatorname{Ind}_{H_i}^G T_{H_i}\}_{i=1,\ldots,m}$ is a linearly independent set in the space of rational characters. Hence $m \leq q$. $\quad\square$

CHAPTER 3

Representations of compact groups

In which we move out of the province of finite groups to explore the larger country of compact groups. We are relieved to see that a lot of methods remain efficient. Not to mention unitary representations.

1. Compact groups

Let G be a group which is also a topological space. We say that G is a *topological group* if both the multiplication from $G \times G$ to G and the inverse from G to G are continuous maps. Naturally, we say that G is compact (respectively, locally compact) if it is a compact (resp., locally compact) topological space.

Examples.

- The circle
$$S^1 = \{z \in \mathbb{C} \mid |z| = 1\}.$$

- The torus $T^n = S^1 \times \cdots \times S^1$.
  Note that in general, the direct product of two compact groups is compact.
- The unitary group
$$U_n = \{X \in \mathrm{GL}_n(\mathbb{C}) \mid \bar{X}^t X = 1_n\}.$$

  To see that U_n is compact, note that a matrix $X = (x_{ij}) \in U_n$ satisfies the equations $\sum_{j=1}^n |x_{ij}|^2 = 1$ for $i = 1, \ldots, n$. Hence U_n is a closed subset of the product of n spheres of dimension $(2n - 1)$.
- The special unitary group
$$SU_n = \{X \in U_n \mid \det X = 1\}.$$

- The orthogonal group
$$O_n = \{X \in \mathrm{GL}_n(\mathbb{R}) \mid X^t X = 1_n\}.$$

- The special orthogonal group
$$SO_n = \{X \in O_n \mid \det X = 1\}.$$

© Springer Nature Switzerland AG 2018
C. Gruson and V. Serganova, *A Journey Through Representation Theory*,
Universitext, https://doi.org/10.1007/978-3-319-98271-7_3

1.1. Haar measure. A measure dg on a locally compact group G is called right-invariant if, for every integrable function F on G and every h in G:

$$\int_G F(gh)\, dg = \int_G F(g)\, dg.$$

Similarly, a measure $d'g$ on G is called left-invariant if for every integrable function F on G and every h in G:

$$\int_G F(hg)\, d'g = \int_G F(g)\, d'g.$$

THEOREM 1.1. *Let G be compact group. There exists a unique right-invariant measure dg on G such that*

$$\int_G dg = 1.$$

In the same way there exists a unique left-invariant measure $d'g$ such that

$$\int_G d'g = 1.$$

Moreover, $dg = d'g$.

DEFINITION 1.2. The measure dg is called the *Haar measure* on G.

We do not give the proof of this theorem in its general form. In the sketch of the proof below (just after Exercise 1.3), we assume general knowledge of submanifolds and of the notion of vector bundle. All the examples we consider here are smooth submanifolds in $\mathrm{GL}_k(\mathbb{R})$ or $\mathrm{GL}_k(\mathbb{C})$.

EXERCISE 1.3. Assume that G is a subgroup of $\mathrm{GL}_k(\mathbb{R})$ or $\mathrm{GL}_k(\mathbb{C})$, and that G is a closed submanifold, i.e. locally it is given by equations $f_1 = \cdots = f_m = 0$ for some smooth functions $f_1, \ldots, f_m$, such that the rank r of the Jacobian matrix is locally constant. Then, G is a smooth submanifold in GL_k of codimension r, i.e. at every point of $g \in G$ one can find functions $h_1, \ldots, h_r$ such that G is defined by $h_1 = \cdots = h_r = 0$ in a small neighbourhood of g and $dh_1(g), \ldots, dh_r(g)$ are linearly independent. Consider the map $m_g : G \to G$ given by left (right) multiplication by $g \in G$. Then its differential $(m_g)_* : T_e G \to T_g G$ is an isomorphism between tangent spaces at e and g.

To define the invariant measure, we just need to define a volume form on the tangent space at identity, $T_e G$, and then use the right (left) multiplication to define it on the whole group. More precisely, let $\gamma \in \Lambda^{\mathrm{top}} T_e^* G$. Then the map

$$g \mapsto \gamma_g := m_{g^{-1}}^*(\gamma),$$

where $m_{g^{-1}}^*$ is the induced differential map $\Lambda^{\mathrm{top}} T_e^* G \to \Lambda^{\mathrm{top}} T_g^* G$, is a section of the bundle $\Lambda^{\mathrm{top}} T^* G$. This section is a right (left) invariant differential form of maximal degree on the group G, i.e. an invariant volume form. If G is compact one can normalize γ to satisfy $\int_G \gamma = 1$.

REMARK 1.4. If G is locally compact but not compact, there are still left-invariant and right-invariant measures on G, each is unique up to scalar multiplication, but the left-invariant ones are not necessarily proportional to the right-invariant ones. We speak of a left-Haar measure or a right-Haar measure.

1.2. Continuous representations. Consider a vector space V over $\mathbb{C}$ equipped with a topology such that addition and multiplication by a scalar are continuous. We always assume that a topological vector space satisfies the following conditions

(1) for any $v \in V \setminus 0$ there exists a neighbourhood of 0 which does not contain v;

(2) there is a basis of convex neighbourhoods of zero.

Topological vector spaces satisfying the above conditions are called *locally convex*. We do not intend to explore the theory of such spaces. All we need to know is the fact that there exists a non-zero continuous linear functional on a locally convex space.

DEFINITION 1.5. A representation $\rho : G \to \mathrm{GL}(V)$ is called *continuous* if the map $G \times V \to V$ given by $(g, v) \mapsto \rho_g v$ is continuous. Two continuous representations are *equivalent* or *isomorphic* if there exists a bicontinuous invertible intertwining operator between them. In this chapter we consider only continuous representations.

A representation $\rho : G \to \mathrm{GL}(V)$, $V \neq \{0\}$ is called *topologically irreducible* if the only G-invariant closed subspaces of V are V and 0.

1.3. Unitary representations. Recall that a Hilbert space is a vector space over $\mathbb{C}$ equipped with a positive definite Hermitian form $\langle , \rangle$, which is complete with respect to the topology defined by the norm

$$\|v\| = \langle v, v \rangle^{1/2}.$$

We will use the following facts about Hilbert spaces:

(1) A Hilbert space V has an orthonormal topological basis, i.e. an orthonormal system of vectors $\{e_i\}_{i \in I}$ such that $\bigoplus_{i \in I} \mathbb{C}e_i$ is dense in V. Two Hilbert spaces are isomorphic if and only if their topological orthonormal bases have the same cardinality.

(2) If V^* denotes the space of all continuous linear functionals on V, then we have an isomorphism $V^* \simeq V$ given by $v \mapsto \langle v, \cdot \rangle$.

DEFINITION 1.6. A continuous representation $\rho : G \to \mathrm{GL}(V)$ is called *unitary* if V is a Hilbert space and

$$\langle v, w \rangle = \langle \rho_g v, \rho_g w \rangle$$

for any $v, w \in V$ and $g \in G$. If $U(V)$ denotes the group of all unitary operators in V, then ρ defines a homomorphism $G \to U(V)$.

The following is an important example of unitary representation.

Regular representation. Let G be a compact group and $L^2(G)$ be the space of all complex-valued functions φ on G such that

$$\int |\varphi(g)|^2 dg$$

exists. Then $L^2(G)$ is a Hilbert space with respect to the Hermitian form

$$\langle \varphi, \psi \rangle = \int_G \bar{\varphi}(g)\,\psi(g)\,dg.$$

Moreover, the representation R of G in $L^2(G)$ given by

$$R_g \varphi(h) = \varphi(hg)$$

is continuous and the Hermitian form is G-invariant. This representation is called the *regular representation of G*.

1.4. Linear operators in a Hilbert space. We will recall certain facts about linear operators in a Hilbert space. We only sketch the proofs hiding technical details in exercises. The enthusiastic reader is encouraged to supply those details and the less enthusiastic reader can find those details in textbooks on the subject, for instance in [18].

DEFINITION 1.7. A linear operator T in a Hilbert space is called *bounded* if there exists $C > 0$ such that for any $v \in V$ we have $\|Tv\| \leq C\|v\|$.

EXERCISE 1.8. Let $\mathcal{B}(V)$ denote the set of all bounded operators in a Hilbert space V.
(a) Check that $\mathcal{B}(V)$ is an algebra over $\mathbb{C}$ with multiplication given by composition.
(b) Show that $T \in \mathcal{B}(V)$ if and only if the map $T : V \to V$ is continuous.
(c) Introduce the norm on $\mathcal{B}(V)$ by setting

$$\|T\| = \sup_{\|v\|=1} \|Tv\|.$$

Check that $\|T_1 T_2\| \leq \|T_1\|\|T_2\|$ and $\|T_1 + T_2\| \leq \|T_1\| + \|T_2\|$ for all $T_1, T_2 \in \mathcal{B}(V)$ and that $\mathcal{B}(V)$ is complete in the topology defined by this norm. Thus, $\mathcal{B}(V)$ is a Banach algebra.

THEOREM 1.9. *Let $T \in \mathcal{B}(V)$ be invertible. Then T^{-1} is also bounded.*

PROOF. Consider the unit ball

$$B := \{x \in V \mid \|x\| < 1\}.$$

For any $k \in \mathbb{N}$ denote by S_k the closure of $T(kB) = kT(B)$ and let $U_k = V \setminus S_k$. Note that

$$V = \bigcup_{k \in \mathbb{N}} kB.$$

Since T is invertible, it is surjective, and therefore

$$\bigcup_{k\in\mathbb{N}} S_k = V.$$

We claim that there exists k such that U_k is not dense. Indeed, otherwise there exists a sequence of embedded closed balls $B_k \subset U_k$, $B_{k+1} \subset B_k$, which has a common point by completeness of V. This contradicts to the fact that the intersection of all U_k is empty. Then S_k contains a ball $x + \varepsilon B$ for some $x \in V$ and $\varepsilon > 0$. It is not hard to see that for any $r > \frac{2}{\varepsilon}$ contains B.

Now we will prove the inclusion $B \subset T(2rB)$ for r as above. Indeed, let $y \in B \subset S_r$. There exists $x_1 \in rB$ such that $\|y - Tx_1\| < \frac{1}{2}$. Note that $y - Tx_1 \in \frac{1}{2}B \subset \frac{1}{2}S_r$. Then one can find $x_2 \in \frac{r}{2}B$ such that $\|y - Tx_1 - Tx_2\| < \frac{1}{4}$. Proceeding in this way we can construct a sequence $\{x_n \in \frac{1}{2^{n-1}}B\}$ such that $\|y - T(x_1 + \cdots + x_n)\| < \frac{1}{2^n}$. Consider $w = \sum_{i=1}^{\infty} x_i$, which is well defined due to completeness of V. Then $w \in 2rB$ and $Tw = y$. That implies $B \subset T(2rB)$.

Now we have $T^{-1}B \subset 2rB$ and hence $\|T^{-1}\| \leq 2r$. $\qquad\square$

Bounded operators have a nice spectral theory, see [18] for instance.

DEFINITION 1.10. Let T be bounded. The *spectrum* $\sigma(T)$ of T is the subset of complex numbers λ such that $T - \lambda\operatorname{Id}$ is not invertible.

In a finite-dimensional Hilbert space $\sigma(T)$ is the set of eigenvalues of T. In the infinite-dimensional case a point of the spectrum is not necessarily an eigenvalue. We need the following fundamental result.

THEOREM 1.11. *If the operator T is bounded, then $\sigma(T)$ is a non-empty closed bounded subset of $\mathbb{C}$.*

PROOF. The main idea is to consider the resolvent $\mathcal{R}(\lambda) = (T - \lambda\operatorname{Id})^{-1}$ as a function of λ. If T is invertible, then we have the decomposition

$$\mathcal{R}(\lambda) = T^{-1}(\operatorname{Id} + T^{-1}\lambda + T^{-2}\lambda^2 + \dots),$$

which converges for $|\lambda| < \frac{1}{\|T^{-1}\|}$. Thus, $\mathcal{R}(\lambda)$ is analytic in a neighbourhood of 0. Using the shift $\mathcal{R}(\lambda) \to \mathcal{R}(\lambda + c)$, we obtain that $\mathcal{R}(\lambda)$ is analytic on its domain which is $\mathbb{C} \setminus \sigma(T)$. The domain of $\mathcal{R}(\lambda)$ is an open set. Hence $\sigma(T)$ is closed.

Furthermore, we can write the series for $\mathcal{R}(\lambda)$ at infinity:

$$(3.1) \qquad \mathcal{R}(\lambda) = -\lambda^{-1}(\operatorname{Id} + \lambda^{-1}T + \lambda^{-2}T^2 + \dots).$$

This series converges for $|\lambda| > \|T\|$. Therefore $\sigma(\lambda)$ is a subset of the disc $|\lambda| \leq \|T\|$. Hence $\sigma(T)$ is bounded.

Finally, (3.1) also implies $\lim_{\lambda\to\infty} \mathcal{R}(\lambda) = 0$. Suppose that $\sigma(T) = \emptyset$, then $\mathcal{R}(\lambda)$ is analytic and bounded. By Liouville's theorem, $\mathcal{R}(\lambda)$ is constant, which is impossible. $\qquad\square$

DEFINITION 1.12. For any linear operator T in a Hilbert space V, we denote by T^* the adjoint operator. Since $V^* \simeq V$, we can consider T^* as a linear operator in V such that for any $x, y \in V$

$$\langle x, Ty \rangle = \langle T^*x, y \rangle .$$

An operator T is *self-adjoint* if $T^* = T$. A self-adjoint operator T defines a Hermitian form on V, $\langle x, y \rangle_T = \langle x, Ty \rangle$. We call T (semi)positive if this form is (semi)positive definite. For any operator X, the operator X^*X is semipositive self-adjoint.

EXERCISE 1.13. (a) If T is bounded, then T^* is bounded and $\sigma(T^*)$ is the complex conjugate of $\sigma(T)$.

(b) If T is bounded self-adjoint, then $\sigma(T) \subset \mathbb{R}$.

LEMMA 1.14. *Let T be a self-adjoint operator in a Hilbert space. Then $\|T^2\| = \|T\|^2$.*

PROOF. For any bounded operator A, the Cauchy–Schwarz inequality implies that for all $v \in V$

$$|\langle Av, v \rangle| \le \|Av\| \|v\| \le \|A\| \|v\|^2.$$

For a self-adjoint T we have

$$\langle T^2v, v \rangle = \|Tv\|^2.$$

Therefore

$$\|T^2\| \ge \sup_{\|v\|=1} \langle T^2v, v \rangle = \sup_{\|v\|=1} \|Tv\|^2 = \|T\|^2.$$

On the other hand $\|T^2\| \le \|T\|^2$. Hence $\|T^2\| = \|T\|^2$. $\qquad\qquad \square$

LEMMA 1.15. *Let T be a self-adjoint operator in a Hilbert space V such that $\sigma(T) = \{\mu\}$ is a single point. Then $T = \mu \operatorname{Id}$.*

PROOF. Without loss of generality, we may assume $\mu = 0$. Then the series (3.1) converges for all $\lambda \ne 0$. Therefore, by the root test we have

$$\lim_{n \to \infty} \sup \|T^n\|^{1/n} = 0.$$

By Lemma 1.14, if $n = 2^k$, then $\|T^n\| = \|T\|^n$. This implies $\|T\| = 0$ and thus $T = 0$. $\qquad\qquad \square$

EXERCISE 1.16. Let X be a self-adjoint bounded operator.

(a) If $f \in \mathbb{R}[x]$ is a polynomial with real coefficients, then $\sigma(f(X)) = f(\sigma(X))$.

(b) Let $f : \mathbb{R} \to \mathbb{R}$ be a continuous function. Show that one can define $f(X)$ by approximating f by polynomials f_n over the interval $|x| \le \|X\|$ and setting $f(X) = \lim_{n \to \infty} f_n(X)$ and that the result does not depend on the choice of approximation.

(c) For a continuous function f we still have $\sigma(f(X)) = f(\sigma(X))$.

DEFINITION 1.17. An operator T in a Hilbert space V is called *compact* if the closure of the image $T(S)$ of the unit sphere $S = \{x \in V \mid \|x\| = 1\}$ is compact.

Clearly, any compact operator is bounded.

EXERCISE 1.18. Let $\mathcal{C}(V)$ be the subset of all compact operators in a Hilbert space V.

(a) Show that $\mathcal{C}(V)$ is a closed ideal in $\mathcal{B}(V)$.

(b) Let $\mathcal{F}(V)$ be the ideal in $\mathcal{B}(V)$ of all operators with finite-dimensional image. Prove that $\mathcal{C}(V)$ is the closure of $\mathcal{F}(V)$.

LEMMA 1.19. *Let A be a compact self-adjoint operator in V. Then*

$$\lambda := \sup_{u \in S} \langle Au, u \rangle$$

is either zero or an eigenvalue of A.

PROOF. Consider the hermitian form $x \mapsto \lambda \langle x, x \rangle - \langle Ax, x \rangle$ on V. It is semi-positive therefore the Cauchy–Schwarz inequality gives

$$(3.2) \qquad |\lambda \langle x, y \rangle - \langle Ax, y \rangle|^2 \le (\lambda \langle x, x \rangle - \langle Ax, x \rangle)(\lambda \langle y, y \rangle - \langle Ay, y \rangle)$$

Let (x_n) be a sequence in S such that $\langle Ax_n, x_n \rangle$ converges to λ. Since A is a compact operator, after extracting a subsequence we may assume that Ax_n converges to $z \in V$. By the inequality 3.2, we get that $\langle \lambda x_n - Ax_n, y \rangle$ tends to 0 uniformly in $y \in S$. Hence, $\|\lambda x_n - Ax_n\|$ tends to 0. Therefore, (x_n) converges to $\frac{1}{\lambda} z$ and z is a eigenvector for A with eigenvalue λ, if $\lambda > 0$. $\qquad \square$

1.5. Schur's lemma for unitary representations.

THEOREM 1.20. *Let $\rho : G \to U(V)$ a topologically irreducible unitary representation of G and $T \in \mathcal{B}(V)$ be a bounded intertwining operator. Then $T = \lambda \operatorname{Id}$ for some $\lambda \in \mathbb{C}$.*

PROOF. First, by Theorem 1.11, the spectrum $\sigma(T)$ is not empty. Therefore by adding a suitable scalar operator we may assume that T is not invertible. Note that T^* is also an intertwiner, and therefore $S = TT^*$ is an intertwiner as well. Moreover, S is not invertible. If $\sigma(S) = \{0\}$, then $S = 0$ by Lemma 1.15. Then we claim that $\operatorname{Ker} T \ne 0$. Indeed, if T is injective, then $\operatorname{Im} T^* \subset \operatorname{Ker} T = 0$. That implies $T^* = T = 0$. Since $\operatorname{Ker} T$ is a closed G-invariant subspace of V, we obtain $T = 0$.

Now we assume that $\sigma(S)$ consists of more than one point. We will use Exercise 1.16. One can always find two continuous functions $f, g : \mathbb{R} \to \mathbb{R}$ such that $fg(\sigma(S)) = 0$, but $f(\sigma(S)) \ne 0$ and $g(\sigma(S)) \ne 0$. Then Exercise 1.16(c) together with Lemma 1.15 implies $f(S)g(S) = 0$. Both $f(S)$ and $g(S)$ are non-zero intertwiners. At least one of $\operatorname{Ker} f(S)$ and $\operatorname{Ker} g(S)$ is a proper non-zero G-invariant subspace of V. Contradiction. $\qquad \square$

COROLLARY 1.21. *Let $\rho : G \to U(V)$ and $\rho' : G \to U(V')$ be two topologically irreducible unitary representations and $T : V \to V'$ be a continuous intertwining operator. Then either $T = 0$ or there exists $c > 0$ such that $cT : V \to V'$ is an isometry of Hilbert spaces.*

PROOF. Let $T \neq 0$. By Theorem 1.20 we have $T^*T = TT^* = \lambda \operatorname{Id}$ for some positive real λ. Set $c = \lambda^{-1/2}$ and $U = cT$. Then $U^* = U^{-1}$, hence U is an isometry. $\square$

COROLLARY 1.22. *Every topologically irreducible unitary representation of an abelian topological group G is one-dimensional.*

1.6. Irreducible unitary representations of compact groups.

PROPOSITION 1.23. *Every non-zero unitary representation of a compact group G contains a non-zero finite-dimensional invariant subspace.*

PROOF. Let $\rho : G \to \operatorname{GL}(V)$ be an irreducible unitary representation. Choose $v \in V$, $\|v\| = 1$. Define an operator $T : V \to V$ by the formula

$$Tx = \langle v, x \rangle v.$$

One can check easily that T is a semipositive self-adjoint operator of rank 1. Define the operator

$$Qx = \int_G \rho_g T \left(\rho_g^{-1} x \right) dg.$$

EXERCISE 1.24. Check $Q : V \to V$ is a compact semipositive intertwining operator.

Lemma 1.19 implies that Q has a positive eigenvalue λ. Consider $W = \operatorname{Ker} (Q - \lambda \operatorname{Id})$. Then W is an invariant subspace of V. Note that for any orthonormal system of vectors $e_1, \ldots, e_n \in W$,

$$\sum_{i=1}^{n} \langle e_i, T e_i \rangle \leq 1.$$

Hence

$$\sum_{i=1}^{n} \langle e_i, Q e_i \rangle = \sum_{i=1}^{n} \int_G \langle \rho_g e_i, T \rho_g e_i \rangle \leq 1.$$

That implies $\lambda n \leq 1$. Hence $\dim W \leq \frac{1}{\lambda}$. $\square$

COROLLARY 1.25. *Every irreducible unitary representation of a compact group G is finite-dimensional.*

LEMMA 1.26. *Every topologically irreducible representation of a compact group G is isomorphic to a subrepresentation of the regular representation in $L^2(G)$.*

PROOF. Let $\rho : G \to \operatorname{GL}(V)$ be irreducible. Pick a non-zero continuous linear functional φ on V and define the map $\Phi : V \to L^2(G)$ which sends v to the matrix coefficient $f_{v,\varphi}(g) = \langle \varphi, \rho_g v \rangle$. The continuity of ρ and φ implies that $f_{v,\varphi}$ is a continuous function on G, therefore $f_{v,\varphi} \in L^2(G)$. Furthermore

$$R_g f_{v,\varphi}(h) = f_{v,\varphi}(hg) = \langle \varphi, \rho_{hg} v \rangle = \langle \varphi, \rho_h \rho_g v \rangle = f_{\rho_g v, \varphi}(h).$$

Hence Φ is a continuous intertwining operator and the irreducibility of ρ implies $\operatorname{Ker} \Phi = 0$. The bicontinuity assertion follows from Corollary 1.25. $\qquad \square$

COROLLARY 1.27. *Every topologically irreducible representation of a compact group G is equivalent to some unitary representation.*

COROLLARY 1.28. *Every irreducible continuous representation of a compact group G is finite-dimensional.*

THEOREM 1.29. *If $\rho : G \to GL(V)$ is a unitary representation, then for any closed invariant subspace $W \subset V$ there exists a closed invariant subspace $U \subset V$ such that $V = U \oplus W$.*

PROOF. Consider $U = W^{\perp}$. $\qquad \square$

DEFINITION 1.30. *Let $\widehat{G}$ denotes the set of isomorphism classes of irreducible unitary representations of G. This set is called the* unitary dual *of G.*

LEMMA 1.31. *Let V be a unitary representation of a compact group G. Then it has a unique dense semisimple G-submodule, namely $\oplus_{\rho \in \widehat{G}} \operatorname{Hom}_G(V_\rho, V) \otimes V_\rho$.*

PROOF. Let $M = \oplus_{\rho \in \widehat{G}} \operatorname{Hom}_G(V_\rho, V) \otimes V_\rho$, and $\bar{M}$ denote the closure of M. We claim that $\bar{M} = V$. Indeed, if $\bar{M}^{\perp}$ is not zero, then it contains an irreducible finite-dimensional subrepresentation by Proposition 1.23, but any such representation is contained in M.

On the other hand, if N is a dense semisimple submodule of V, then N must contain all finite-dimensional irreducible subrepresentations of V. Therefore $N = M$. $\qquad \square$

2. Orthogonality relations and Peter–Weyl Theorem

2.1. Matrix coefficients. Let $\rho : G \to \operatorname{GL}(V)$ be a unitary representation of a compact group G. The function $G \to \mathbb{C}$ defined by the formula

$$f_{v,w}(g) = \langle w, \rho_g v \rangle .$$

for v, w in V is called a *matrix coefficient* of the representation ρ.

Since ρ is unitary,

$$(3.3) \qquad\qquad f_{v,w}(g^{-1}) = \bar{f}_{w,v}(g).$$

THEOREM 2.1. *For every irreducible unitary representation $\rho : G \to \operatorname{GL}(V)$,*

$$\langle f_{v,w}, f_{v',w'} \rangle = \frac{1}{\dim \rho} \langle v, v' \rangle \langle w', w \rangle .$$

Moreover, the matrix coefficients of two non-isomorphic representations of G are orthogonal in $L^2(G)$.

PROOF. Take v and v' in V. Define $T \in \operatorname{End}_{\mathbb{C}}(V)$ by

$$Tx := \langle v, x \rangle v'$$

and let

$$Q = \int_G \rho_g T \rho_g^{-1} dg.$$

As follows from Schur's lemma, since ρ is irreducible, Q is a scalar multiplication. Since

$$\operatorname{Tr} Q = \operatorname{Tr} T = \langle v, v' \rangle,$$

we obtain

$$Q = \frac{\langle v, v' \rangle}{\dim \rho} \operatorname{Id}.$$

Hence

$$\langle w', Qw \rangle = \frac{1}{\dim \rho} \langle v, v' \rangle \langle w', w \rangle.$$

On the other hand,

$$\langle w', Qw \rangle = \int_G \langle w', \langle v, \rho_g^{-1} w \rangle \rho_g v' \rangle dg = \int_G f_{w,v}\left(g^{-1}\right) f_{v',w'}(g)\, dg =$$

$$= \int_G \bar{f}_{v,w}(g) f_{v',w'}(g)\, dg = \langle f_{v,w}, f_{v',w'} \rangle.$$

If $f_{v,w}$ and $f_{v',w'}$ are matrix coefficients of two non-isomorphic representations, then $Q = 0$, and the calculation is even simpler. $\quad\square$

COROLLARY 2.2. *Let $\rho : G \to GL(V)$ and $\sigma : G \to GL(W)$ be two irreducible unitary representations, then $\langle \chi_\rho, \chi_\sigma \rangle = 1$ if ρ is isomorphic to σ and $\langle \chi_\rho, \chi_\sigma \rangle = 0$ otherwise.*

PROOF. Let $v_1, \ldots, v_n$ be an orthonormal basis in V and $w_1, \ldots, w_m$ be an orthonormal basis in W. Then

$$\langle \chi_\rho, \chi_\sigma \rangle = \sum_{i=1}^{n} \sum_{j=1}^{m} \langle f_{v_i, v_i}, f_{w_j, w_j} \rangle.$$

Therefore the statement follows from Theorem 2.1. $\quad\square$

LEMMA 2.3. *Let $\rho : G \to GL(V)$ be an irreducible unitary representation of G. Then the map $V \to \operatorname{Hom}_G(V, L^2(G))$ defined by*

$$w \mapsto \varphi_w, \quad \varphi_w(v) := f_{v,w} \text{ for all } v, w \in V$$

is an isomorphism of vector spaces.

PROOF. It is easy to see that $\varphi_w \in \mathrm{Hom}_G(V, L^2(G))$. Moreover, the value of $\varphi_w(w)$ at e equals $\langle w, w \rangle$. Hence $\varphi_w \neq 0$ if $w \neq 0$. Thus, the map is injective. To check surjectivity note that $\mathrm{Hom}_G(V, L^2(G))$ is the subspace of functions $f : G \times V \to \mathbb{C}$ satisfying the condition

$$f(gh, v) = f(g, \rho_h v) \quad \text{for all } v \in V,\, g, h \in G.$$

For any such f there exists $w \in V$ such that $f(e, v) = \langle w, v \rangle$. The above condition implies $f(g, v) = \langle w, \rho_g v \rangle$, i.e. $f = \varphi_w$. $\qquad\square$

THEOREM 2.4. (Peter–Weyl) *Matrix coefficients of all irreducible unitary representations span a dense subspace in $L^2(G)$ for a compact group G.*

PROOF. We apply Lemma 1.31 to the regular representation of G. Let $\rho \in \widehat{G}$. Lemma 2.3 implies that $V_\rho \otimes \mathrm{Hom}_G(V_\rho, L^2(G))$ coincides with the space of all matrix coefficients of ρ. Hence the span of matrix coefficients is the unique dense semisimple G-submodule in $L^2(G)$. $\qquad\square$

2.2. Convolution algebra. For a group G we denote by $L^1(G)$ the set of all complex-valued functions φ on G such that

$$\|\varphi\|_1 := \int_G |\varphi(g)|\,dg$$

is finite.

DEFINITION 2.5. The *convolution product* of two continuous complex-valued functions φ and ψ on G is defined by the formula:

$$(3.4) \qquad (\varphi * \psi)(g) := \int_G \varphi(h)\psi(h^{-1}g)\,dh.$$

EXERCISE 2.6. The following properties are easily checked:

(1) $\|\varphi * \psi\|_1 \leq \|\varphi\|_1 \|\psi\|_1$
(2) The convolution product extends uniquely as a continuous bilinear map $L^1(G) \times L^1(G) \to L^1(G)$.
(3) The convolution is an associative product.
(4) Let V be a unitary representation of G, show that we can see it as a $L^1(G)$-module by setting $\varphi.v := \int_G \varphi(g) gv\,dg$.

COROLLARY 2.7. *Let G be a compact group and R denote the representation of $G \times G$ in $L^2(G)$ given by the formula*

$$R_{s,t} f(x) = f(s^{-1}xt).$$

Then

$$L^2(G) \cong \widehat{\bigoplus_{\rho \in \widehat{G}}} V_\rho^* \boxtimes V_\rho,$$

where the direct sum is in the sense of Hilbert spaces.

Moreover, this isomorphism is actually an isomorphism of algebras (without unit) between $L^2(G)$ equipped with the convolution and $\widehat{\bigoplus}_{\rho \in \widehat{G}} End(V_\rho)$.

PROOF. For any $\rho \in \widehat{G}$ consider the map $\Phi_\rho : V_\rho^* \boxtimes V_\rho \to L^2(G)$ defined by

$$\Phi_\rho(v \otimes w)(g) = \langle v, \rho_g w \rangle \,.$$

It is easy to see that Φ_ρ defines an embedding of the irreducible $G \times G$-representation $\rho^* \boxtimes \rho$ in $L^2(G)$. Moreover, by orthogonality relation $\langle \operatorname{Im} \Phi_\rho, \operatorname{Im} \Phi_\sigma \rangle = 0$ if ρ and σ are not isomorphic. The direct sum $\bigoplus_{\rho \in \widehat{G}} \operatorname{Im} \Phi_\rho$ coincides with the span of all matrix coefficients of all irreducible representations of G. Hence it is dense in $L^2(G)$. That implies the first statement. The final statement is clear by applying item (4) of Exercise 2.6. □

REMARK 2.8. A finite group G is a compact group in discrete topology and $L^2(G)$ with the convolution product is the group algebra $\mathbb{C}(G)$. Therefore Theorem 1.13 of Chapter 2 is a particular case of Corollary 2.7 when the ground field is $\mathbb{C}$.

COROLLARY 2.9. *The characters of irreducible representations form an orthonormal basis in the subspace of class function in $L^2(G)$.*

PROOF. Let $\mathcal{C}(G)$ denote the subspace of class functions in $L^2(G)$, it is clearly the centre of $L^2(G)$. On the other hand, the centre of $End(V_\rho)$ is $\mathbb{C}$ and its image in $L^2(G)$ is $\mathbb{C}\chi_\rho$ (χ_ρ denotes as usual the character of ρ). The assertion is a direct consequence of Corollary 2.7. □

EXERCISE 2.10. Let $r : G \to U(V)$ be a unitary representation of a compact group G and ρ be an irreducible representation with character χ_ρ. Then the linear operator

$$P_\rho(x) := \dim \rho \int_G \chi_\rho(g^{-1}) r_g x dg$$

is a projector onto the corresponding isotypic component.

EXERCISE 2.11. Let E be a faithful finite-dimensional representation of a compact group G. Show that all irreducible representations of G appear in $T(E) \otimes T(E^*)$ as subrepresentations. Hint: Note that G is a subgroup in $GL(E)$. Using the Weierstrass theorem prove that the matrix coefficients of E and E^* generate a dense subalgebra in $L^2(G)$ (with usual pointwise multiplication).

3. Examples

3.1. The circle. Let $\mathbb{T} = S^1 = \{z \in \mathbb{C} \mid |z| = 1\}$, if $z \in S^1$, one can write $z = e^{i\theta}$ with θ in $\mathbb{R}/2\pi\mathbb{Z}$. The Haar measure on S^1 is equal to $\frac{d\theta}{2\pi}$. All irreducible representations of S^1 are one-dimensional since S^1 is abelian. They are given by the characters $\chi_n : S^1 \to \mathbb{C}^*$, $\chi_n(\theta) = e^{in\theta}$, $n \in \mathbb{Z}$. Hence $\widehat{S^1} = \mathbb{Z}$ and

$$L^2(S^1) = \widehat{\bigoplus}_{n \in \mathbb{Z}} \mathbb{C} e^{in\theta}.$$

This is a representation-theoretic explanation of the theorem of Parseval, meaning that every square integrable periodic function is the sum (with respect to the L^2 norm) of its Fourier series.

3.2. The group SU_2. Consider the compact group $G = SU_2$. Then G consists of all matrices

$$\begin{pmatrix} a & b \\ -\bar{b} & \bar{a} \end{pmatrix},$$

satisfying the relations $|a|^2 + |b|^2 = 1$. Thus, as a topological space, SU_2 is isomorphic to the 3-dimensional sphere S^3.

EXERCISE 3.1. Check that SU_2 is isomorphic to the multiplicative subgroup of quaternions with norm 1 by identifying the quaternion $a + bi + cj + dk = a + bi + (c + di)j$ with the matrix $\begin{pmatrix} a + bi & c + di \\ -c + di & a - bi \end{pmatrix}$.

To find the irreducible representations of SU_2, consider the polynomial ring $\mathbb{C}[x, y]$, with the action of SU_2 given by the formula

$$\rho_g(x) = \bar{a}x + \bar{b}y, \quad \rho_g(y) = -bx + ay, \text{ if } g = \begin{pmatrix} a & b \\ -\bar{b} & \bar{a} \end{pmatrix}.$$

Let ρ_n be the representation of G in the space $\mathbb{C}_n[x, y]$ of homogeneous polynomials of degree n. The monomials $x^n, x^{n-1}y, \ldots, y^n$ form a basis of $\mathbb{C}_n[x, y]$. Therefore $\dim \rho_n = n + 1$. We claim that all ρ_n are irreducible and that every irreducible representation of SU_2 is isomorphic to ρ_n for some $n \geq 0$. We will show this by checking that the characters χ_n of ρ_n form an orthonormal basis in the Hilbert space of class functions on G.

Recall that every unitary matrix is diagonal in some orthonormal basis, therefore every conjugacy class of SU_2 intersects the diagonal subgroup. Moreover, $\begin{pmatrix} z & 0 \\ 0 & \bar{z} \end{pmatrix}$ and $\begin{pmatrix} \bar{z} & 0 \\ 0 & z \end{pmatrix}$ are conjugate. Hence the set of conjugacy classes can be identified with the quotient of S^1 by the equivalence relation $z \sim \bar{z}$. Let $z = e^{i\theta}$, then

$$(3.5) \qquad \chi_n(z) = z^n + z^{n-2} + \cdots + z^{-n} = \frac{z^{n+1} - z^{-n-1}}{z - z^{-1}} = \frac{\sin(n+1)\theta}{\sin\theta}.$$

First, let us compute the Haar measure for G.

EXERCISE 3.2. Let $G = SU_2$.

(a) Using Exercise 3.1 show that the action of $G \times G$ given by the multiplication on the right and on the left coincides with the standard action of $SO(4)$ on S^3. Use it to prove that $SO(4)$ is isomorphic to the quotient of $G \times G$ by the two element subgroup $\{(1, 1), (-1, -1)\}$.

(b) Prove that the Haar measure on G is proportional to the standard volume form on S^3 invariant under the action of the orthogonal group SO_4.

More generally: let us compute the volume form on the n-dimensional sphere $S^n \subset \mathbb{R}^{n+1}$ which is invariant under the action SO_{n+1}. We use the spherical coordinates in $\mathbb{R}^{n+1}$,

$$x_1 = r\cos\theta, x_2 = r\sin\theta\cos\varphi_1, x_3 = r\sin\theta\sin\varphi_1\cos\varphi_2,$$

$$\cdots$$

$$x_{n-1} = r\sin\theta\sin\varphi_1\sin\varphi_2\ldots\sin\varphi_{n-2}\cos\varphi_{n-1},$$
$$x_n = r\sin\theta\sin\varphi_1\sin\varphi_2\ldots\sin\varphi_{n-2}\sin\varphi_{n-1},$$

where $r > 0$, $\theta, \varphi_1, \ldots, \varphi_{n-2}$ vary in $[0, \pi]$ and $\varphi_{n-1} \in [0, 2\pi]$. The Jacobian relating spherical and Euclidean coordinates is equal to

$$r^n \sin^{n-1}\theta \sin^{n-2}\varphi_1\ldots\sin\varphi_{n-2},$$

thus when we restrict to the sphere $r = 1$ we obtain the volume form

$$\sin^{n-1}\theta \sin^{n-2}\varphi_1\ldots\sin\varphi_{n-2}d\theta d\varphi_1\ldots d\varphi_{n-1},$$

which is SO_{n+1}-invariant. It is not normalized.

Let us return to the case $G = SU_2 \simeq S^3$. After normalization the invariant volume form is

$$\frac{1}{2\pi^2}\sin^2\theta\sin\varphi_1 d\theta d\varphi_1 d\varphi_2.$$

The conjugacy class $C(\theta)$ of all matrices with eigenvalues $e^{i\theta}, e^{-i\theta}$ ($\theta \in [0, \pi]$) is the set of points in S^3 with spherical coordinates $(1, \theta, \varphi_1, \varphi_2)$: indeed, the minimal polynomial on $\mathbb{R}$ of the quaternion with those coordinates is

$$t^2 - 2t\cos\theta + 1$$

which is also the characteristic polynomial of the corresponding matrix in SU_2, so it belongs to $C(\theta)$.

Hence, one gets that, for a class function ψ on G

$$\int_G \psi(g)\,dg = \frac{1}{2\pi^2}\int_0^\pi \psi(\theta)\sin^2\theta d\theta \int_0^\pi \sin\varphi_1 d\varphi_1 \int_0^{2\pi} d\varphi_2 = \frac{2}{\pi}\int_0^\pi \psi(\theta)\sin^2\theta d\theta.$$

EXERCISE 3.3. Prove that the functions χ_n form an orthonormal basis of the space $L^2([0, \pi])$ with the measure $\frac{2}{\pi}\sin^2\theta d\theta$ and hence of the space of class functions on G.

3.3. The orthogonal group $G = SO_3$. Recall that SU_2 can be realized as the set of quaternions with norm 1. Consider the representation γ of SU_2 in the space of quaternions $\mathbb{H}$ defined by the formula $\gamma_g(\alpha) = g\alpha g^{-1}$. One can see that the 3-dimensional space $\mathbb{H}_{\text{im}}$ of pure imaginary quaternions is invariant and $(\alpha, \beta) = \text{Re}(\alpha\bar{\beta})$ is an invariant positive definite scalar product on $\mathbb{H}_{\text{im}}$. Therefore γ defines a homomorphism $\gamma\colon SU_2 \to SO_3$.

EXERCISE 3.4. Check that $\operatorname{Ker} \gamma = \{1, -1\}$ and that γ is surjective. Hence $SO_3 \cong SU_2/\{1, -1\}$. Thus, we can see that as a topological space SO_3 is a 3-dimensional sphere with opposite points identified, or the real 3-dimensional projective space.

Therefore every representation of SO_3 can be lifted to the representations of SU_2, and a representation of SU_2 factors to the representation of SO_3 if and only if it is trivial on -1. One can check easily that $\rho_n(-1) = 1$ if and only if n is even. Thus, any irreducible representation of SO_3 is isomorphic to ρ_{2m} for some $m > 0$ and $\dim \rho_{2m} = 2m + 1$. Below we give an independent realization of irreducible representations of SO_3.

3.4. Harmonic analysis on the sphere. Consider the sphere S^2 in $\mathbb{R}^3$ defined by the equation

$$x_1^2 + x_2^2 + x_3^2 = 1.$$

The action of SO_3 on S^2 induces a representation of SO_3 in the space $L^2(S^2)$ of complex-valued square integrable functions on S^2. This representation is unitary. We would like to decompose it into a sum of irreducible representations of SO_3. We first note that the space $\mathbb{C}[S^2]$ obtained by restriction of the polynomial functions $\mathbb{C}[x_1, x_2, x_3]$ to S^2 is the invariant dense subspace in $L^2(S^2)$. Indeed, it is dense in the space of continuous functions on S^2 by the Weierstrass theorem and the latter space is dense in $L^2(S^2)$.

Let us introduce the following differential operators in $\mathbb{R}^3$:

$$e := -\frac{1}{2}\left(x_1^2 + x_2^2 + x_3^2\right), \quad h := x_1 \frac{\partial}{\partial x_1} + x_2 \frac{\partial}{\partial x_2} + x_3 \frac{\partial}{\partial x_3} + \frac{3}{2}, \quad f := \frac{1}{2}\left(\frac{\partial^2}{\partial x_1^2} + \frac{\partial^2}{\partial x_2^2} + \frac{\partial^2}{\partial x_3^2}\right),$$

note that e, f, and h commute with the action of SO_3 and satisfy the relations

$$[e, f] = h, \quad [h, e] = 2e, \quad [h, f] = -2f,$$

where $[a, b] = ab - ba$.

Let P_n be the space of homogeneous polynomials of degree n and $H_n = \operatorname{Ker} f \cap P_n$. The polynomials of H_n are called *harmonic* polynomials since they are annihilated by the Laplace operator f. For any $\varphi \in P_n$ we have

$$h(\varphi) = \left(n + \frac{3}{2}\right)\varphi.$$

If $\varphi \in H_n$, then

$$fe(\varphi) = ef(\varphi) - h(\varphi) = -\left(n + \frac{3}{2}\right)\varphi,$$

and by induction

$$fe^k(\varphi) = efe^{k-1}(\varphi) - he^{k-1}(\varphi) = -\left(nk + k(k-1) + \frac{3k}{2}\right)e^{k-1}\varphi.$$

In particular, this implies that

(3.6) $$fe^k(H_n) = e^{k-1}(H_n).$$

We will prove now that

(3.7) $$P_n = H_n \oplus e(H_{n-2}) \oplus e^2(H_{n-4}) + \ldots$$

by induction on n. Indeed, by the induction assumption

$$P_{n-2} = H_{n-2} \oplus e(H_{n-4}) + \ldots.$$

Furthermore, (3.6) implies $fe(P_{n-2}) = P_{n-2}$. Hence $H_n \cap eP_{n-2} = 0$. On the other hand, $f: P_n \to P_{n-2}$ is surjective, and therefore $\dim H_n + \dim P_{n-2} = \dim P_n$. Therefore

(3.8) $$P_n = H_n \oplus eP_{n-2},$$

which implies (3.7). Note that after restriction to S^2, the operator e acts as the multiplication on $\frac{-1}{2}$.

Hence (3.7) implies that

$$\mathbb{C}[S^2] = \bigoplus_{n \geq 0} H_n.$$

To calculate the dimension of H_n use (3.8)

$$\dim H_n = \dim P_n - \dim P_{n-2} = \frac{(n+1)(n+2)}{2} - \frac{n(n-1)}{2} = 2n+1.$$

Finally, we will prove that the representation of SO_3 in H_n is irreducible and isomorphic to ρ_{2n}. Consider the subgroup $D \subset SO_3$ consisting of all rotations about x_3-axis. Then D is the image under $\gamma: SU_2 \to SO_3$ of a diagonal subgroup of SU_2. Let R_θ denote the rotation by the angle θ.

EXERCISE 3.5. Let V_{2n} be the space of the representation ρ_{2n}. Check that the set of eigenvalues of R_θ in the representation V_{2n} equals $\{e^{k\theta i} \mid -n \leq k \leq n\}$.

Let $\varphi_n = (x_1 + ix_2)^n$. It is easy to see that $\varphi_n \in H_n$ and $R_\theta(\varphi_n) = e^{n\theta i}\varphi_n$. By Exercise 3.5 this implies that H_n contains a subrepresentation isomorphic to ρ_{2k} for some $k \geq n$. By comparing dimensions we see that this implies $H_n = V_{2n}$. Thus, we obtain the following decompositions

$$\mathbb{C}[S^2] = \bigoplus_{n \in \mathbb{N}} H_n, \quad L^2(S^2) = \widehat{\bigoplus_{n \in \mathbb{N}}} H_n.$$

Now, we are able to prove the following geometrical theorem.

THEOREM 3.6. *A convex centrally symmetric solid in $\mathbb{R}^3$ is uniquely determined by the areas of the plane cross sections through the origin.*

PROOF. A convex solid B can be defined by an even continuous function on S^2. Indeed, for each unit vector v let

$$\varphi(v) = \sup\left\{t^2 \in \mathbb{R} \mid tv \in B\right\}.$$

Define a linear operator T in the space of all even continuous functions on S^2 by the formula

$$T\varphi(v) = \frac{1}{2}\int_0^{2\pi} \varphi(w)\, d\theta,$$

where w runs over the set of unit vectors orthogonal to v, and θ is the angular parameter on the circle $S^2 \cap v^\perp$. Check that $T\varphi(v)$ is the area of the cross section by the plane $v^\perp$. We have to prove that T is invertible.

Obviously T commutes with the SO_3-action. Therefore T can be diagonalized by using Schur's lemma and the decomposition

$$L^2(G)_{even} = \widehat{\bigoplus_{n \in \mathbb{N}}} H_{2n}.$$

Indeed, T acts on H_{2n} as the scalar operator $\lambda_n Id$. We have to check that $\lambda_n \neq 0$ for all n. Consider again $\varphi_{2n} \in H_{2n}$. Then $\varphi_{2n}(1,0,0) = 1$ and

$$T\varphi_{2n}(1,0,0) = \frac{1}{2}\int_0^{2\pi} (iy)^{2n}\, d\theta = \frac{(-1)^n}{2}\int_0^{2\pi} \sin^{2n}\theta\, d\theta,$$

here we take the integral over the circle $x_2^2 + x_3^2 = 1$, and assume $x_2 = \sin\theta$, $x_3 = \cos\theta$. Since $T\varphi = \lambda_n\varphi$, we obtain

$$\lambda_n = \frac{(-1)^n}{2}\int_0^{2\pi} \sin^{2n}\theta\, d\theta \neq 0.$$

$\square$

CHAPTER 4

Results about unitary representations

In which we take a quick glimpse at unitary representations of locally compact groups and meet some deep results in analysis. Unfortunately, a rendez-vous with algebra prevents us from going too far in this direction. Nevertheless, we cannot resist the temptation to prove Stone–von Neumann theorem and construct unitary representations of $SL_2(\mathbb{R})$ before leaving the premises.

In this chapter, we consider unitary representations of groups which are locally compact but not compact. We do not intend to go very far in this deep subject, but we want to give three examples in order to show how the structure of the dual of the group changes.

1. Unitary representations of $\mathbb{R}^n$ and Fourier transform

1.1. Unitary dual of a locally compact abelian group. Let G be a locally compact abelian group. Then by Corollary 1.22 of Chapter 3, every unitary representation of G is one-dimensional. Therefore the unitary dual $\widehat{G}$ (see Definition 1.30 in Chapter 3) is the set of continuous homomorphisms $\rho : G \to S^1$. Moreover, $\widehat{G}$ is an abelian group with multiplication defined by the tensor product.

For example, as we have seen in Section 3.1 Chapter 3, if $G = S^1$ is the circle, then $\widehat{G}$ is isomorphic to $\mathbb{Z}$. In general, if G is compact, then $\widehat{G}$ is discrete. If G is not compact, we define a topology on $\widehat{G}$ by the uniform convergence on compact sets, and then the natural homomorphism $s : G \to \widehat{\widehat{G}}$, defined by $s(g)(\rho) = \rho(g)$, is an isomorphism of topological groups. This fact is usually called the *Pontryagin duality*.

Let us concentrate on the case when $G = V$ is a real vector space of finite dimension n. Let us fix an invariant volume form dx on V. The unitary dual of V is isomorphic to the usual dual V^* via the identification

$$\rho_\xi(x) = e^{2i\pi<\xi,x>} \text{ for all } x \in V, \xi \in V^*,$$

where $< \xi, x >$ is the duality evaluation.

We immediately see that, in contrast with the compact case, ρ_ξ is no longer in $L^2(V)$. We have an analogue of the formula giving the projector P_ξ from $L^2(V)$ onto the irreducible representation ρ_ξ, as in Exercise 2.10 Chapter 3: for $f \in L^2(V)$,

© Springer Nature Switzerland AG 2018
C. Gruson and V. Serganova, *A Journey Through Representation Theory*,
Universitext, https://doi.org/10.1007/978-3-319-98271-7_4

$y \in V$ and $\xi \in V^*$, let us set

$$P_\xi(f)(y) := \int_V f(x+y)e^{-2i\pi<\xi,x>}dx = (\int_V f(z)e^{-2i\pi<\xi,z>}dz)\rho_\xi(y) :$$

the coefficient $\int_V f(z)e^{-2i\pi<\xi,z>}dz$ is nothing but the value $\hat{f}(\xi)$ of the *Fourier transform* $\hat{f}$. However, the integral defining $\hat{f}$ has a meaning for $f \in L^1(V)$, but not necessarily for $f \in L^2(V)$. In this section, we explain how to overcome this difficulty, see the theorem of Plancherel (Theorem 1.12).

We also would like to claim that every $f \in L^2(V)$ is in a certain sense the "sum of its projections", which leads to the equality

$$f(x) = \int_V \hat{f}(\xi)e^{2i\pi<\xi,x>}d\xi.$$

This formula expresses the involutivity of the Fourier transform, see Theorem 1.7 below.

1.2. General facts on the Fourier transform. Let $L^1(V)$ be the set of integrable complex-valued functions on V.

DEFINITION 1.1. Let $f \in L^1(V)$, the *Fourier transform* of f is the function on V^*

$$\hat{f}(\xi) := \int_V f(x)e^{-2i\pi<\xi,x>}dx.$$

REMARK 1.2.   (1) One checks that $\lim_{\xi\to\infty} \hat{f}(\xi) = 0$ and that $\hat{f}$ is continuous on V^*.
(2) Nevertheless, there is no reason for $\hat{f}$ to belong to $L^1(V^*)$ (check on the characteristic function over an interval in $\mathbb{R}$).
(3) The Fourier transform of the convolution product of two functions (see Definition 2.5 Chapter 3) is the product of the Fourier transforms of the two factors.
(4) (Adjunction formula for Fourier transforms) Let $f \in L^1(V)$ and $\varphi \in L^1(V^*)$, then

$$\int_V f(x)\hat{\varphi}(x)dx = \int_{V^*} \hat{f}(\xi)\varphi(\xi)d\xi.$$

EXERCISE 1.3. Let $\gamma \in GL(V)$, show that the Fourier transform of the function $\gamma.f$ defined by $(\gamma.f)(x) = f(\gamma^{-1}(x))$ is equal to $\det(\gamma)^t\gamma^{-1}.\hat{f}$.

Let us consider the *generalized Wiener algebra* $\mathcal{W}(V)$ consisting of integrable functions on V whose Fourier transform is integrable on V^*.

PROPOSITION 1.4. *The subspace $\mathcal{W}(V) \subset L^1(V)$ is a dense subset (for the L^1-norm).*

PROOF. Let Q be a positive definite quadratic form on V, denote by B its polarization and by Q^{-1} the quadratic form on V^* whose polarization is B^{-1}. Let $Disc(Q)$ denote the discriminant of Q in a basis of V of volume 1. The proof of this Proposition is done after the following Lemma and Exercise.

LEMMA 1.5. *The Fourier transform of the function* $\phi : x \mapsto e^{-\pi Q(x)}$ *on* V *is the function* $\xi \mapsto Disc(Q)^{-1/2} e^{-\pi Q^{-1}(\xi)}$ *on* V^*.

PROOF. (of Lemma 1.5) One can reduce this lemma to the case $n = 1$ by using an orthogonal basis for Q and Fubini's theorem. We just need to compute the Fourier transform of the function $\varepsilon(x) := x \mapsto e^{-\pi x^2}$ on the line $\mathbb{R}$.

One has

$$\hat{\varepsilon}(\xi) = \int_{\mathbb{R}} e^{-\pi x^2 - 2i\pi\xi x} dx = e^{-\pi\xi^2} \int_{\mathbb{R}} e^{-\pi(x+i\xi)^2} dx.$$

By complex integration, the integral factor in the far right-hand side does not depend on ξ and its value for $\xi = 0$ is the Gauss integral $\int_{\mathbb{R}} e^{-\pi x^2} dx = 1$. Hence the lemma. $\square$

To finish the proof of the Proposition let us take Q such that $Disc(Q) = 1$. Lemma 1.5 implies that ϕ belongs to $\mathcal{W}(V)$. For every $\lambda \in \mathbb{R}_{>0}$, we set $\phi_\lambda(x) := \lambda^n \phi(\lambda x)$.

EXERCISE 1.6. Check that $\phi_\lambda(x)$ is a positive-valued function and $\int_V \phi_\lambda(x)dx=1$. Prove that, when λ tends to infinity, $\phi_\lambda(x)$ converges uniformly to 0 in the complement of any neighbourhood of $0 \in V$.

Now take any function $f \in L^1(V)$. By Remark 1.2 the convolution product $f_\lambda := f * \phi_\lambda$ belongs to $\mathcal{W}(V)$. By Exercise 1.6, f_λ converges to f for the L^1-norm. Hence the Proposition. $\square$

THEOREM 1.7. *(Fourier inversion formula) If* $f \in \mathcal{W}(V)$ *then*

$$\hat{\hat{f}}(x) = f(-x), \forall x \in V.$$

PROOF. By Proposition 1.4 the subset of continuous bounded functions is dense in $\mathcal{W}(V)$. Hence it suffices to prove the statement for continuous bounded f. We use a slight extension of the adjunction formula (Remark 1.2, (1.2)): let $\lambda \in \mathbb{R}_{>0}$, then for all $f \in L^1(V)$ and $\varphi \in L^1(V^*)$,

$$(4.1) \quad \int_V f(\lambda x)\hat{\varphi}(x)dx = \int_{V^*} \hat{f}(\xi)\varphi(\lambda\xi)d\xi = \iint_{V\times V^*} f(x)\varphi(\xi)e^{-2i\pi\lambda<\xi,x>}dxd\xi.$$

If λ goes to 0, the function $x \mapsto f(\lambda x)$ tends to $f(0)$ and remains bounded by $\|f\|_\infty := \sup |f|$. By dominated convergence, we obtain the equality

$$(4.2) \quad f(0)\hat{\hat{\varphi}}(0) = \hat{f}(0)\varphi(0).$$

Using the function ϕ of Lemma 1.5 for φ, we know that $\hat{\hat{\phi}} = \phi$, thus $\hat{f}(0) = f(0)$.

We use the action of the additive group V on $\mathcal{W}(V)$ (translation) given by

$$\tau_y(f) : (x \mapsto f(x - y))$$

and on $\mathcal{W}(V^*)$ (multiplication) given by

$$\mu_y(\varphi) : \xi \mapsto e^{-2i\pi<\xi,y>}\varphi(\xi)$$

for all $y \in V$.

We apply the translation τ_y to f, and Exercise 1.8 (see just below) shows that $\widehat{\tau_y(f)} = \tau_{-y}\hat{f}$, hence the result. $\qquad\square$

EXERCISE 1.8. Check that
(1) $\widehat{\tau_y(f)} = \mu_y(\hat{f})$ for all $f \in L^1(V)$,
(2) $\widehat{\mu_y(\varphi)} = \tau_{-y}(\hat{\varphi})$ for all $\varphi \in L^1(V^*)$.

REMARK 1.9. The Fourier inversion formula is equivalent to the following statement

$$(4.3) \qquad\qquad\qquad \hat{\bar{\hat{f}}} = f$$

where $\bar{f}$ denotes the complex conjugate of f.

COROLLARY 1.10. *The space $\mathcal{W}(V)$ is a dense subspace in $L^2(V)$.*

PROOF. By Theorem 1.7 and Remark 1.2, $\mathcal{W}(V)$ is a subset of the set $\mathcal{C}^0(V)$ of continuous functions on V which tend to 0 at infinity. For details see [29]. Therefore $\mathcal{W}(V)$ is included in $L^2(V)$. The proof of Proposition 1.4 can be adapted to prove the density of $\mathcal{W}(V)$ in $L^2(V)$ (using that $\phi \in L^2(V)$). $\qquad\square$

COROLLARY 1.11. *The Fourier transform is an injective map from $L^1(V)$ to $\mathcal{C}^0(V^*)$.*

PROOF. We first notice that $\mathcal{W}(V)$ is dense in $\mathcal{C}^0(V)$ by the same argument as in the proof of Proposition 1.4.

Hence, if $f \in L^1(V)$ is such that $\hat{f} = 0$, to show that $f = 0$ it is sufficient to prove that

$$\int_V f(x)g(x)dx = 0$$

for any $g \in \mathcal{W}(V)$. By Theorem 1.7, g is the Fourier transform of $\xi \mapsto \hat{g}(-\xi)$ and by Remark 1.2(1.2), so

$$\int_V f(x)g(x)dx = \int_{V^*} \hat{f}(\xi)\hat{g}(-\xi)d\xi = 0.$$

$\qquad\square$

THEOREM 1.12. *(Plancherel) The Fourier transform extends to an isometry from $L^2(V)$ to $L^2(V^*)$.*

PROOF. Since $\mathcal{W}(V)$ is dense in $L^2(V)$, all we have to show is that for f and g in $\mathcal{W}(V)$,

$$(4.4) \qquad \int_V \bar{f}(x)g(x)dx = \int_{V^*} \bar{\hat{f}}(\xi)\hat{g}(\xi)d\xi.$$

By Remark 1.2 (1.2), the right-hand side is equal to

$$\int_V \bar{\hat{\hat{f}}}(x)g(x)dx.$$

But, by Remark 1.9 (4.3), $\hat{\hat{f}} = \check{f}$.   □

REMARK 1.13. By Plancherel's theorem, the Fourier transform maps $\mathcal{W}(V)$ to $\mathcal{W}(V^*)$ exchanging the roles of usual and convolution products.

1.3. The link between Fourier series and Fourier transform on $\mathbb{R}$. Let f be a function over the interval $[-\frac{1}{2}, \frac{1}{2}]$. The Fourier series of f is

$$(4.5) \qquad \sum_{n \in \mathbb{Z}} c_n(f)e^{2i\pi nx}$$

where

$$c_n := \int_{-\frac{1}{2}}^{\frac{1}{2}} f(t)e^{-2i\pi nt}dt.$$

Now, for $\lambda \in \mathbb{R}_{>0}$, if g is a function defined over the interval $[-\frac{\lambda}{2}, \frac{\lambda}{2}]$, changing the variable by $y := \lambda x$, the corresponding Fourier series is written

$$(4.6) \qquad \sum_{n \in \mathbb{Z}} \left(\int_{-\frac{\lambda}{2}}^{\frac{\lambda}{2}} \frac{1}{\lambda} g(u)e^{-2i\pi n\frac{u}{\lambda}}du \right) e^{2i\pi n\frac{y}{\lambda}}.$$

We consider that, formally, g is the sum of its Fourier series on $[-\frac{\lambda}{2}, \frac{\lambda}{2}]$.

Now if we consider g as a function defined on $\mathbb{R}$ with compact support by extending by 0 outside the interval $[-\frac{\lambda}{2}, \frac{\lambda}{2}]$, we may interpret the n-th Fourier coefficient as $\frac{1}{\lambda}\hat{g}(\frac{n}{\lambda})$ and the Fourier series as the sum

$$(4.7) \qquad \frac{1}{\lambda}\sum_{n \in \mathbb{Z}} \hat{g}(\frac{n}{\lambda})e^{2i\pi \frac{ny}{\lambda}}.$$

Formally, this series is exactly the Riemann sum, corresponding to the partition of $\mathbb{R}$ associated to the intervals $[\frac{n}{\lambda}, \frac{n+1}{\lambda}]$, of the infinite integral $\int_{\mathbb{R}} \hat{g}(t)e^{2i\pi nt}dt$.

If now g is compactly supported and λ tends to $+\infty$, this formal expression of the sum suggests the equality

$$g(t) = \int_{\mathbb{R}} \hat{g}(u)e^{2i\pi tu}du.$$

2. Heisenberg groups and the Stone–von Neumann theorem

2.1. The Heisenberg group and some examples of its unitary representations. Let V be a real vector space of finite even dimension $n = 2g$ together with a non-degenerate symplectic form $\omega : (x, y) \mapsto (x|y)$. Let $\mathbb{T} = S^1$ be the group of complex numbers of modulus 1. We define the *Heisenberg group* H as the set theoretical product $\mathbb{T} \times V$ with the composition law

$$(t, x)(t', x') := (tt' e^{i\pi(x|x')}, x + x').$$

The centre of H is $\mathbb{T}$, imbedded in H by $t \mapsto (t, 0)$. There is a non-split exact sequence

$$1 \to \mathbb{T} \to H \to V \to 0.$$

The commutator map $(g, h) \mapsto ghg^{-1}h^{-1}$ of elements of H naturally factorises as the map $V \times V \to \mathbb{T}$

$$(x, y) \mapsto e^{2i\pi(x|y)}.$$

EXERCISE 2.1. Show that the formula

$$r_{(t,x)}f(y) := tf(y - x)e^{i\pi(x|y)} \text{ for all } t \in \mathbb{T}, \, x, y \in V, \, f \in L^2(V)$$

defines a unitary representation r of the group H in the space $L^2(V)$.

DEFINITION 2.2. Consider two maximal isotropic subspaces of V with trivial intersection. If we denote one of them by W, then the second one can be identified by ω with the dual space W^*. Since the restriction of ω to both W and W^* is zero, the map $x \mapsto (1, x)$ from V to H induces group homomorphisms on both W and W^*. The *Schrödinger representation* σ of H in the Hilbert space $L^2(W)$ is defined by

$$\sigma_{(t,w+\eta)}f(x) = tf(x - w)e^{2i\pi(\eta|x)} \text{ for all } t \in \mathbb{T}, \, x, w \in W, \, \eta \in W^*, \, f \in L^2(W).$$

EXERCISE 2.3. Prove that σ is an irreducible unitary representation of H. To show irreducibility, it is sufficient to check that any bounded operator T in $L^2(W)$, commuting with the action of H, is a scalar multiplication. First, since T commutes with the action of W^*, it commutes also with multiplication by any continuous function with compact support. Making use of partitions of unity, show that this implies that T is the multiplication by some bounded measurable function g on W. Moreover, since T commutes with the action of W, the function g is invariant under translations, hence is a constant function.

2.2. The Stone–von Neumann theorem. The aim of this subsection is to show

THEOREM 2.4. *(Stone–von Neumann) Let ρ be an irreducible unitary representation of H such that $\rho_t = t\,\mathrm{Id}$ for all $t \in \mathbb{T}$. Then ρ is isomorphic to the Schrödinger representation.*

Let $\mathcal{H}$ be a Hilbert space together with an action ρ of the Heisenberg group H. We assume that the hypotheses of the theorem are satisfied by $(\rho, \mathcal{H})$. To simplify the notations, we identify $x \in V$ with $(1, x) \in H$, although this identification is not a group homomorphism. We set $\rho(x) := \rho_{(1,x)}$ for all $x \in V$. Then the condition that ρ is a representation is equivalent to

$$(4.8) \qquad \rho(x)\rho(y) = e^{i\pi(x|y)}\rho(x + y).$$

We denote by $\mathcal{A}$ the minimal closed subalgebra of the algebra $\mathcal{B}(\mathcal{H})$ of bounded operators on $\mathcal{H}$, which contains the image ρ_H. Let $\mathcal{C}_c^0(V)$ be the space of compactly supported continuous complex-valued functions on V. For every $\varphi \in \mathcal{C}_c^0(V)$, set

$$T_\varphi := \int_V \varphi(x)\rho(x)dx,$$

where dx is the Lebesgue measure on V. It is easy to see that $T_\varphi \in \mathcal{A}$. We have

$$T_\varphi T_\psi = \iint_{V \times V} \varphi(x)\psi(y)\rho(x)\rho(y)dxdy$$

$$T_\varphi T_\psi = \iint_{V \times V} \varphi(x)\psi(y)e^{i\pi(x|y)}\rho(x + y)dxdy$$

$$T_\varphi T_\psi = \iint_{V \times V} \varphi(x)\psi(u - x)e^{i\pi(x|u-x)}\rho(u)dxdu$$

$$T_\varphi T_\psi = T_{\varphi * \psi},$$

where $\varphi * \psi$ is defined by the formula

$$(4.9) \qquad \varphi * \psi(u) = \int_V \varphi(x)\psi(u - x)e^{i\pi(x|u-x)}dx.$$

Since clearly $\|T_\varphi\| \le \|\varphi\|_1 (= \int_V |\varphi(x)|dx)$, we get the following statements:

- The map $\varphi \mapsto T_\varphi$ extends by continuity to $L^1(V)$, the space of integrable complex-valued functions on V.
- The product $(\varphi, \psi) \mapsto \varphi * \psi$ extends to a product $L^1(V) \times L^1(V) \to L^1(V)$
- The formula (4.9) remains valid for φ and ψ in $L^1(V)$ for almost every $u \in V$.

LEMMA 2.5. *The map $\varphi \mapsto T_\varphi$ is injective on $L^1(V)$.*

PROOF. Denote by N the kernel of this map. We notice the equality

$$\rho(y)T_\varphi\rho(-y) = \int_V \varphi(x)\rho(y)\rho(x)\rho(-y)dx = \int_V \varphi(x)e^{2i\pi(y|x)}\rho(x)dx.$$

It shows that if $\varphi(x)$ is in N then $\varphi(x)e^{2i\pi(y|x)}$ lies in N for every $y \in V$. For a, b in $\mathcal{H}$, consider the matrix coefficient function

$$\chi_{a,b}(x) = <\rho(x)a, b>,$$

where $<,>$ is the scalar product on $\mathcal{H}$. It is a continuous bounded function of $x \in V$. Moreover, for any x, there exists at least one coefficient function which doesn't vanish at x.

If φ belongs to N, we have

$$\int_V \varphi(x)\chi_{a,b}(x)dx = 0,$$

and therefore

$$\int_V \varphi(x)\chi_{a,b}(x)e^{2i\pi(x|y)}dx = 0$$

for all $y \in V$. This means that the Fourier transform of the function $\varphi\chi_{a,b} \in L^1(V)$ is identically zero, hence $\varphi\chi_{a,b} = 0$ for all a, b, therefore $\varphi = 0$. $\square$

We will also use the following equality:

(4.10) $T_\varphi^* = T_{\varphi^*},$

with $\varphi^*(x) := \overline{\varphi}(-x)$.

Our ultimate goal is to construct a continuous intertwiner $\tau : L^2(V) \to \mathcal{H}$. The following observation is crucial for this construction.

LEMMA 2.6. (a) For all $f \in \mathcal{C}_c^0(V)$ and $h \in H$, we have $T_{r_h f} = \rho_h T_f$.

(b) For any $u \in \mathcal{H}$, the map $\pi_u := \mathcal{C}_c^0(V) \to \mathcal{H}$ defined by $f \mapsto T_f u$ is H-equivariant.

PROOF. It is sufficient to check (a) for $h = (1, y)$ with $y \in V$. Then using (4.8) and making the substitution $z = x - y$, we obtain

$$T_{r_h f} = \int_V f(x - y)\rho(x)e^{i\pi(y|x)}dx = \int_V f(x - y)\rho(y)\rho(x - y)dx =$$

$$= \rho(y) \int_V f(z)\rho(z)dz = \rho(y)T_f.$$

(b) follows immediately from (a). $\square$

Thus we have an equivariant map $\pi_u : \mathcal{C}_c^0(V) \to \mathcal{H}$. It remains to show that for a suitable choice of $u \in \mathcal{H}$ we are able to extend π_u to a continuous map $L^2(V) \to \mathcal{H}$.

LEMMA 2.7. Let φ be a continuous bounded function on V which lies in the intersection $L^1(V) \cap L^2(V)$. Assume that T_φ is an orthogonal projection onto a line $\mathbb{C}\varepsilon_\varphi$ for some vector ε_φ in $\mathcal{H}$ of norm 1. Then the map $\pi_{\varepsilon_\varphi} : \mathcal{C}_c^0(V) \to \mathcal{H}$ extends to a continuous linear H-equivariant map $\tau : L^2(V) \to \mathcal{H}$.

PROOF. Observe that for any $f \in L^2(V)$ the convolution $f * \varphi$ lies in $L^1(V)$. Hence we can use

$$T_f \varepsilon_\varphi = T_f T_\varphi \varepsilon_\varphi = T_{f*\varphi}\varepsilon_\varphi.$$

$\square$

The next step is to look for a function φ such that T_φ is an orthogonal projector of rank 1.

LEMMA 2.8. *Let $P \in \mathcal{B}(\mathcal{H})$ be a non-zero self-adjoint bounded operator. Then P is a scalar multiple of an orthogonal projector of rank 1 if and only if, for any $x \in V$, we have*

$$(4.11) \qquad\qquad P\rho(x)P \in \mathbb{C}P.$$

PROOF. We first note that if P is a multiple of an orthogonal projector of rank 1, then clearly $P\rho(x)P \in \mathbb{C}P$ for all $x \in V$.

Let us assume now that P satisfies the latter condition. First, we have $P^2 = \lambda P$ for some non-zero λ. Hence after normalization we can assume $P^2 = P$. Hence P is a projector. It is an orthogonal projector since P is self-adjoint.

It remains to show that P has rank 1. Let u be a non-zero vector in $P(\mathcal{H})$ and M be the span of $\rho(x)u$ for all $x \in V$. The assumption on P implies that M is included in $\mathbb{C}u \oplus \operatorname{Ker} P$. The fact that $\mathcal{H}$ is irreducible implies that M is dense in $\mathcal{H}$. Hence we have $\mathcal{H} = \mathbb{C}u \oplus \operatorname{Ker} P$. Thus P has rank 1.   □

LEMMA 2.9. *Let φ be an element in $L^1(V)$ such that $\varphi = \varphi^*$. Then T_φ is a multiple of an orthogonal projection on a line if and only if, for all $u \in V$, the function*

$$x \mapsto \varphi(u+x)\varphi(u-x)$$

is its own Fourier transform.

REMARK 2.10. A priori, the Fourier transform is defined on the dual V^* of V, but those spaces are identified viaC the symplectic form ω.

We will refer to the characterization of φ from Lemma 2.9 as the *functional equation*.

PROOF. We use Lemma 2.8. Let $v \in V$. We compute

$$T_\varphi \rho(v) T_\varphi = \iint_{V \times V} \varphi(x)\varphi(y)\rho(x)\rho(v)\rho(y)\,dx\,dy$$

$$= \iint_{V \times V} \varphi(x)\varphi(y) e^{i\pi((x|v)+(x|y)+(v|y))} \rho(x+v+y)\,dx\,dy$$

$$= \iint_{V \times V} \varphi(x)\varphi(z-v-x) e^{i\pi((x|v)+(x|z)+(v|z))} \rho(z)\,dx\,dz.$$

For almost every value of z, this operator is T_ψ for

$$\psi(v,z) = \int_V \varphi(x)\varphi(z-v-x) e^{i\pi((x|v)+(x|z)+(v|z))}\,dx$$

by Fubini's theorem (consider ψ as a function of the variable v when z is fixed). The relation (4.11) is equivalent to the fact that for every v, $\psi = C(v)\varphi$. So (4.11) is equivalent to:

$$\int_V \varphi(x)\varphi(z - v - x)e^{i\pi((x|v)+(x|z)+(v|z))}dx = C(v)\varphi(z).$$

In the left-hand side, we set $x = -y$ and use $\varphi^* = \varphi$. Then we obtain

$$\int_V \overline{\varphi}(y)\overline{\varphi}(v - z - y)e^{-i\pi((y|v)+(y|z)+(z|v))}dy = C(v)\varphi(z) = \overline{C}(z)\overline{\varphi}(v).$$

Hence

(4.12)
$$\frac{\varphi(z)}{\overline{C}(z)} = \frac{\overline{\varphi}(v)}{C(v)}$$

so that $\frac{\varphi(z)}{\overline{C}(z)}$ does not depend on z, moreover it is equal to its complex conjugate: hence it belongs to $\mathbb{R}$. We set $D = \frac{\varphi(z)}{\overline{C}(z)}$, and get $C(z) = D\varphi(-z)$.

Finally,

$$\int_V \varphi(x)\varphi(z - v - x)e^{i\pi((x|v)+(x|z)+(v|z))}dx = D\varphi(-v)\varphi(z).$$

Now we set $t := x - \frac{1}{2}(z - v)$ and we get

$$\int_V \varphi(\frac{z - v}{2} + t)\varphi(\frac{z - v}{2} - t)e^{i\pi(t|z+v)}dt = D\varphi(-v)\varphi(z).$$

The left-hand side is precisely the value at $\frac{z+v}{2}$ of the Fourier transform of $t \mapsto \varphi(\frac{z-v}{2} + t)\varphi(\frac{z-v}{2} - t)$, the Fourier inversion formula implies $D^2 = 1$ and D is a positive real number as can be seen by setting $z = v$ in (4.12), hence the Lemma. $\square$

In order to find a non-trivial solution of the functional equation, we choose a positive definite quadratic form Q on V and denote by $B : V \to V^*$ the morphism induced by the polarization of Q. We recall (Lemma 1.5) that the Fourier transform of the function $z \mapsto e^{-\pi Q(z)}$ on V is the function $w \mapsto Disc(Q)^{-\frac{1}{2}}e^{-\pi Q^{-1}(w)}$ on V^*. Let $\Omega : V \to V^*$ be the isomorphism induced by the symplectic form ω.

LEMMA 2.11. *The function* $x \mapsto \psi(x) = e^{-\pi Q(x)}$ *is its own Fourier transform if and only if*

$$(\Omega^{-1}B)^2 = -Id_V.$$

PROOF. Straightforward computation. $\square$

LEMMA 2.12. *The function*

$$\varphi(x) := e^{-\pi\frac{Q(x)}{2}}$$

satisfies the functional equation of Lemma 2.9.

PROOF. This is easily shown using the fact that $\varphi(u+x)\varphi(u-x) = \varphi^2(u)\varphi^2(x)$. $\square$

Now by application of Lemma 2.7 we obtain a bounded H-invariant linear operator $\tau : L^2(V) \to \mathcal{H}$. Consider the dual operator $\tau^* : \mathcal{H} \to L^2(V)$. The composition $\tau\tau^*$ is a bounded intertwiner in $\mathcal{H}$. Hence Theorem 1.20 in Chapter 3 implies that $\tau\tau^* = \lambda\,\mathrm{Id}_\mathcal{H}$ for some positive real λ, since $\tau\tau^*$ is positive and self-adjoint.

Next we will show that $\lambda = 1$.

LEMMA 2.13. We have $\tau^*(\varepsilon_\varphi) = \varphi$ and $\tau\tau^* = \mathrm{Id}_\mathcal{H}$.

PROOF. Consider the operator $Y : L^2(V) \to L^2(V)$ defined by

$$Y(f) := \varphi * f * \varphi.$$

Applying Lemma 2.5 and the relations $\varphi * \varphi = \varphi$ and $T_\varphi T_f T_\varphi \in \mathbb{C}T_\varphi$, we get that Y is the orthogonal projection on the line $\mathbb{C}\varphi$. Hence $Y(f) = \langle \varphi, f \rangle_{L^2(V)}\, \varphi$.

If f is orthogonal to $\tau^*(\varepsilon_\varphi)$, then

$$\langle \tau^*(\varepsilon_\varphi), f \rangle_{L^2(V)} = \langle \varepsilon_\varphi, \tau(f) \rangle_\mathcal{H} = \langle \varepsilon_\varphi, T_f(\varepsilon_\varphi) \rangle_\mathcal{H} = 0.$$

This is equivalent to $T_\varphi T_f T_\varphi = T_{Y(f)} = 0$. Hence f is orthogonal to φ. We obtain that $\tau^*(\varepsilon_\varphi) = c\varphi$ for some $c \in \mathbb{C}$. But

$$c = \langle c\varphi, \varphi \rangle_{L^2(V)} = \langle \tau^*(\varepsilon_\varphi), \varphi \rangle_{L^2(V)} = \langle \varepsilon_\varphi, \tau(\varphi) \rangle_\mathcal{H} = \langle \varepsilon_\varphi, \varepsilon_\varphi \rangle_\mathcal{H} = 1.$$

The first assertion is proven.

Now

$$\langle \tau\tau^*(\varepsilon_\varphi), \varepsilon_\varphi \rangle_\mathcal{H} = \langle \tau^*(\varepsilon_\varphi), \tau^*(\varepsilon_\varphi) \rangle_{L^2(V)} = \langle \varphi, \varphi \rangle_{L^2(V)} = 1.$$

Hence the second assertion. $\square$

Thus, we have shown that an arbitrary irreducible unitary representation $\mathcal{H}$ is equivalent to the subrepresentation of $L^2(V)$ generated by $\varphi(x) = e^{-\pi\frac{Q(x)}{2}}$. Therefore, the Stone–von Neumann theorem is proven.

2.3. Fock representation. Let us continue with a lovely avatar of the representation $\mathcal{H}$, the *Fock representation*. We would like to characterize the image $\tau^*(\mathcal{H})$ inside $L^2(V)$.

Just before Lemma 2.11, we chose a quadratic form Q on V such that $(\Omega^{-1}B)^2 = -Id_V$, and this equips V with a structure of complex vector space of dimension g for which $\Omega^{-1}B$ is the scalar multiplication by the imaginary unit i. We denote by J this complex structure and by V_J the corresponding complex space.

Furthermore $B + i\Omega : V_J \to V_J^*$ is a sesquilinear isomorphism, we denote by A the corresponding Hermitian form on V_J.

In this context, for a given $x \in V$ we have:

$$(4.13) \quad r_{(1,x)}\varphi(y) = \varphi(y-x)e^{i\pi(x|y)} = e^{-\pi\frac{Q(y-x)}{2}+i\pi(x|y)} = e^{-\pi\left(\frac{Q(x)+Q(y)}{2}-A(x,y)\right)}$$

which is the product of $\varphi(y)$ with a holomorphic function $f(y) = e^{-\pi(\frac{Q(x)}{2} - A(x,y))}$.

The *Fock representation* associated to the complex structure J is the subspace $\mathcal{F}_J \subset L^2(V)$ consisting of the products $f\varphi$ where φ was defined before and f is a holomorphic function on V_J. We have just proven that this space is stable under the H-action. Moreover, it is closed in $L^2(V)$ since holomorphy is preserved under uniform convergence on compact sets. Let us choose complex coordinates $z = (z_1, \ldots, z_g)$ in V_J so that the Hermitiam product has the form $A(w, z) = \sum \bar{w}_i z_i$. The scalar product $(\cdot, \cdot)_F$ in $\mathcal{F}_J$ is given by

$$(f\varphi, g\varphi)_F = \int_V \bar{f}(z)g(z)e^{-\pi|z|^2} d\bar{z}dz,$$

where $|z|^2 = \sum_{i=1}^g |z_i|^2$. If $\mathbf{m} = (m_1, \ldots, m_g) \in \mathbb{N}^g$ we denote by $z^{\mathbf{m}}$ the monomial function $z_1^{m_1} \ldots z_g^{m_g}$. Any analytic function $f(z)$ can be represented by a convergent series

(4.14) $$f(z) = \sum_{\mathbf{m} \in \mathbb{N}^g} a_{\mathbf{m}} z^{\mathbf{m}}.$$

EXERCISE 2.14. Check that if $f(z)\varphi \in \mathcal{F}_J$ then the series

$$f(z)\varphi = \sum_{\mathbf{m} \in \mathbb{N}^g} a_{\mathbf{m}} z^{\mathbf{m}} \varphi$$

is convergent in the topology defined by the norm in $\mathcal{F}_J$. Furthermore, prove that $\{z^{\mathbf{m}} \mid \mathbf{m} \in \mathbb{N}^g\}$ is an orthogonal topological basis of $\mathcal{F}_J$.

LEMMA 2.15. *The image $\tau^*(\mathcal{H})$ is equal to $\mathcal{F}_J$. Hence the representation of H in $\mathcal{F}_J$ is irreducible.*

PROOF. Recall that $\tau^*(\varepsilon_\varphi) = \varphi$. Therefore taking into account (4.13) it is sufficient to show that the set $\{e^{-\pi A(x,y)}\varphi(y) \mid x \in V\}$ is dense in $\mathcal{F}_J$. Let $f\varphi \in \mathcal{F}_J$. Assume that

$$(f(y)\varphi(y), e^{-\pi A(x,y)})_F = 0 \quad \text{for all } x \in V.$$

In the z-coordinates it amounts to saying that

$$F(w) = \int_V \bar{f}(z)e^{\sum_{i=1}^g w_i z_i} e^{-\pi|z|^2} d\bar{z}dz$$

is identically zero. Note that then the partial derivative

$$\frac{\partial F}{\partial w_i} = \int_V z_i \bar{f}(z)e^{\sum_{i=1}^g w_i z_i} e^{-\pi|z|^2} d\bar{z}dz,$$

is also zero. Hence for every monomial $z^{\mathbf{m}}$ and $w \in V_J$ we have

$$\int_V z^{\mathbf{m}} \bar{f}(z)e^{\sum_{i=1}^g w_i z_i} e^{-\pi|z|^2} d\bar{z}dz = 0.$$

Consider the Taylor series (4.14). By Exercise 2.14 we have for any $w \in V_J$

$$\int_V f(z)\bar{f}(z)e^{\sum_{i=1}^g w_i z_i}e^{-\pi|z|^2}\,d\bar{z}dz = 0,$$

in particular,

$$\int_V f(z)\bar{f}(z)d\bar{z}dz = 0,$$

which implies $f(z) = 0$. Hence the set $\{e^{-\pi A(x,y)}\varphi(y) \mid x \in V\}$ is dense in $\mathcal{F}_J$. $\qquad\square$

EXERCISE 2.16. Check that $f * \varphi \in \mathcal{F}_J$ for any $f \in L^2(V)$. Therefore the map $f \mapsto f * \varphi$ from $L^2(V)$ to $L^2(V)$ is an orthogonal projection onto $\mathcal{F}_J$.

2.4. Unitary dual of H. Now it is not hard to classify unitary irreducible representations of the Heisenberg group H. If ρ is an irreducible representation of H in a Hilbert space $\mathcal{H}$, then by Theorem 1.20 Chapter 3, for every $t \in \mathbb{T}$ we have $\rho_t = \chi_t \operatorname{Id}_{\mathcal{H}}$ for some character $\chi \in \widehat{\mathbb{T}}$. In other words, using the description of $\widehat{\mathbb{T}}$, $\rho_t = t^n \operatorname{Id}_{\mathcal{H}}$ for some $n \in \mathbb{Z}$. Hence we have defined the map $\Phi : \widehat{H} \to \mathbb{Z} = \widehat{\mathbb{T}}$.

We know that the fiber $\Phi^{-1}(1) = \{\sigma\}$ is a single point due to the Stone–von Neumann theorem. We claim that for any $n \neq 0$ the fiber $\Phi^{-1}(n)$ is also a single point. Indeed, consider a linear transformation γ of V such that $\langle \gamma(x)|\gamma(y)\rangle = n\langle x|y\rangle$. Then we can define a homomorphism $\tilde{\gamma} : H \to H$ by setting $\tilde{\gamma}(t, x) = (t^n, \gamma(x))$. We have the exact sequence of groups

$$1 \to \mathbb{Z}/n\mathbb{Z} \to H \xrightarrow{\tilde{\gamma}} H \to 1.$$

If ρ lies in the fiber over n, then $\operatorname{Ker}\rho \subset \operatorname{Ker}\tilde{\gamma}$. Hence $\rho = \tilde{\gamma} \circ \rho'$, where ρ' lies in $\Phi^{-1}(1)$. Thus, $\rho \simeq \tilde{\gamma} \circ \sigma$.

Finally, $\Phi^{-1}(0)$ consists of all representations which are trivial on $\mathbb{T}$. Therefore $\Phi^{-1}(0)$ coincides with the unitary dual of $V = H/\mathbb{T}$ and hence it is isomorphic to V^*.

3. Representations of SL$_2$ (ℝ)

In this section we give a construction of all unitary irreducible representations of the group $SL_2(\mathbb{R})$, up to isomorphism. We do not provide a proof that our list is complete and refer to [18] for this.

3.1. Geometry of $SL_2(\mathbb{R})$. In this section we use the notation

$$G = \text{SL}_2(\mathbb{R}) = \{g \in \text{GL}_2(\mathbb{R}) \mid \det g = 1\}.$$

EXERCISE 3.1. (a) Since $G = \{\begin{pmatrix} a & b \\ c & d \end{pmatrix} \mid ad - bc = 1\}$, topologically G can be described as a non-compact 3-dimensional quadric in $\mathbb{R}^4$.

(b) Describe conjugacy classes in G.

(c) The only proper non-trivial normal closed subgroup of G is the center $\{1, -1\}$.

Let us start with the following observation.

LEMMA 3.2. Let $\rho : G \to GL(V)$ be a unitary finite-dimensional representation of G. Then ρ is trivial.

PROOF. Let $g = \begin{pmatrix} 1 & 1 \\ 0 & 1 \end{pmatrix}$. Then g^k is conjugate to g for every non-zero integer k. Hence $\operatorname{Tr} \rho_g = \operatorname{Tr} \rho_{g^k}$. Note that ρ_g is unitary and hence diagonalizable in V. Let $\lambda_1, \ldots, \lambda_n$ be the eigenvalues of ρ_g (taken with muliplicities). Then for any $k \neq 0$ we have

$$\lambda_1 + \cdots + \lambda_n = \lambda_1^k + \cdots + \lambda_n^k.$$

Hence $\lambda_1 = \cdots = \lambda_n = 1$. Then $g \in \operatorname{Ker} \rho$. By Exercise 3.1(c) we have $G = \operatorname{Ker} \rho$. $\square$

Let K be the subgroup of matrices

$$g_\theta = \begin{pmatrix} \cos \theta & \sin \theta \\ -\sin \theta & \cos \theta \end{pmatrix}.$$

The group K is a maximal compact subgroup of G, and clearly K is isomorphic to $\mathbb{T} = S^1$. If $\rho : G \to GL(V)$ is a unitary representation of G in a Hilbert space then the restricted K-representation $\operatorname{Res}_K \rho$ splits into the sum of 1-dimensional representations of K. In particular, one can find $v \in V$ such that, for some n, $\rho_{g_\theta}(v) = e^{in\theta}v$. We define the matrix coefficient function $f : G \to \mathbb{C}$ by the formula

$$f(g) = \langle v, \rho_g v \rangle.$$

Then f satisfies the condition

$$f(gg_\theta) = e^{in\theta} f(g).$$

One can consider f as a section of a line bundle on the space G/K (if $n = 0$, then f is a function). Thus, it is clear that the space G/K is an important geometric object, on which the representations of G are "realized". To be a trifle more precise, consider the quotient $(G \times \mathbb{C})/K$ where K acts on G by right multiplication and on $\mathbb{C}$ by $e^{in\theta}$. It is a topological line bundle on G/K, and one can see f as a section of this bundle.

Consider the Lobachevsky plane

$$H = \{z = x + iy \in \mathbb{C} \mid y > 0\}$$

equipped with the Riemannian metric defined by the formula $\frac{dx^2 + dy^2}{y^2}$ and the corresponding volume form $\frac{dxdy}{y^2}$. Then G coincides with the group of rigid motions of H preserving orientation. The action of the matrix $\begin{pmatrix} a & b \\ c & d \end{pmatrix} \in G$ on H is given by the formula

$$z \mapsto \frac{az + b}{cz + d}.$$

EXERCISE 3.3. Check that G acts transitively on H, preserves the metric and the volume form. Moreover, the stabilizer of $i \in H$ coincides with K. Thus, we identify H with G/K.

3.2. Discrete series. The discrete series are the representations with matrix coefficients in $L^2(G)$. For $n \in \mathbb{Z}_{>1}$, let $\mathcal{H}_n^+$ be the space of holomorphic densities on H, i.e. the set of formal expressions $\varphi(z)(dz)^{n/2}$, where $\varphi(z)$ is a holomorphic function on H satisfying the condition that the integral

$$\int_H |\varphi|^2 y^{n-2} dz d\bar{z}$$

is finite. Define a representation of G in $\mathcal{H}_n^+$ by the formula

$$\rho_g\left(\varphi(z)(dz)^{n/2}\right) = \varphi\left(\frac{az+b}{cz+d}\right) \frac{1}{(cz+d)^n} (dz)^{n/2},$$

and a Hermitian product on $\mathcal{H}_n$ by the formula

$$(4.15) \qquad \left\langle \varphi(dz)^{n/2}, \psi(dz)^{n/2} \right\rangle = \int_H \bar{\varphi}\psi y^{n-2} dz d\bar{z},$$

for $n > 1$. For $n = 1$ the product is defined by

$$(4.16) \qquad \left\langle \varphi(dz)^{n/2}, \psi(dz)^{n/2} \right\rangle = \int_{-\infty}^{\infty} \bar{\varphi}\psi dx,$$

in this case $\mathcal{H}_1^+$ consists of all densities which converge to L^2-functions on the boundary (real line).

EXERCISE 3.4. Check that this Hermitian product is invariant.

To show that $\mathcal{H}_n^+$ is irreducible it is convenient to consider the Poincaré model of the Lobachevsky plane using the conformal map

$$w = \frac{z-i}{z+i},$$

that maps H to the unit disk $|w| < 1$. Then the group G acts on the unit disk by linear fractional maps $w \mapsto \frac{aw+b}{\bar{b}w+\bar{a}}$ for all complex a, b satisfying $|a|^2 - |b|^2 = 1$, and K is defined by the condition $b = 0$. If $a = e^{i\theta}$, then $\rho_{g_\theta}(w) = e^{2i\theta}w$. The invariant volume form is $\frac{dw d\bar{w}}{1 - \bar{w}w}$.

It is clear that $w^k (dw)^{n/2}$ for all $k \geq 0$ form an orthogonal basis in $\mathcal{H}_n^+$, each vector $w^k (dw)^{n/2}$ is an eigenvector with respect to K, namely

$$\rho_{g_\theta}\left(w^k (dw)^{n/2}\right) = e^{(2k+n)i\theta} w^k (dw)^{n/2}.$$

It is now easy to check that $\mathcal{H}_n^+$ is irreducible. Indeed, every invariant closed subspace M in $\mathcal{H}_n^+$ has a topological basis consisting of eigenvectors of K, in other words $w^k (dw)^{n/2}$ for all positive k must form a topological basis of M. Without loss of generality assume that M contains $(dw)^{n/2}$, then by applying ρ_g one can get that $\frac{1}{(bw+a)^n}(dw)^{n/2}$, and in Taylor series for $\frac{1}{(bw+a)^n}$ all elements of the basis appear with non-zero coefficients. That implies $w^k(dw)^{n/2} \in M$ for all $k \geq 0$, hence $M = \mathcal{H}_n^+$.

One can construct another series $\mathcal{H}_n^-$ by considering holomorphic densities in the lower half-plane $\operatorname{Re} z < 0$.

EXERCISE 3.5. Check that all representations in the discrete series $\mathcal{H}_n^{\pm}$ are pairwise non-isomorphic.

3.3. Principal series. The representations of the principal series are parameterized by a continuous parameter $s \in \mathbb{R}i \, (s \neq 0)$. Consider the action of G on the real line by linear fractional transformations $x \mapsto \frac{ax+b}{cx+d}$. Let $\mathcal{P}_s^+$ denote the space of densities $\varphi(x)(dx)^{\frac{1+s}{2}}$ with G-action given by

$$\rho_g\left(\varphi(x)(dx)^{\frac{1+s}{2}}\right) = \varphi\left(\frac{ax+b}{cx+d}\right)|cx+d|^{-s-1}.$$

The Hermitian product given by

$$\langle \varphi, \psi \rangle = \int_{-\infty}^{\infty} \bar{\varphi}\psi \, dx \tag{4.17}$$

is invariant. The property of invariance justifies the choice of weight for the density as $(dx)^{\frac{1+s}{2}}(dx)^{\frac{1+\bar{s}}{2}} = dx$. To check that the representation is irreducible one can move the real line to the unit circle as in the example of discrete series and then use $e^{ik\theta}(d\theta)^{\frac{1+s}{2}}$ as an orthonormal basis in $\mathcal{P}_s^+$. Note that the eigenvalues of ρ_{g_θ} in this case are $e^{2ki\theta}$ for all integer k.

The second principal series $\mathcal{P}_s^-$ can be obtained if instead of densities we consider the pseudo-densities which are transformed by the law

$$\rho_g\left(\varphi(x)(dx)^{\frac{1+s}{2}}\right) = \varphi\left(\frac{ax+b}{cx+d}\right)|cx+d|^{-s-1}\operatorname{sgn}(cx+d)\,dx^{\frac{1+s}{2}}.$$

3.4. Complementary series. Those are representations which do not appear in the regular representation $L^2(G)$. They can be realized as the representations in $\mathcal{C}_s$ of all densities $\varphi(x)(dx)^{\frac{1+s}{2}}$ for real $0 < s < 1$ and have an invariant Hermitian product

$$\langle \varphi, \psi \rangle = \int_{-\infty}^{\infty}\int_{-\infty}^{\infty} \bar{\varphi}(x)\psi(y)|x-y|^{s-1}\,dx\,dy. \tag{4.18}$$

CHAPTER 5

On algebraic methods

et... plusieurs ratons laveurs. (J. Prévert, L'inventaire)

In which we put together all the facts in algebra coming to our mind for further use. We apologize for the kaleidoscopic outcome.

1. Introduction

This chapter deals with modules over a general ring R: in particular, we do not assume that R is an algebra over some field. We may consider R-modules which are not finitely generated. We want to be able to consider infinite direct sums and products; since by definition the direct sum is the set of families with finite support, the injection in the product is not an isomorphism any more.

The R-modules are difficult to classify in general: the semisimple ones have the nice property that every submodule appears as a direct factor, but there are other modules, which are not completely reducible though they are not irreducible. Homological algebra provides a powerful tool towards the classification of R-modules.

Many results have a nice formulation as long as we accept the Axiom of choice and make use of Zorn's Lemma: for instance, "every vector space has a basis", or "every non-zero ring contains a maximal left-ideal". Let us just recall the statement:

LEMMA 1.1. *(Zorn) Let X be a non-empty partially ordered set, assume that every totally ordered subset of X admits an upper bound. Then X has a maximal element.*

2. Semisimple modules and density theorem

2.1. Semisimplicity. Let R be a unital ring. We will use indifferently the terms R-module and module whenever the context is clear.

DEFINITION 2.1. An R-module M is *semisimple* if for any submodule $N \subset M$ there exists a submodule N' of M such that $M = N \oplus N'$.

Recall that a non-zero R-module M is *simple* if any submodule of M is either M or 0. Clearly, a simple module is semisimple.

EXERCISE 2.2. Show that if M is a semisimple R-module and if N is a quotient of M, then N is isomorphic to some submodule of M.

© Springer Nature Switzerland AG 2018
C. Gruson and V. Serganova, *A Journey Through Representation Theory*,
Universitext, https://doi.org/10.1007/978-3-319-98271-7_5

LEMMA 2.3. *Every submodule, every quotient of a semisimple R-module is semisimple.*

PROOF. Let N be a submodule of a semisimple module M, and let P be a submodule of N. By semisimplicity of M, there exists a submodule $P' \subset M$ such that $M = P \oplus P'$, then there exists an R-invariant projector $p : M \to P$ with kernel P'. The restriction of p to N defines the projector $N \to P$, and the kernel of this projector is the complement of P in N. Apply Exercise 2.2 to complete the proof.   □

For what comes next, it is essential that the ring R is unital. Indeed it is necessary to have this property to ensure that R has a maximal left ideal and this can be proven using Zorn's Lemma.

LEMMA 2.4. *Any semisimple R-module contains a simple submodule.*

PROOF. Let M be semisimple, $m \in M$. Then Rm is isomorphic to R/J for some left ideal $J \subset R$. Let I be a maximal left ideal in R containing J. Then Rm is semisimple by Lemma 2.3 and $Rm = Im \oplus N$. We claim that N is simple. Indeed, every submodule of Rm is of the form Km for some left ideal $K \subset R$. If N' is a submodule of N, then $N' \oplus Im = Km$ and hence $I \subset K$. But, by maximality of I, $K = I$ or R, therefore $N' = 0$ or N.   □

LEMMA 2.5. *Let M be an R-module. The following conditions are equivalent:*

(1) *M is semisimple;*
(2) *$M = \sum_{i \in I} M_i$ for a family of simple submodules M_i of M indexed by a set I, i.e. M is generated by the union of the M_is.*
(3) *$M = \bigoplus_{j \in J_0} M_j$ for a family of simple submodules M_j of M indexed by a set J_0.*

PROOF. (1) $\Rightarrow$ (2) Let $\{M_i\}_{i \in I}$ be the collection of all simple submodules of M. We want to show that $M = \sum_{i \in I} M_i$. Let $N = \sum_{i \in I} M_i$ and assume that N is a proper submodule of M. Then $M = N \oplus N'$ by the semisimplicity of M. By Lemma 2.4, N' contains a simple submodule which cannot be contained in the family $\{M_i\}_{i \in I}$. Contradiction.

Let us prove (2) $\Rightarrow$ (3). We consider all possible families $\{M_j\}_{j \in J}$ of simple submodules of M such that $\sum_{j \in J} M_j = \oplus_{j \in J} M_j$. First, we note the set of such families satisfies the conditions of Zorn's lemma, namely that any totally ordered subset of such families has a maximal element, where the order is the inclusion order. To check this just take the union of all sets in a totally ordered subset. Hence there is a maximal subset J_0 and the corresponding maximal family $\{M_j\}_{j \in J_0}$ which satisfies $\sum_{j \in J_0} M_j = \oplus_{j \in J} M_j$. We claim that the sum $M = \oplus_{j \in J_0} M_j$. Indeed, otherwise we would have a simple submodule M_k which does not belong to $\oplus_{j \in J_0} M_j$. But then $\sum_{j \in J_0 \cup k} M_j = \oplus_{j \in J_0 \cup k} M_j$ which contradicts maximality of J_0.

Finally, let us prove (3) $\Rightarrow$ (1). Let $N \subset M$ be a submodule and $S \subset J_0$ be a maximal subset such that $N \cap (\oplus_{j \in S} M_j) = 0$ (Zorn's lemma once more). Let $M' = N \oplus (\oplus_{j \in S} M_j)$. We claim that $M' = M$. Indeed, otherwise there exists $k \in J_0$ such that M_k does not belong to M'. Then $M_k \cap M' = 0$ by simplicity of M_k, and therefore $N \cap (\oplus_{j \in S \cup k} M_j) = 0$. Contradiction. $\square$

EXERCISE 2.6. If R is a field, after noticing that an R-module is a vector space, show that every simple R-module is one-dimensional, and therefore, through the existence of bases, show that every module is semisimple.

EXERCISE 2.7. If $R = \mathbb{Z}$, then some R-modules are not semisimple, for instance $\mathbb{Z}$ itself.

LEMMA 2.8. *Let M be a semisimple module. Then M is simple if and only if $\operatorname{End}_R (M)$ is a division ring.*

PROOF. In one direction this is Schur's lemma. In the opposite direction let $M = M_1 \oplus M_2$ for some proper submodules M_1 and M_2 of M. Let p_1, p_2 be the canonical projections onto M_1 and M_2, respectively. Then $p_1 \circ p_2 = 0$ and therefore p_1, p_2 cannot be invertible. $\square$

2.2. Jacobson density theorem. Let M be any R-module. Set $K := \operatorname{End}_R (M)$, then set $S := \operatorname{End}_K (M)$. There exists a natural homomorphism $R \to S$. In general it is neither surjective nor injective. In the case when M is semisimple it is very close to being surjective.

THEOREM 2.9. *(Jacobson density theorem). Assume that M is semisimple. Then for any $m_1, \ldots, m_n \in M$ and $s \in S$ there exists $r \in R$ such that $rm_i = sm_i$ for all $i = 1, \ldots, n$.*

PROOF. First let us prove the statement for $n = 1$. We just have to show that $Rm_1 = Sm_1$. The inclusion $Rm_1 \subset Sm_1$ is obvious. We will prove the inverse inclusion. The semisimplicity of M implies $M = Rm_1 \oplus N$ for some R-submodule N of M. Let p be the projector $M \to N$ with kernel Rm_1. Then $p \in K$ and therefore $p \circ s = s \circ p$ for every $s \in S$. Hence $\operatorname{Ker} p$ is S-invariant. So $Sm_1 \subset Rm_1$.

For arbitrary n we use the following lemma.

LEMMA 2.10. *Let $\widehat{K} := \operatorname{End}_R (M^{\oplus n})$ and $\widehat{S} := \operatorname{End}_{\widehat{K}} (M^{\oplus n})$. Then $\widehat{K}$ is isomorphic to the matrix ring $\operatorname{Mat}_n (K)$ and $\widehat{S}$ is isomorphic to S. The latter isomorphism is given by the diagonal action*

$$s (m_1, \ldots, m_n) = (sm_1, \ldots, sm_n).$$

EXERCISE 2.11. Adapt the proof of Lemma 1.12 Chapter 2 to check the above lemma. $\square$

COROLLARY 2.12. *Let M be a semisimple R-module, which is finitely generated over K. Then the natural map $R \to \operatorname{End}_K(M)$ is surjective.*

PROOF. Let $m_1, \ldots, m_n$ be generators of M over K, apply Theorem 2.9. $\square$

COROLLARY 2.13. *Let R be an algebra over a field k, and $\rho: R \to \operatorname{End}_k(V)$ be an irreducible finite-dimensional representation of R. Then*

- *There exists a division ring D containing k such that $\rho(R) \cong \operatorname{End}_D(V)$.*
- *If k is algebraically closed, then $D = k$ and therefore ρ is surjective.*

PROOF. Apply Schur's lemma. $\square$

EXERCISE 2.14. Let V be an infinite-dimensional vector space over $\mathbb{C}$ and R be the ring of linear operators in V of the form $k\operatorname{Id} + F$ for all linear operators F with finite-dimensional image. Check that R is a ring and V is a simple R-module. Then $K = \mathbb{C}$, S is the ring of all linear operators in V and R is dense in S but R does not coincide with S.

3. Wedderburn–Artin theorem

A ring R is called *semisimple* if every R-module is semisimple. For example, a group algebra $k(G)$, for a finite group G such that $\operatorname{char} k$ does not divide $|G|$, is semisimple by Maschke's Theorem 3.3 in Chapter 1.

LEMMA 3.1. *A ring R is semisimple if and only if as a module over itself R is isomorphic to a finite direct sum of minimal left ideals.*

PROOF. Consider R as an R-module. By definition the simple submodules of R are exactly the minimal left ideals of R. Hence since R is semisimple we can write R as a direct sum $\oplus_{i \in I} L_i$ of minimal left ideals L_i. It remains to show that this direct sum is finite. Indeed, let $l_i \in L_i$ be the image of the identity element 1 under the projection $R \to L_i$. But R as a module is generated by 1. Therefore $l_i \neq 0$ for all $i \in I$. Hence I is finite. Converse statement follows from the fact that every module is a quotient of a free module. $\square$

COROLLARY 3.2. *A direct product of finitely many semisimple rings is semisimple.*

EXERCISE 3.3. Let D be a division ring, and $R = \operatorname{Mat}_n(D)$ be a matrix ring over D.

(a) Let L_i be the subset of R consisting of all matrices which have zeros everywhere outside the i-th column. Check that L_i is a minimal left ideal of R and that $R = L_1 \oplus \cdots \oplus L_n$. Therefore R is semisimple.

(b) Show that L_i and L_j are isomorphic R-modules and that any simple R-module is isomorphic to L_i.

(c) Using Corollary 2.12 show that $F := \operatorname{End}_R(L_i)$ is isomorphic to D^{op}, and that R is isomorphic to $\operatorname{End}_F(L_i)$.

By the above exercise and Corollary 3.2 a direct product $\mathrm{Mat}_{n_1}(D_1) \times \cdots \times \mathrm{Mat}_{n_k}(D_k)$ of finitely many matrix rings is semisimple. In fact any semisimple ring is of this form.

THEOREM 3.4. *(Wedderburn–Artin) Let R be a semisimple ring. Then there exist division rings $D_1, \ldots, D_k$ such that R is isomorphic to a finite product of matrix rings*

$$\mathrm{Mat}_{n_1}(D_1) \times \cdots \times \mathrm{Mat}_{n_k}(D_k).$$

Furthermore, $D_1, \ldots, D_k$ are unique up to isomorphism and this presentation of R is unique up to permutation of the factors.

PROOF. Take the decomposition of Lemma 3.1 and combine isomorphic factors together. Then the following decomposition holds

$$R = L_1^{\oplus n_1} \oplus \cdots \oplus L_k^{\oplus n_k},$$

where L_i is not isomorphic to L_j if $i \neq j$. Set $J_i = L_i^{\oplus n_i}$. We claim that J_i is actually a two-sided ideal. Indeed Lemma 1.10 of Chapter 2 and simplicity of L_i imply that $L_i r$ is isomorphic to L_i for any $r \in R$ such that $L_i r \neq 0$. Thus, $L_i r \subset J_i$.

Now we will show that each J_i is isomorphic to a matrix ring. Let $F_i := \mathrm{End}_{J_i}(L_i)$. The natural homomorphism $J_i \to \mathrm{End}_{F_i}(L_i)$ is surjective by Corollary 2.12. This homomorphism is also injective since $rL_i = 0$ implies $rJ_i = 0$ for any $r \in R$. Then, since J_i is a unital ring $r = 0$. On the other hand, F_i is a division ring by Schur's lemma. Therefore we have an isomorphism $J_i \simeq \mathrm{End}_{F_i}(L_i)$. By Exercise 1.7 in Chapter 2 L_i is a free F_i-module. Moreover, L_i is finitely generated over F_i as J_i is a sum of finitely many left ideals. Thus, by Exercise 3.3 (c), J_i is isomorphic to $\mathrm{Mat}_{n_i}(D_i)$ where $D_i = F_i^{op}$.

The uniqueness of presentation follows easily from Krull–Schmidt theorem (Theorem 4.19) which we prove in the next section. Indeed, let $S_1, \ldots, S_k$ be a complete list of non-isomorphic simple R-modules. Then both D_i and n_i are defined intrinsically, since $D_i^{op} \simeq \mathrm{End}_R(S_i)$ and n_i is the multiplicity of the indecomposable module S_i in R.   □

4. Jordan-Hölder theorem and indecomposable modules

Let R be a unital ring.

4.1. Artinian and Noetherian modules.

DEFINITION 4.1. We say that an R-module M is *Noetherian* or satisfies the ascending chain condition (ACC for short) if every increasing sequence

$$M_1 \subset M_2 \subset \ldots$$

of submodules of M stabilizes.

Similarly, we say that M is Artinian or satisfies the descending chain condition (DCC) if every decreasing sequence

$$M_1 \supset M_2 \supset \cdots$$

of submodules of M stabilizes.

EXERCISE 4.2. Consider $\mathbb{Z}$ as a module over itself. Show that it is Noetherian but not Artinian.

EXERCISE 4.3. (a) A submodule or a quotient of a Noetherian (respectively, Artinian) module is always Noetherian (resp. Artinian).

(b) Let

$$0 \to N \to M \to L \to 0$$

be an exact sequence of R-modules. Assume that both N and L are Noetherian (respectively, Artinian), then M is also Noetherian (respectively, Artinian).

EXERCISE 4.4. Let M be a semisimple module. Prove that M is Noetherian if and only if it is Artinian.

4.2. Jordan-Hölder theorem.

DEFINITION 4.5. Let M be an R-module. A finite sequence of submodules of M

$$M = M_0 \supset M_1 \supset \cdots \supset M_k = 0$$

such that M_i/M_{i+1} is a simple module for all $i = 0, \ldots, k-1$ is called a *Jordan-Hölder series* of M.

LEMMA 4.6. *An R-module M has a Jordan-Hölder series if and only if M is both Artinian and Noetherian.*

PROOF. Let M be an R-module which is both Artinian and Noetherian. Then it is easy to see that there exists a finite sequence of properly included submodules of M

$$M = M_0 \supset M_1 \supset \cdots \supset M_k = 0$$

which cannot be refined. Then M_i/M_{i+1} is a simple module for all $i = 0, \ldots, k-1$.

Conversely, assume that M has a Jordan-Hölder series

$$M = M_0 \supset M_1 \supset \cdots \supset M_k = 0.$$

We prove that M is both Noetherian and Artinian by induction on k. If $k = 1$, then M is simple and hence both Noetherian and Artinian. For $k > 1$ consider the exact sequence

$$0 \to M_1 \to M \to M/M_1 \to 0$$

and use Exercise 4.3 (b). $\qquad\square$

We say that two Jordan-Hölder series of M

$$M = M_0 \supset M_1 \supset \cdots \supset M_k = 0$$

and

$$M = N_0 \supset N_1 \supset \cdots \supset N_l = 0$$

are *equivalent* if $k = l$ and for some permutation s of indices $1, \ldots, k-1$ we have $M_i/M_{i+1} \cong N_{s(i)}/N_{s(i)+1}$.

THEOREM 4.7. *Let M be an R-module which is both Noetherian and Artinian. Let*

$$M = M_0 \supset M_1 \supset \cdots \supset M_k = 0$$

and

$$M = N_0 \supset N_1 \supset \cdots \supset N_l = 0$$

be two Jordan-Hölder series of M. Then they are equivalent.

PROOF. First note that if M is simple, then the statement is trivial. We will prove that if the statement holds for any proper submodule of M then it is also true for M. If $M_1 = N_1$, then the statement is obvious. Otherwise, $M_1 + N_1 = M$, hence we have two isomorphisms $M/M_1 \cong N_1/(M_1 \cap N_1)$ and $M/N_1 \cong M_1/(M_1 \cap N_1)$, like the second isomorphism theorem for groups. Now let

$$M_1 \cap N_1 \supset K_1 \supset \cdots \supset K_s = 0$$

be a Jordan-Hölder series for $M_1 \cap N_1$. This gives us two new Jordan-Hölder series of M

$$M = M_0 \supset M_1 \supset M_1 \cap N_1 \supset K_1 \supset \cdots \supset K_s = 0$$

and

$$M = N_0 \supset N_1 \supset N_1 \cap M_1 \supset K_1 \supset \cdots \supset K_s = 0.$$

These series are obviously equivalent. By our assumption on M_1 and N_1 the first series is equivalent to $M = M_0 \supset M_1 \supset \cdots \supset M_k = 0$, and the second one is equivalent to $M = N_0 \supset N_1 \supset \cdots \supset N_l = \{0\}$. Hence the original series are also equivalent. $\square$

Thus, we can now give two definitions:

DEFINITION 4.8. First, we define the *length $l(M)$* of an R-module M which satisfies ACC and DCC as the length of any Jordan-Hölder series of M. Note that we can easily see that if N is a proper submodule of M, then $l(N) < l(M)$.

Furthermore, this gives rise to the notion of *finite length R-module*.

REMARK 4.9. Note that in the case of *infinite series* with simple quotients, we may have many non-equivalent series. For example, consider $\mathbb{Z}$ as a $\mathbb{Z}$-module. Then the series

$$\mathbb{Z} \supset 2\mathbb{Z} \supset 4\mathbb{Z} \supset \ldots$$

is not equivalent to

$$\mathbb{Z} \supset 3\mathbb{Z} \supset 9\mathbb{Z} \supset \ldots.$$

4.3. Indecomposable modules and Krull–Schmidt theorem. A module M is *indecomposable* if it is not zero and $M = M_1 \oplus M_2$ implies $M_1 = 0$ or $M_2 = 0$.

EXAMPLE 4.10. Every simple module is indecomposable. Furthermore, if a semisimple module M is indecomposable then M is simple.

DEFINITION 4.11. An element $e \in R$ is called an *idempotent* if $e^2 = e$.

LEMMA 4.12. *An R-module M is indecomposable if and only if every idempotent in $\mathrm{End}_R(M)$ is either 1 or 0.*

PROOF. If M is decomposable, then $M = M_1 \oplus M_2$ for some proper submodules M_1 and M_2. Then the projection $e : M \to M_1$ with kernel M_2 is an idempotent in $\mathrm{End}_R M$, which is neither 0 nor 1. Conversely, any non-trivial idempotent $e \in \mathrm{End}_R M$ gives rise to a decomposition $M = \mathrm{Ker}\, e \oplus \mathrm{Im}\, e$. □

EXERCISE 4.13. Show that $\mathbb{Z}$ is an indecomposable module over itself, although it is not simple.

LEMMA 4.14. *Let M and N be indecomposable R-modules, $\alpha \in \mathrm{Hom}_R(M, N)$, $\beta \in \mathrm{Hom}_R(N, M)$ be such that $\beta \circ \alpha$ is an isomorphism. Then α and β are isomorphisms.*

PROOF. We claim that $N = \mathrm{Im}\, \alpha \oplus \mathrm{Ker}\, \beta$. Indeed, since $\mathrm{Im}\, \alpha \cap \mathrm{Ker}\, \beta \subset \mathrm{Ker}\, \beta \circ \alpha$, we have $\mathrm{Im}\, \alpha \cap \mathrm{Ker}\, \beta = 0$. Furthermore, for any $x \in N$ set $y := \alpha \circ (\beta \circ \alpha)^{-1} \circ \beta(x)$ and $z = x - y$. Then $\beta(y) = \beta(x)$. One can write $x = y + z$, where $z \in \mathrm{Ker}\, \beta$ and $y \in \mathrm{Im}\, \alpha$.

Since N is indecomposable, $\mathrm{Im}\, \alpha = N$, $\mathrm{Ker}\, \beta = 0$; hence N is isomorphic to M. □

LEMMA 4.15. *Let M be an indecomposable R-module of finite length and $\varphi \in \mathrm{End}_R(M)$, then either φ is an isomorphism or φ is nilpotent.*

PROOF. Since M is of finite length and $\mathrm{Ker}\, \varphi^n, \mathrm{Im}\, \varphi^n$ are submodules, there exists $n > 0$ such that $\mathrm{Ker}\, \varphi^n = \mathrm{Ker}\, \varphi^{n+1}$, $\mathrm{Im}\, \varphi^n = \mathrm{Im}\, \varphi^{n+1}$. Then $\mathrm{Ker}\, \varphi^n \cap \mathrm{Im}\, \varphi^n = 0$. The latter implies that the exact sequence

$$0 \to \mathrm{Ker}\, \varphi^n \to M \to \mathrm{Im}\, \varphi^n \to 0$$

splits. Thus, $M = \mathrm{Ker}\, \varphi^n \oplus \mathrm{Im}\, \varphi^n$. Since M is indecomposable, either $\mathrm{Im}\, \varphi^n = 0$, $\mathrm{Ker}\, \varphi^n = M$ or $\mathrm{Ker}\, \varphi^n = 0$, $\mathrm{Im}\, \varphi^n = M$. In the former case φ is nilpotent. In the latter case φ^n is an isomorphism and hence φ is also an isomorphism. □

LEMMA 4.16. *Let M be as in Lemma 4.15 and $\varphi, \varphi_1, \varphi_2 \in \mathrm{End}_R(M)$ such that $\varphi = \varphi_1 + \varphi_2$. Then if φ is an isomorphism, at least one of φ_1 and φ_2 is also an isomorphism.*

PROOF. Without loss of generality we may assume that $\varphi = \mathrm{id}$ (otherwise multiply by φ^{-1}). In this case φ_1 and φ_2 commute. If both φ_1 and φ_2 are nilpotent, then $\varphi_1 + \varphi_2$ is nilpotent, but this is impossible as $\varphi_1 + \varphi_2 = \mathrm{id}$. □

COROLLARY 4.17. *Let M be as in Lemma 4.15. Let $\varphi = \varphi_1 + \cdots + \varphi_k \in \mathrm{End}_R(M)$. If φ is an isomorphism then φ_i is an isomorphism at least for one i.*

EXERCISE 4.18. Let M be of finite length. Show that M has a decomposition

$$M = M_1 \oplus \cdots \oplus M_k,$$

where all M_i are indecomposable.

THEOREM 4.19. *(Krull–Schmidt) Let M be an R-module of finite length. Consider two decompositions*

$$M = M_1 \oplus \cdots \oplus M_k \quad \text{and} \quad M = N_1 \oplus \cdots \oplus N_l$$

such that all M_i and N_j are indecomposable. Then $k = l$, and there exists a permutation s of indices $1, \ldots, k$ such that M_i is isomorphic to $N_{s(i)}$.

PROOF. We prove the statement by induction on k. The case $k = 1$ is clear since in this case M is indecomposable.

Let

$$p_i^{(1)} : M \to M_i, \quad p_j^{(2)} : M \to N_j$$

denote the natural projections, and

$$q_i^{(1)} : M_i \to M, \quad q_j^{(2)} : N_j \to N$$

denote the injections. We have

$$\sum_{j=1}^{l} q_j^{(2)} \circ p_j^{(2)} = \mathrm{id}_M,$$

hence

$$\sum_{j=1}^{l} p_1^{(1)} \circ q_j^{(2)} \circ p_j^{(2)} \circ q_1^{(1)} = \mathrm{id}_{M_1}.$$

By Corollary 4.17 there exists j such that $p_1^{(1)} \circ q_j^{(2)} \circ p_j^{(2)} \circ q_1^{(1)}$ is an isomorphism. After permuting indices we may assume that $j = 1$. Then Lemma 4.14 implies that $p_1^{(2)} \circ q_1^{(1)}$ is an isomorphism between M_1 and N_1. Set

$$M' := M_2 \oplus \cdots \oplus M_k, \quad N' := N_2 \oplus \cdots \oplus N_l.$$

Since M_1 intersects trivially $N' = \mathrm{Ker}\, p_1^{(2)}$ we have $M = M_1 \oplus N'$. But we also have $M = M_1 \oplus M'$. Therefore M' is isomorphic to N'. By induction assumption the statement holds for $M' \simeq N'$. Hence the statement holds for M. □

EXERCISE 4.20. Let $R = \mathbb{C}[x, y, z]/(x^2 + y^2 + z^2 - 1)$. Consider a homomorphism $\varphi : R \oplus R \oplus R \to R$ defined by

$$\varphi(a, b, c) = xa + yb + zc.$$

Show that φ is surjective and the kernel of φ is not isomorphic to a free R-module of rank 2. On the other hand

$$R \oplus R \oplus R \simeq R \oplus \operatorname{Ker} \varphi,$$

hence in this case the Krull–Schmidt theorem does not hold.

5. A bit of homological algebra

Homology groups initially come from topology, and they compute some important invariants like the genus of a Riemann surface (which dates back to Riemann of course) and more generally Betti numbers of manifolds.

For a general exposition of the topic of homological algebra, see, for instance, [26], [19], [28] or [37].

Let R be a unital ring.

5.1. Complexes.
Let $C_\bullet = \oplus_{i \geq 0} C_i$ be a graded R-module. An R-morphism f from $C_\bullet$ to $D_\bullet$ is of *degree* k ($k \in \mathbb{Z}$) if f maps C_i to D_{i+k} for all i. An R-*differential* on $C_\bullet$ is an R-morphism d from $C_\bullet$ to $C_\bullet$ of degree -1 such that $d^2 = 0$.

An R-module $C_\bullet$ together with a differential d is called a *complex*.

We usually represent $C_\bullet$ the following way:

$$\ldots \xrightarrow{d} C_i \xrightarrow{d} \ldots \xrightarrow{d} C_1 \xrightarrow{d} C_0 \to 0.$$

REMARK 5.1. It will be convenient to look at similar situations for an R-morphism δ of degree $+1$ on a graded R-module such that $\delta^2 = 0$. In this case, we will use upper indices C^i (instead of C_i) and represent the complex the following way:

$$0 \to C^0 \xrightarrow{\delta} C^1 \xrightarrow{\delta} \ldots \xrightarrow{\delta} C^i \xrightarrow{\delta} \ldots$$

EXERCISE 5.2. (Koszul complex) The following example is very important.

Let V be a finite-dimensional vector space over a field k and denote by V^* its dual space. By $S(V) = \bigoplus S^i(V)$ and $\Lambda(V) = \bigoplus \Lambda^i(V)$ we denote the symmetric and the exterior algebras of V, respectively.

Choose a basis $e_1, \ldots, e_n$ of V and let $f_1, \ldots, f_n$ be the dual basis in V^*, i.e. $f_i(e_j) = \delta_{ij}$. For any $x \in V^*$ we define the linear derivation $\partial_x : S(V) \to S(V)$ given by $\partial_x(v) := x(v)$ for $v \in V$ and extend it to the whole $S(V)$ via the Leibniz relation

$$\partial_x(u_1 u_2) = \partial_x(u_1) u_2 + u_1 \partial_x(u_2) \quad \text{for all} \quad u_1, u_2 \in S(V).$$

Now set $C^k := S(V) \otimes \Lambda^k(V)$ and $C^\bullet := S(V) \otimes \Lambda(V)$. Define $\delta : C^\bullet \to C^\bullet$ by

$$\delta(u \otimes w) := \sum_{j=1}^{n} d_{f_j}(u) \otimes (e_j \wedge w) \quad \text{for all} \quad u \in S(V), w \in \Lambda(V).$$

(a) Show that δ does not depend on the choice of basis in V.

(b) Prove that $\delta^2 = 0$, and therefore $(C^\bullet, \delta)$ is a complex. It is called the *Koszul complex*.

(c) Let $p(w)$ denote the parity of the degree of w if w is homogeneous in $\Lambda(V)$. For any $x \in V^*$ define the linear map $\partial_x : \Lambda(V) \to \Lambda(V)$ by setting $\partial_x(v) := x(v)$ for all $v \in V$ and extend it to the whole $\Lambda(V)$ by the $\mathbb{Z}_2$-graded version of the Leibniz relation

$$\partial_x(w_1 \wedge w_2) = \partial_x(w_1) \wedge w_2 + (-1)^{p(w_1)} w_1 \wedge \partial_x(w_2) \quad \text{for all} \quad w_1, w_2 \in \Lambda(V).$$

Check that one can construct a differential d of degree -1 on the Koszul complex by

$$d(u \otimes w) := \sum_{j=1}^{n} (ue_j) \otimes \partial_{f_j}(w) \quad \text{for all} \quad u \in S(V), w \in \Lambda(V).$$

5.2. Homology and Cohomology. Since in any complex $d^2 = 0$, we have $\mathrm{Im}\, d \subset \mathrm{Ker}\, d$ (in every degree). The complex $(C_\bullet, d)$ is *exact* if $\mathrm{Im}\, d = \mathrm{Ker}\, d$. The key notion of homological algebra is defined below. This notion expresses how far a given complex is from being exact.

DEFINITION 5.3. Let $(C_\bullet, d)$ be a complex of R-modules (with d of degree -1). Its i-th *homology*, $H_i(C_\bullet)$, is the quotient

$$H_i(C_\bullet) = (\mathrm{Ker}\, d \cap C_i) / (\mathrm{Im}\, d \cap C_i).$$

A complex $(C_\bullet, d)$ is exact if and only if $H_i(C_\bullet) = 0$ for all $i \geq 0$.

If $(C^\bullet, \delta)$ is a complex with a differential δ of degree $+1$ we use the term *cohomology* instead of homology and we consistently use upper indices in the notation:

$$H^i(C^\bullet) = \left(\mathrm{Ker}\, \delta \cap C^i\right) / \left(\mathrm{Im}\, \delta \cap C^i\right).$$

DEFINITION 5.4. Given two complexes $(C_\bullet, d)$ and $(C'_\bullet, d')$, a homomorphism $f : C_\bullet \to C'_\bullet$ of R-modules of degree 0 which satisfies the relation $f \circ d = d' \circ f$ is called a *morphism of complexes*.

EXERCISE 5.5. Let $f : C_\bullet \to C'_\bullet$ be a morphism of complexes. Check that $f(\mathrm{Ker}\, d) \subset \mathrm{Ker}\, d'$ and $f(\mathrm{Im}\, d) \subset \mathrm{Im}\, d'$. Therefore f induces a homomorphism

$$f_* : H_i(C_\bullet) \to H_i(C'_\bullet)$$

between homology groups of the complexes.

Let $(C_\bullet, d)$, $(C'_\bullet, d')$, $(C''_\bullet, d'')$ be complexes and $f : C'_\bullet \to C''_\bullet$ and $g : C_\bullet \to C'_\bullet$ be morphisms such that the sequence

$$0 \to C_i \xrightarrow{g} C'_i \xrightarrow{f} C''_i \to 0$$

is exact for all $i \geq 0$.

EXERCISE 5.6. (Snake Lemma) One can define a homomorphism $\delta : H_i(C_\bullet'') \to H_{i-1}(C_\bullet)$ as follows. Let $x \in \operatorname{Ker} d'' \cap C_i''$ and y be an arbitrary element in the preimage $f^{-1}(x) \subset C_i'$. Check that $d'(y)$ lies in the image of g. Pick up $z \in g^{-1}(d'(y)) \subset C_{i-1}$. Show that $z \in \operatorname{Ker} d$. Moreover, show that for a different choice of $y' \in f^{-1}(x) \subset C_i'$ and of $z' \in g^{-1}(d'(y')) \subset C_{i-1}$ the difference $z - z'$ lies in the image of $d : C_i \to C_{i-1}$. Thus, $x \mapsto z$ gives a well-defined map $\delta : H_i(C_\bullet'') \to H_{i-1}(C_\bullet)$.

Why is it called "snake lemma"? Look at the following diagram

$$
\begin{array}{ccccc}
C_i & \xrightarrow{\ g\ } & C_i' & \xrightarrow{\ f\ } & C_i'' \\
\ \downarrow{\scriptstyle d} & & \ \downarrow{\scriptstyle d'} & & \ \downarrow{\scriptstyle d''} \\
C_{i-1} & \xrightarrow{\ g\ } & C_{i-1}' & \xrightarrow{\ f\ } & C_{i-1}''
\end{array}
$$

In this diagram $\delta = g^{-1} \circ d' \circ f^{-1}$ goes from the upper right to the lower left corners.

THEOREM 5.7. (Long exact sequence). The following sequence

$$
\xrightarrow{\delta} H_i(C_\bullet) \xrightarrow{g_*} H_i(C_\bullet') \xrightarrow{f_*} H_i(C_\bullet'') \xrightarrow{\delta} H_{i-1}(C_\bullet) \xrightarrow{g_*} \dots
$$

is an exact complex.

We skip the proof of this theorem. The enthusiastic reader might verify it as an exercise or read the proof in [26] or [37].

5.3. Homotopy.

DEFINITION 5.8. Consider complexes $(C_\bullet, d)$, $(C_\bullet', d')$ of R-modules and let $f, g : C_\bullet \to C_\bullet'$ be morphisms. We say that f and g are *homotopically equivalent* if there exists a map $h : C_\bullet \to C_\bullet'$ of degree 1 such that

$$
f - g = h \circ d + d' \circ h.
$$

LEMMA 5.9. If f and g are homotopically equivalent then $f_* = g_*$.

PROOF. Let $\varphi := f - g$ and $x \in C_i$ such that $dx = 0$. Then

$$
\varphi(x) = h(dx) + d'(hx) = d'(hx) \in \operatorname{Im} d'.
$$

Hence $f_* - g_* = 0$. □

We say that complexes $C_\bullet$ and $C_\bullet'$ are *homotopically equivalent* if there exist $f : C_\bullet \to C_\bullet'$ and $g : C_\bullet' \to C_\bullet$ such that $f \circ g$ is homotopically equivalent to $\operatorname{id}_{C'}$ and $g \circ f$ is homotopically equivalent to id_C. Lemma 5.9 implies that homotopically equivalent complexes have isomorphic homology.

Let $(C_\bullet, d)$ be a complex of R-modules and M be an R-module. Then we have a complex of abelian groups

$$
0 \to \operatorname{Hom}_R(C_0, M) \xrightarrow{\delta} \operatorname{Hom}_R(C_1, M) \xrightarrow{\delta} \dots \xrightarrow{\delta} \operatorname{Hom}_R(C_i, M) \xrightarrow{\delta} \dots,
$$

where $\delta : \mathrm{Hom}_R(C_i, M) \to \mathrm{Hom}_R(C_{i+1}, M)$ is defined by

$$(5.1) \qquad \delta(\varphi)(x) := \varphi(dx) \quad \text{for all} \quad \varphi \in \mathrm{Hom}_R(C_i, M) \quad \text{and} \quad x \in C_{i+1}.$$

Note that the differential δ on $\mathrm{Hom}_R(C_\bullet, M)$ has degree 1. The following Lemma will be used in the next section. The proof is straightforward, and we leave it to the reader.

LEMMA 5.10. *Let $(C_\bullet, d)$ and $(C'_\bullet, d')$ be homotopically equivalent complexes and M be an arbitrary R-module. Then the complexes $(\mathrm{Hom}_R(C_\bullet, M), \delta)$ and $(\mathrm{Hom}_R(C'_\bullet, M), \delta')$ are also homotopically equivalent.*

The following lemma is useful for calculating cohomology.

LEMMA 5.11. *Let $(C_\bullet, d)$ be a complex of R-modules and $h : C_\bullet \to C_\bullet$ be a map of degree 1. Set $f := d \circ h + h \circ d$. Then f is a morphism of complexes. Furthermore, if $f : C_i \to C_i$ is an isomorphism for all $i \geq 0$, then $C_\bullet$ is exact.*

PROOF. First, f has degree 0 and since $d^2 = 0$ we have

$$d \circ f = d \circ h \circ d = f \circ d.$$

Thus, f is a morphism of complexes.

Now let f be an isomorphism. Then $f_* : H_i(C_\bullet) \to H_i(C_\bullet)$ is also an isomorphism for all i. On the other hand, f is homotopically equivalent to 0. Hence, by Lemma 5.9, $f_* = 0$. Therefore $H_i(C_\bullet) = 0$ for all i. $\qquad\square$

EXERCISE 5.12. Recall the Koszul complex $(C^\bullet, \delta)$ from Exercise 5.2. Assume the field k has characteristic zero. Show that $H^i(C^\bullet) = 0$ for $i \geq 0$ and $H^0(C^\bullet) = k$.

Hint. For every $m \geq 0$ consider the subcomplex $C^\bullet_m$ with graded components

$$C^i_m := S^{m-i}(V) \otimes \Lambda^i(V).$$

Check that $d(C^i_m) \subset C^{i-1}_m$, $\delta(C^i_m) \subset C^{i+1}_m$ and that the relation

$$d \circ \delta + \delta \circ d = m\,\mathrm{id}$$

holds on $C^\bullet_m$. Then use Lemma 5.11 and the decomposition $C^\bullet = \bigoplus_{m \geq 0} C^\bullet_m$.

6. Projective modules

Let R be a unital ring.

6.1. Projective modules. An R-module P is *projective* if for any surjective morphism $\varphi : M \to N$ of R-modules and any morphism $\psi : P \to N$ there exists a morphism $f : P \to M$ such that $\psi = \varphi \circ f$.

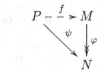

EXAMPLE 6.1. A free R-module F is projective. Indeed, let $\{e_i\}_{i \in I}$ be a set of generators of F. Define $f : F \to M$ by $f(e_i) = \varphi^{-1}(\psi(e_i))$.

LEMMA 6.2. *Let P be an R-module, the following conditions are equivalent*

(1) *P is projective;*
(2) *There exists a free module F such that F is isomorphic to $P \oplus P'$;*
(3) *Any exact sequence of R-modules*

$$0 \to N \to M \to P \to 0$$

splits.

PROOF. $(1) \Rightarrow (3)$
Consider the exact sequence

$$0 \to N \to M \xrightarrow{\varphi} P \to 0.$$

Set $\psi = \mathrm{id}_P$. Since φ is surjective and P is projective, there exists $f : P \to M$ such that $\psi = \mathrm{id}_P = \varphi \circ f$.

$(3) \Rightarrow (2)$ Every module is a quotient of a free module. Therefore we just have to apply (3) to the exact sequence

$$0 \to N \to F \to P \to 0$$

for a free module F.

$(2) \Rightarrow (1)$ Choose a free module F such that $F = P \oplus P'$. Let $\varphi : M \to N$ be a surjective morphism of R-modules and ψ a morphism $\psi : P \to N$. Now extend ψ to $\tilde{\psi} : F \to N$ such that the restriction of $\tilde{\psi}$ to P (respectively, P') is ψ (respectively, zero). There exists $f : F \to M$ such that $\varphi \circ f = \tilde{\psi}$. After restriction to P we get

$$\varphi \circ f_{|P} = \tilde{\psi}_{|P} = \psi.$$

$\square$

EXERCISE 6.3. Recall that a ring A is called a principal ideal domain if A is commutative, has no zero divisors and every ideal of A is principal, i.e. generated by a single element.

(a) Let F be a free A-module. Show that every submodule of F is free. For finitely generated F this can be done by induction on the rank of F. In the infinite case it is necessary to use transfinite induction, see [28](651).

(b) Let P be a projective A-module. Show that P is free.

EXERCISE 6.4. An *injective* R-module is a module I such that, for any injective homomorphism $i : X \to Y$ and any homomorphism $\varphi : X \to I$, there exists a homomorphism $\psi : Y \to I$ such that $\varphi = \psi \circ i$.

(a) Show that a module I is injective if and only if every exact sequence of R-modules,

$$0 \to I \to E \to E/I \to 0,$$

splits.

(Notice the analogy with projective modules.) However, a free module is not injective in general. Thus, checking that every R-module has an injective resolution is harder, see for instance [23].

(b) Let A be an algebra over a field and P be a projective right A-module. Then P^* is a left A-module with action of A given by

$$\langle a\varphi, x \rangle = \langle \varphi, ax \rangle$$

for all $a \in A$, $x \in P$ and $\varphi \in P^*$. Show that P^* is injective.

EXERCISE 6.5. Check that $\mathbb{Q}$ is an injective $\mathbb{Z}$-module.

6.2. Projective cover.

DEFINITION 6.6. Let M be an R-module. A submodule N of M is *small* if for any submodule $L \subset M$ such that $L + N = M$, we have $L = M$.

EXERCISE 6.7. Let $f : P \to M$ be a surjective morphism of modules such that $\mathrm{Ker}\, f$ is a small submodule of P. Assume that $f = f \circ \gamma$ for some homomorphism $\gamma : P \to P$. Show that γ is surjective.

DEFINITION 6.8. Let M be an R-module. A *projective cover* of M is a projective R-module P equipped with a surjective morphism $f : P \to M$ such that $\mathrm{Ker}\, f \subset P$ is small.

LEMMA 6.9. *Let* $f : P \to M$ *and* $g : Q \to M$ *be two projective covers of* M. *Then there exists an isomorphism* $\varphi : P \to Q$ *such that* $g \circ \varphi = f$.

PROOF. The existence of φ such that $g \circ \varphi = f$ follows immediately from projectivity of P. Similarly, we obtain the existence of a homomorphism $\psi : Q \to P$ such that $f \circ \psi = g$. Therefore we have $g \circ \varphi \circ \psi = g$. By Exercise 6.7 $\varphi \circ \psi$ is surjective. This implies surjectivity of $\varphi : P \to Q$. Since Q is projective we have an isomorphism $P \simeq Q \oplus \mathrm{Ker}\, \varphi$. Since $\mathrm{Ker}\, \varphi \subset \mathrm{Ker}\, f$, we have $P = Q + \mathrm{Ker}\, f$. Recall that $\mathrm{Ker}\, f \subset P$ is a small. Hence $P = Q$ and $\mathrm{Ker}\, \varphi = 0$. Thus φ is an isomorphism. $\qquad\square$

Let R be a ring and $e \in R$ be an idempotent. Then the left R-module Re is a direct summand of a free module R and hence is projective.

LEMMA 6.10. *The endomorphism algebra* $\mathrm{End}_R(Re)$ *is isomorphic to* $eR^{op}e$.

PROOF. The right multiplication by e defines the projector $\pi : R \to Re$, and we have $\mathrm{End}_R(Re) = \pi\, \mathrm{End}_R(R)\pi$, the statement follows from Lemma 1.10 Chapter 2. $\qquad\square$

6.3. Projective resolutions.

DEFINITION 6.11. Let M be an R-module. A complex $(P_\bullet, d)$ of R-modules

$$\ldots \xrightarrow{d} P_i \xrightarrow{d} \ldots \xrightarrow{d} P_1 \xrightarrow{d} P_0 \to 0$$

such that P_i is projective for all $i \geq 0$, $H_0(P_\bullet) = M$ and $H_i(P_\bullet) = 0$ for all $i \geq 1$, is called a *projective resolution* of M.

It is sometimes useful to see a projective resolution as the exact complex

$$\cdots \to P_i \to \cdots \to P_1 \to P_0 \xrightarrow{p} M \to 0,$$

where $p : P_0 \to M$ is the lift of the identity map between $H_0(P_\bullet)$ and M.

EXERCISE 6.12. Show that for every R-module M, there exists a resolution of M

$$\cdots \to F_i \to \cdots \to F_1 \to F_0 \to 0$$

such that all F_i are free. Such a resolution is called a free resolution.

This exercise immediately implies:

PROPOSITION 6.13. *Every R-module has a projective resolution.*

EXAMPLE 6.14. Let $R = k[x_1, ..., x_n]$ be a polynomial ring over a field k. Consider the simple R-module $M := R/(x_1, \ldots, x_n)$. One can use the Koszul complex, introduced in Exercise 5.2, to construct a projective resolution of M. First, we identify R with the symmetric algebra $S(V)$ of the vector space $V = k^n$. Let P_i denote the free R-module $R \otimes \Lambda^i(V)$ and recall $d : P_i \to P_{i-1}$ from Exercise 5.2 (c). Then $H_0(P_\bullet) = M$ and $H_i(P_\bullet) = 0$ for $i \geq 1$. Hence $(P_\bullet, d)$ is a free resolution of M.

LEMMA 6.15. *Let $(P_\bullet, d)$ and $(P_\bullet, d')$ be two projective resolutions of an R-module M. Then there exists a morphism of complexes $f : P_\bullet \to P'_\bullet$ such that $f_* : H_0(P_\bullet) \to H_0(P'_\bullet)$ induces the identity id_M. Moreover, f is unique up to homotopy equivalence.*

PROOF. We use an induction procedure to construct a morphism $f_i : P_i \to P'_i$. For $i = 0$, we denote by $p : P_0 \to M$ and $p' : P'_0 \to M$ the natural projections. Since P_0 is projective there exists a morphism $f_0 : P_0 \to P'_0$ such that $p' \circ f_0 = p$:

$$
\begin{array}{ccccccc}
\longrightarrow & P_1 & \xrightarrow{d} & P_0 & \xrightarrow{p} & M & \longrightarrow 0 \\
& & & \downarrow f_0 & & \downarrow \mathrm{id} & \\
\longrightarrow & P'_1 & \xrightarrow{d'} & P'_0 & \xrightarrow{p'} & M & \longrightarrow 0
\end{array}
$$

then we have $f(\operatorname{Ker} p) \subset \operatorname{Ker} p'$. We construct $f_1 : P_1 \to P'_1$ using the following commutative diagram:

$$
\begin{array}{ccccc}
\longrightarrow & P_1 & \xrightarrow{d} & \operatorname{Ker} p & \longrightarrow 0 \\
& \downarrow f_1 & & \downarrow f_0 & \\
\longrightarrow & P'_1 & \xrightarrow{d'} & \operatorname{Ker} p' & \longrightarrow 0.
\end{array}
$$

The existence of f_1 follows from projectivity of P_1 and surjectivity of d'.

We repeat the procedure to construct $f_i : P_i \to P_i$ for all i.

Suppose now that f and g are two morphisms satisfying the assumptions of the lemma. Let us prove that f and g are homotopically equivalent. Let $\varphi = f - g$. We have to prove the existence of maps $h_i : P_i \to P'_{i+1}$ such that $h_i \circ d = d' \circ h_{i+1}$. Let us explain how to construct h_0 and h_1 using the following diagram

$$
\begin{array}{ccccccccc}
\longrightarrow & P_2 & \xrightarrow{d} & P_1 & \xrightarrow{d} & P_0 & \xrightarrow{p} & M & \longrightarrow 0 \\
& \downarrow{\varphi_2} & \swarrow_{h_1} & \downarrow{\varphi_1} & \swarrow_{h_0} & \downarrow{\varphi_0} & & \downarrow{0} & \\
\longrightarrow & P'_2 & \xrightarrow{d'} & P'_1 & \xrightarrow{d'} & P'_0 & \xrightarrow{p'} & M & \longrightarrow 0.
\end{array}
$$

Since the morphism $\varphi_* : H_0(P_\bullet) \to H_0(P'_\bullet)$ is zero, we get $p' \circ \varphi_0 = 0$, and hence $\operatorname{Im} \varphi_0 \subset \operatorname{Im} d'$. Recall that P_0 is projective, therefore there exists $h_0 : P_0 \to P'_1$ such that $d' \circ h_0 = \varphi_0$.

To construct the map h_1, set $\psi := \varphi_1 - h_0 \circ d$. The relation

$$d' \circ h_0 \circ d = \varphi_0 \circ d = d' \circ \varphi_1$$

implies $d' \circ \psi = 0$. Since $H_1(P'_\bullet) = 0$, the image of ψ belongs to $d'(P'_2)$, and by projectivity of P_1 there exists a morphism $h_1 : P_1 \to P'_2$ such that

$$d' \circ h_1 = \psi = \varphi_1 - h_0 \circ d.$$

The construction of h_i for $i > 1$ is similar to the one for $i = 1$. The collection of the maps h_i gives the homotopy equivalence. $\qquad\square$

The following proposition expresses in what sense a projective resolution is unique.

PROPOSITION 6.16. *Let M be an R-module, and $(P_\bullet, d)$, $(P'_\bullet, d')$ be two projective resolutions of M. Then $(P_\bullet, d)$ and $(P'_\bullet, d')$ are homotopically equivalent.*

PROOF. By Lemma 6.15 there exist $f : P_\bullet \to P'_\bullet$ and $g : P'_\bullet \to P_\bullet$ such that $g \circ f$ is homotopically equivalent to $\mathrm{id}_{P_\bullet}$ and $f \circ g$ is homotopically equivalent to $\mathrm{id}_{P'_\bullet}$. $\qquad\square$

6.4. Extensions.

DEFINITION 6.17. Let M and N be two R-modules and $P_\bullet$ be a projective resolution of M. Consider the complex of abelian groups

$$0 \to \operatorname{Hom}_R (P_0, N) \xrightarrow{\delta} \operatorname{Hom}_R (P_1, N) \xrightarrow{\delta} \dots,$$

where δ is defined by (5.1). We define the i-th *extension group* $\operatorname{Ext}^i_R (M, N)$ as the i-th cohomology group of this complex. Lemma 5.10 ensures that $\operatorname{Ext}^i_R (M, N)$ does not depend on the choice of a projective resolution of M.

EXERCISE 6.18. Check that $\operatorname{Ext}^0_R(M, N) = \operatorname{Hom}_R(M, N)$.

Let us give an interpretation of $\operatorname{Ext}^1_R(M, N)$. Consider an exact sequence of R-modules

(5.2) $0 \to N \xrightarrow{\alpha} Q \xrightarrow{\beta} M \to 0$

and a projective resolution

(5.3) $\cdots \xrightarrow{d} P_2 \xrightarrow{d} P_1 \xrightarrow{d} P_0 \xrightarrow{\varphi} M \to 0$

of M. Then by projectivity of $P_\bullet$ there exist $\psi \in \operatorname{Hom}_R(P_0, Q)$ and $\gamma \in \operatorname{Hom}_R(P_1, N)$ which make the following diagram

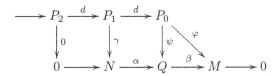

commutative. Let δ be the differential of degree $+1$ in Definition 6.17. The commutativity of this diagram implies that $\gamma \circ d = 0$ and hence $\delta(\gamma) = 0$. The choice of ψ and γ is not unique. If we choose another pair $\psi' \in \operatorname{Hom}_R(P_0, Q)$ and $\gamma' \in \operatorname{Hom}_R(P_1, N)$, then there exists $\theta \in \operatorname{Hom}_R(P_0, N)$ such that $\psi' - \psi = \alpha \circ \theta$ as in the diagram below

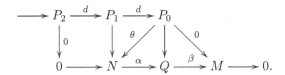

Furthermore, $\gamma' - \gamma = \theta \circ d$ or, equivalently, $\gamma' - \gamma = \delta(\theta)$. Thus, we can associate the class $[\gamma] \in \operatorname{Ext}^1_R(M, N)$ to the exact sequence (5.2).

Conversely, if we start with resolution (5.3) and a class $[\gamma] \in \operatorname{Ext}^1_R(M, N)$, we may consider some lift $\gamma \in \operatorname{Hom}_R(P_1, N)$. Then we can associate the following short exact sequence with $[\gamma]$

$$0 \to P_1/\operatorname{Ker} \gamma \to P_0/d(\operatorname{Ker} \gamma) \to M \to 0.$$

The reader may check that this exact sequence splits if and only if $[\gamma] = 0$.

EXAMPLE 6.19. Let R be $\mathbb{C}[x]$. Since $\mathbb{C}$ is algebraically closed, every simple R-module is one-dimensional over $\mathbb{C}$ and isomorphic to $\mathbb{C}_\lambda := \mathbb{C}[x]/(x - \lambda)$. It is easy to check that

$$0 \to \mathbb{C}[x] \xrightarrow{d} \mathbb{C}[x] \to 0,$$

where $d(1) = x - \lambda$ is a projective resolution of $\mathbb{C}_\lambda$. We can compute $\operatorname{Ext}^\bullet(\mathbb{C}_\lambda, \mathbb{C}_\mu)$. It amounts to calculating the cohomology of the complex

$$0 \to \mathbb{C} \xrightarrow{\delta} \mathbb{C} \to 0$$

where δ is the multiplication by $\mu - \lambda$. Hence

$$\operatorname{Ext}_R^0(\mathbb{C}_\lambda, \mathbb{C}_\mu) = \operatorname{Ext}_R^1(\mathbb{C}_\lambda, \mathbb{C}_\mu) = \begin{cases} 0 \text{ if } \lambda \neq \mu \\ \mathbb{C} \text{ if } \lambda = \mu \end{cases}$$

EXAMPLE 6.20. Let $R = \mathbb{C}[x] / (x^2)$. Then R has only one simple module (up to isomorphism), which we denote $\mathbb{C}_0$. Then

$$\ldots \xrightarrow{d} R \xrightarrow{d} R \to 0,$$

where $d(1) = x$ is a projective resolution for $\mathbb{C}_0$ and $\operatorname{Ext}^i(\mathbb{C}_0, \mathbb{C}_0) = \mathbb{C}$ for all $i \geq 0$.

7. Representations of Artinian rings

7.1. Idempotents, nilpotent ideals and Jacobson radical. A (left or right) ideal N of a ring R is called *nilpotent* if there exists $p > 0$ such that $N^p = 0$. The smallest such p is called the *degree of nilpotency* of N. The following lemma is sometimes called "lifting of an idempotent".

LEMMA 7.1. *Let N be a left (or right) nilpotent ideal of R and take $r \in R$ such that $r^2 \equiv r \mod N$. Then there exists an idempotent $e \in R$ such that $e \equiv r \mod N$.*

PROOF. Let N be a left ideal. We prove the statement by induction on the degree of nilpotency $d(N)$. The case $d(N) = 1$ is trivial. Let $d(N) > 1$. Set $n = r^2 - r$, then n belongs to N and $rn = nr$. Therefore we have

$$(r + n - 2rn)^2 \equiv r^2 + 2rn - 4r^2n \mod N^2.$$

We set $s = r + n - 2rn$. Then we have

$$s^2 \equiv s \mod N^2, \quad s \equiv r \mod N.$$

Since $d(N^2) < d(N)$, the induction assumption ensures that there exists an idempotent $e \in R$ such that $e \equiv s \mod N^2$, hence $e \equiv r \mod N$. $\square$

For an R-module M let

$$\operatorname{Ann} M = \{x \in R \mid xM = 0\}.$$

DEFINITION 7.2. The *Jacobson radical* $\operatorname{rad} R$ of a ring R is the intersection of $\operatorname{Ann} M$ for all simple R-modules M.

EXERCISE 7.3. (a) Prove that $\operatorname{rad} R$ is the intersection of all maximal left ideals of R as well as the intersection of all maximal right ideals.
(b) Show that x belongs to $\operatorname{rad} R$ if and only if $1 + rx$ is invertible for any $r \in R$.
(c) Show that if N is a nilpotent left ideal of R, then N is contained in $\operatorname{rad} R$.

LEMMA 7.4. *Let $e \in \operatorname{rad} R$ such that $e^2 = e$. Then $e = 0$.*

PROOF. By Exercise 7.3 (b) we have that $1 - e$ is invertible. But $e(1 - e) = 0$ and therefore $e = 0$. $\square$

7.2. The Jacobson radical of an Artinian ring.

DEFINITION 7.5. A ring R is *Artinian* if it satisfies the descending chain condition for left ideals.

A typical example of Artinian ring is a finite-dimensional algebra over a field. It follows from the definition that any left ideal in an Artinian ring contains a minimal (non-zero) ideal.

LEMMA 7.6. *Let R be an Artinian ring, $I \subset R$ be a left ideal. If I is not nilpotent, then I contains a non-zero idempotent.*

PROOF. Since R is Artinian, one can find a minimal left ideal $J \subset I$ among all non-nilpotent ideals of I. Then $J^2 = J$. We will prove that J contains a non-zero idempotent.

Let $L \subset J$ be some minimal left ideal such that $JL \neq 0$. Then there exists $x \in L$ such that $Jx \neq 0$. By minimality of L we have $Jx = L$. Therefore there exists $r \in J$ such that $rx = x$. Hence $(r^2 - r)x = 0$. Let $N = \{y \in J \mid yx = 0\}$. Note that N is a proper left ideal of J and therefore N is nilpotent. Thus, we have $r^2 \equiv r \mod N$. By Lemma 7.1 there exists an idempotent $e \in R$ such that $e \equiv r \mod N$, and we are done. $\square$

THEOREM 7.7. *If R is Artinian then $\operatorname{rad} R$ is the unique maximal nilpotent ideal of R.*

PROOF. By Exercise 7.3 every nilpotent ideal of R lies in $\operatorname{rad} R$. It remains to show that $\operatorname{rad} R$ is nilpotent. Indeed, otherwise by Lemma 7.6, $\operatorname{rad} R$ contains a non-zero idempotent. This contradicts Lemma 7.4. $\square$

LEMMA 7.8. *An Artinian ring R is semisimple if and only if $\operatorname{rad} R = 0$.*

PROOF. If R is semisimple and Artinian, then by Wedderburn–Artin theorem it is a direct product of matrix rings, which does not have non-trivial nilpotent ideals.

If R is Artinian with trivial radical, then by Lemma 7.6 every minimal left ideal L of R contains an idempotent e such that $L = Re$. Hence R is isomorphic to $L \oplus R(1 - e)$. Therefore R is a direct sum of its minimal left ideals. $\square$

COROLLARY 7.9. *If R is Artinian, then $R/\operatorname{rad} R$ is semisimple.*

PROOF. By Theorem 7.7 the quotient ring $R/\operatorname{rad} R$ does not have non-zero nilpotent ideals. Hence it is semisimple by Lemma 7.8. $\square$

7.3. Modules over Artinian rings.

LEMMA 7.10. *Let R be an Artinian ring and M be an R-module. Then $M/(\operatorname{rad} R)M$ is the maximal semisimple quotient of M.*

PROOF. Since $R/\operatorname{rad} R$ is a semisimple ring and $M/(\operatorname{rad} R)M$ is an $R/\operatorname{rad} R$-module, we obtain that $M/(\operatorname{rad} R)M$ is semisimple. To prove maximality, observe that $\operatorname{rad} R$ acts by zero on any semisimple quotient of M. $\square$

COROLLARY 7.11. *Assume that R is Artinian and M is an R-module. Consider the filtration (called the radical filtration)*

$$M \supset (\mathrm{rad}\,R)\,M \supset (\mathrm{rad}\,R)^2\,M \supset \cdots \supset (\mathrm{rad}\,R)^k\,M = 0,$$

where k is the degree of nilpotency of $\mathrm{rad}\,R$. Then all quotients $(\mathrm{rad}\,R)^i M / (\mathrm{rad}\,R)^{i+1} M$ are semisimple R-modules. In particular, M always has a simple quotient.

PROPOSITION 7.12. *Let R be Artinian. Consider it as a module over itself. Then R is a finite length module. Hence R is a Noetherian ring.*

PROOF. Apply Corollary 7.11 to $M = R$. Then every quotient $(\mathrm{rad}\,R)^i / (\mathrm{rad}\,R)^{i+1}$ is a semisimple Artinian R-module. By Exercise 4.4 $(\mathrm{rad}\,R)^i / (\mathrm{rad}\,R)^{i+1}$ is a Noetherian R-module. Hence R is a Noetherian module over itself. □

Let us apply the Krull–Schmidt theorem to an Artinian ring R considered as a left module over itself. Then R has a decomposition into a direct sum of indecomposable submodules

$$R = L_1 \oplus \cdots \oplus L_n.$$

Recall that $\mathrm{End}_R(R) = R^{\mathrm{op}}$. Therefore the canonical projection on each component L_i is given by multiplication (on the right) by some idempotent element $e_i \in L_i$.

In other words R has a decomposition

(5.4) $$R = Re_1 \oplus \cdots \oplus Re_n.$$

Moreover, $e_i e_j = \delta_{ij} e_i$. Once more by Krull–Schmidt theorem this decomposition is unique up to multiplication by some invertible element on the right.

DEFINITION 7.13. An idempotent $e \in R$ is called *primitive* if it can not be written $e = e' + e''$ for some non-zero idempotents e', e'' such that $e'e'' = e''e' = 0$.

EXERCISE 7.14. Prove that the idempotent $e \in R$ is primitive if and only if Re is an indecomposable R-module.

In the decomposition (5.4) the idempotents $e_1, \ldots, e_n$ are primitive.

LEMMA 7.15. *Assume R is Artinian, $N = \mathrm{rad}\,R$ and $e \in R$ is a primitive idempotent. Then Ne is a unique maximal submodule of Re.*

PROOF. Due to Corollary 7.11 it is sufficient to show that Re/Ne is a simple R-module. Since e is primitive, the left ideal Re is an indecomposable R-module. Assume that Re/Ne is not simple. Then $Re/Ne = Re_1 \oplus Re_2$ for some non-zero idempotent elements e_1 and e_2 in the quotient ring R/N. By Lemma 7.1 there exist idempotents $f_1, f_2 \in R$ such that $f_i \equiv e_i \mod N$. Then the map $Re \to Ref_1 \oplus Ref_2$ is an isomorphism of left R-modules and this contradicts indecomposability of Re. □

THEOREM 7.16. *Assume R is Artinian.*

(1) *Every simple R-module S has a projective cover which is isomorphic to Re for some primitive idempotent $e \in R$.*

(2) *Let P be an indecomposable projective R-module. There exists a primitive idempotent $e \in R$ such that P is isomorphic to Re. Furthermore, P has a unique simple quotient.*

PROOF. Let S be a simple R-module. There exists a surjective homomorphism $f : R \rightarrow S$. Consider the decomposition (5.4). There exists $i \leq n$ such that the restriction of f to Re_i is non-zero. By the simplicity of S the restriction $f : Re_i \rightarrow S$ is surjective. It follows from Lemma 7.15 that Re_i is a projective cover of S.

Let P be an indecomposable projective module. By Lemma 7.10 the quotient $P/(\text{rad } R)P$ is semisimple. Let S be a simple submodule of $P/(\text{rad } R)P$. Then we have a surjection $f : P \rightarrow S$. Let $g : Q \rightarrow S$ be a projective cover of S. There exists a morphism $\varphi : P \rightarrow Q$ such that $f = g \circ \varphi$. Since Q has a unique simple quotient, the morphism φ is surjective. Then P is isomorphic to $Q \oplus \text{Ker } \varphi$. The indecomposability of P implies that P is isomorphic to Q. $\square$

EXAMPLE 7.17. Consider the group algebra $R = \mathbb{F}_3(S_3)$. First let us classify simple and indecomposable projective R-modules.

Let r be a 3-cycle and s be a transposition. Since s and r generate S_3, one can see easily that the elements $r - 1$, $r^2 - 1$, $sr - s$ and $sr^2 - s$ span a nilpotent ideal N, which turns out to be maximal. The quotient R/N is a semisimple R-module with two simple components L_1 and L_2, where L_1 (resp. L_2) is the trivial (resp. the sign) representation of S_3. Set $e_1 = \frac{1+s}{2}$ and $e_2 = \frac{1-s}{2}$. Then e_1, e_2 are primitive idempotents of R such that $1 = e_1 + e_2$ and $e_1 e_2 = 0$. Hence R has exactly two indecomposable projective modules, namely $P_1 = Re_1$ and $P_2 = Re_2$. Those modules can be seen as induced modules

$$Re_1 \cong \text{Ind}_{S_2}^{S_3}(\text{triv}), \quad Re_2 \cong \text{Ind}_{S_2}^{S_3}(\text{sgn}).$$

Thus P_1 is the 3-dimensional permutation representation of S_3, and $P_2 = P_1 \otimes \text{sgn}$.

EXERCISE 7.18. Compute explicitly the radical filtration of P_1 and P_2. Show that

$$P_1/(\text{rad } R)P_1 \simeq L_1, \quad (\text{rad } R)P_1/(\text{rad } R)^2 P_1 \simeq L_2, \quad (\text{rad } R)^2 P_1 \simeq L_1$$

and

$$P_2/(\text{rad } R)P_2 \simeq L_2, \quad (\text{rad } R)P_2/(\text{rad } R)^2 P_2 \simeq L_1, \quad (\text{rad } R)^2 P_2 \simeq L_2.$$

Now we will calculate the extension groups between the simple modules. The above exercise implies the following exact sequences

$$0 \rightarrow L_2 \rightarrow P_2 \rightarrow P_1 \rightarrow L_1 \rightarrow 0, \quad 0 \rightarrow L_1 \rightarrow P_1 \rightarrow P_2 \rightarrow L_2 \rightarrow 0.$$

By gluing these sequences together we obtain a projective resolution for L_1

$$\cdots \rightarrow P_1 \rightarrow P_2 \rightarrow P_2 \rightarrow P_1 \rightarrow P_1 \rightarrow P_2 \rightarrow P_2 \rightarrow P_1 \rightarrow 0$$

and for L_2

$$\cdots \rightarrow P_2 \rightarrow P_1 \rightarrow P_1 \rightarrow P_2 \rightarrow P_2 \rightarrow P_1 \rightarrow P_1 \rightarrow P_2 \rightarrow 0.$$

Using the following obvious relation

$$\mathrm{Hom}(P_i, P_j) = \begin{cases} \mathbb{F}_3, & \text{if } i = j \\ 0, & \text{if } i \neq j \end{cases}$$

we obtain

$$\mathrm{Ext}^p (L_i, L_j) = \begin{cases} 0, & \text{if } p \equiv 1, 2 \mod 4,\, i = j \\ \mathbb{F}_3, & \text{if } p \equiv 0, 3 \mod 4,\, i = j \\ 0, & \text{if } p \equiv 0, 3 \mod 4,\, i \neq j \\ \mathbb{F}_3, & \text{if } p \equiv 1, 2 \mod 4,\, i \neq j \end{cases}$$

EXERCISE 7.19. Let B_n denote the algebra of upper triangular $n \times n$ matrices over a field $\mathbb{F}$. Denote by E_{ij} the elementary matrices. Show that E_{ii} for $i = 1, \ldots, n$, are primitive idempotents of B_n. Furthermore, show that B_n has n simple modules (up to isomorphism) $L_1, \ldots, L_n$ associated with those idempotents and that the dimension of every L_i over $\mathbb{F}$ is 1. Finally check that

$$\mathrm{Ext}^p (L_i, L_j) = \begin{cases} \mathbb{F}, & \text{if } i = j,\, p = 0 \text{ or } i = j + 1,\, p = 1 \\ 0, & \text{otherwise} \end{cases}.$$

8. Abelian categories

An abelian category is a generalization of the categories of modules over a ring. Let us start with definition of an additive category.

DEFINITION 8.1. A category $\mathcal{C}$ is called *additive* if for any pair of objects A, B,

(1) The set of morphisms $\mathrm{Hom}_{\mathcal{C}}(A, B)$ is an abelian group.
(2) There exist an object $A \oplus B$, called a direct sum, and a pair of morphisms $i_A : A \to A \oplus B$ and $i_B : B \to A \oplus B$ such that for any morphisms $\varphi : A \to M$ and $\psi : B \to M$ there exists a unique morphism $\tau : A \oplus B \to M$ such that $\tau \circ i_A = \varphi$ and $\tau \circ i_B = \psi$.
(3) There exist an object $A \times B$ called a direct product and a pair of morphisms $p_A : A \times B \to A$ and $i_B : A \times B \to B$ such that for any morphisms $\alpha : M \to A$ and $\beta : M \to B$ there exists a unique morphism $\theta : M \to A \times B$ such that $p_A \circ \theta = \alpha$ and $p_B \circ \theta = \beta$.
(4) The induced morphism $A \oplus B \to A \times B$ is an isomorphism.

DEFINITION 8.2. An *abelian* category is an additive category $\mathcal{C}$ such that, for every morphism $\varphi \in \mathrm{Hom}_{\mathcal{C}}(A, B)$

(1) There exist an object and a morphism $\mathrm{Ker}\, \varphi \xrightarrow{i} A$ such that for any morphism $\gamma : M \to A$ such that, $\varphi \circ \gamma = 0$, there exists a unique morphism $\delta : M \to \mathrm{Ker}\, \varphi$ such that $\gamma = i \circ \delta$.

(2) There exist an object and morphism $B \xrightarrow{p} \mathrm{Coker}\ \varphi$ such that for any morphism $\tau : B \to M$ such that, $\tau \circ \varphi = 0$, there exists a unique morphism $\sigma : \mathrm{Coker}\ \varphi \to M$ such that $\tau = \sigma \circ p$.

(3) There is an isomorphism $\gamma : \mathrm{Coker}\ i \to \mathrm{Ker}\ p$ such that the composition $A \to \mathrm{Coker}\ i \to \mathrm{Ker}\ p \to B$ equals φ.

EXERCISE 8.3. Let R be a ring, show that the category of finitely generated R-modules is abelian. Show that the category of projective R-modules is additive but not abelian in general. Finally show that the category of projective R-modules is abelian if and only if R is a semisimple ring.

In an abelian category we can define the image of a morphism, a quotient object, exact sequences, projective and injective objects. All the results of Sections 4, 5 and 6 can be generalized to abelian categories. If we want to define extension groups we have to assume the existence of projective covers.

DEFINITION 8.4. Let $\mathcal{C}$ be an abelian category. Its *Grothendieck group* $\mathcal{K}_{\mathcal{C}}$ is the abelian group defined by generators and relations in the following way. For every object M of $\mathcal{C}$ there is one generator $[M]$. For every exact sequence

$$0 \to N \to M \to K \to 0$$

in $\mathcal{C}$ we have the relation $[M] = [K] + [N]$.

EXERCISE 8.5. Let $\mathcal{C}$ be the category of finite-dimensional vector spaces. Show that $\mathcal{K}_{\mathcal{C}}$ is isomorphic to $\mathbb{Z}$.

EXERCISE 8.6. Let G be a finite group and k be a field of characteristic 0. Let $\mathcal{C}$ be the category of finite-dimensional $k(G)$-modules. Then $\mathcal{K}_{\mathcal{C}}$ is isomorphic to the abelian subgroup of $\mathcal{C}(G)$ generated by the characters of irreducible representations. Furthermore, the tensor product equips $\mathcal{K}_{\mathcal{C}}$ with a structure of a commutative ring.

CHAPTER 6

Symmetric groups, Schur–Weyl duality and positive self-adjoint Hopf algebras

This chapter was written with Laurent GRUSON

Though this be madness, yet there is method in it (Hamlet, Act II scene 2)

In which we revisit the province of representations of symmetric groups, encounter Schur-Weyl duality and PSH algebras, and put a bit of order in this mess. Not to mention the partitions, Young tableaux and related combinatorics.

This chapter (from section 3), relies on a book by Andrei Zelevinsky, *Representations of finite classical groups, a Hopf algebra approach*, [38], which gives a very efficient axiomatisation of the essential properties of the representations of symmetric groups and general linear groups over finite fields. In this book lies the first appearance of the notion of *categorification* which has become an ubiquitous tool in representation theory. For the case of symmetric groups, the seminal work is due to Geissinger ([14]).

We will meet graded objects in what follows. Hence the word *degree* will be attached to the grading (and not to the dimension of a representation) throughout this chapter.

1. Representations of symmetric groups

Consider the symmetric group S_n. In this section we classify irreducible representations of S_n over $\mathbb{Q}$. We will see that any irreducible representation over $\mathbb{Q}$ is absolutely irreducible, in other words $\mathbb{Q}$ is a splitting field for S_n. We will realize the irreducible representations of S_n as minimal left ideals in the group algebra $\mathbb{Q}(S_n)$.

DEFINITION 1.1. A *partition* λ of n is a sequence of positive integers $(\lambda_1, \ldots, \lambda_k)$ such that $\lambda_1 \geq \cdots \geq \lambda_k$ and $\lambda_1 + \cdots + \lambda_k = n$. We use the notation $\lambda \vdash n$ when λ is a partition of n. Moreover, the integer k is called the length of the partition λ.

REMARK 1.2. Recall that two permutations lie in the same conjugacy class of S_n if and only if there is a bijection between their sets of cycles which preserves the lengths. Therefore we can parametrize the conjugacy classes in S_n by the partitions of n.

© Springer Nature Switzerland AG 2018
C. Gruson and V. Serganova, *A Journey Through Representation Theory*,
Universitext, https://doi.org/10.1007/978-3-319-98271-7_6

DEFINITION 1.3. To every partition $\lambda = (\lambda_1, \ldots, \lambda_k)$ we associate a table, also denoted λ, consisting of n boxes with rows of length $\lambda_1, \ldots, \lambda_k$, it is called a *Young diagram*. A *Young tableau* $t(\lambda)$ is a Young diagram λ with entries $1, \ldots, n$ in its boxes such that every number occurs in exactly one box. We say that two Young tableaux have the same shape if they are obtained from the same Young diagram. The number of tableaux of shape λ equals $n!$.

EXAMPLE 1.4. Let $n = 7$, $\lambda = (3, 2, 1, 1)$. The corresponding Young diagram is

and a possible example of tableau $t(\lambda)$ is

$$
\begin{array}{|c|c|c|}
\hline
1 & 2 & 3 \\
\hline
4 & 5 \\
\cline{1-2}
6 \\
\cline{1-1}
7 \\
\cline{1-1}
\end{array}
$$

Given a Young tableau $t(\lambda)$, we denote by $P_{t(\lambda)}$ the subgroup of S_n preserving the rows of $t(\lambda)$ and by $Q_{t(\lambda)}$ the subgroup of permutations preserving the columns.

EXAMPLE 1.5. Consider the tableau $t(\lambda)$ from Example 1.4. Then $P_{t(\lambda)}$ is isomorphic to $S_3 \times S_2$, which is the subgroup of S_7 permuting $\{1, 2, 3\}$ and $\{4, 5\}$, and $Q_{t(\lambda)}$ is isomorphic to $S_4 \times S_2$ which permutes $\{1, 4, 6, 7\}$ and $\{2, 5\}$.

EXERCISE 1.6. Check that $P_{t(\lambda)} \cap Q_{t(\lambda)} = \{1\}$ for any tableau $t(\lambda)$.

Introduce the following elements in $\mathbb{Q}(S_n)$:

$$
a_{t(\lambda)} = \sum_{p \in P_t(\lambda)} p, \ b_{t(\lambda)} = \sum_{q \in Q_t(\lambda)} (-1)^q q, \ c_{t(\lambda)} = a_{t(\lambda)} b_{t(\lambda)},
$$

where $(-1)^q$ stands for the sign representation $\varepsilon(q)$.
The element $c_{t(\lambda)}$ is called a *Young symmetrizer*.

THEOREM 1.7. *Let $t(\lambda)$ be a Young tableau.*
(1) *The left ideal $\mathbb{Q}(S_n)c_{t(\lambda)}$ is minimal, therefore it is a simple $\mathbb{Q}(S_n)$-module.*
(2) *Two $\mathbb{Q}(S_n)$-modules $\mathbb{Q}(S_n)c_{t(\lambda)}$ and $\mathbb{Q}(S_n)c_{t'(\mu)}$ are isomorphic if and only if $\mu = \lambda$.*
(3) *Every simple $\mathbb{Q}(S_n)$-module is isomorphic to $V_{t(\lambda)} := \mathbb{Q}(S_n)c_{t(\lambda)}$ for some Young tableau $t(\lambda)$.*
(4) *The representation $V_{t(\lambda)}$ is absolutely irreducible.*

The proof of this Theorem is provided later in this Section.

REMARK 1.8. Note that assertion (3) of the Theorem follows from the first two, since the number of Young diagrams is equal to the number of conjugacy classes (see Remark 1.2).

EXAMPLE 1.9. Consider the partition (of length 1) $\lambda = (n)$. Then the corresponding Young diagram consists of one row with n boxes. For any tableau $t(\lambda)$ we have $P_{t(\lambda)} = S_n$, $Q_{t(\lambda)}$ is trivial and therefore

$$c_{t(\lambda)} = a_{t(\lambda)} = \sum_{s \in S_n} s.$$

The corresponding representation of S_n is trivial.

EXAMPLE 1.10. Consider the partition $\lambda = (1, \ldots, 1)$ whose Young diagram consists of one column with n boxes. Then $Q_{t(\lambda)} = S_n$, $P_{t(\lambda)}$ is trivial and

$$c_{t(\lambda)} = b_{t(\lambda)} = \sum_{s \in S_n} (-1)^s \, s.$$

Therefore the corresponding representation of S_n is the sign representation.

EXAMPLE 1.11. Let us consider the partition $\lambda = (n-1, 1)$ and the Young tableau $t(\lambda)$ which has entries $1, \ldots, n-1$ in the first row and n in the second row. Then $P_{t(\lambda)}$ is isomorphic to S_{n-1} and consists of all permutations which fix n, and $Q_{t(\lambda)}$ is generated by the transposition $(1n)$. We have

$$c_{t(\lambda)} = \left(\sum_{s \in S_{n-1}} s \right) (1 - (1n)).$$

Let E denote the permutation representation of S_n. Let us show that $\mathbb{Q}(S_n)c_{t(\lambda)}$ is the $n-1$ dimensional simple submodule of E. Indeed, $sc_{t(\lambda)} = c_{t(\lambda)}$ for all $s \in P_{t(\lambda)}$, therefore the restriction of $V_{t(\lambda)}$ to $P_{t(\lambda)}$ contains the trivial representation of $P_{t(\lambda)}$. Recall that the permutation representation can be obtained by induction from the trivial representation of S_{n-1}:

$$E = \operatorname{Ind}_{P_{t(\lambda)}}^{S_n} \operatorname{triv}.$$

By Frobenius reciprocity $\mathbb{Q}(S_n)c_{t(\lambda)}$ is a non-trivial submodule of E.

In the rest of this Section we prove Theorem 1.7.

First, let us note that S_n acts simply transitively on the set of Young tableaux of the same shape by permuting the entries, and for any $s \in S_n$ we have

$$a_{st(\lambda)} = sa_{t(\lambda)}s^{-1}, \; b_{st(\lambda)} = sb_{t(\lambda)}s^{-1}, \; c_{st(\lambda)} = sc_{t(\lambda)}s^{-1}.$$

Therefore if we have two tableaux $t(\lambda)$ and $t'(\lambda)$ of the same shape, then

$$\mathbb{Q}(S_n)c_{t(\lambda)} = \mathbb{Q}(S_n)c_{t'(\lambda)}s^{-1}$$

for some $s \in S_n$. Hence $\mathbb{Q}(S_n)c_{t(\lambda)}$ and $\mathbb{Q}(S_n)c_{t'(\lambda)}$ are isomorphic $\mathbb{Q}(S_n)$-modules.

In what follows we denote by V_λ a representative of the isomorphism class of $\mathbb{Q}(S_n)c_{t(\lambda)}$ for some tableau $t(\lambda)$. As we have seen the isomorphism class does not depend on the tableau but only on its shape.

EXERCISE 1.12. Let $t(\lambda)$ be a Young tableau and $s \in S_n$. Show that if s does not belong to the set $P_{t(\lambda)}Q_{t(\lambda)}$, then there exist two entries i, j which lie in the same row of $t(\lambda)$ and in the same column of $st(\lambda)$. In other words, the transposition (ij) lies in the intersection $P_{t(\lambda)} \cap Q_{st(\lambda)}$. *Hint*: Assume the opposite, and check that one can find $s' \in P_{t(\lambda)}$ and $s'' \in Q_{st(\lambda)}$ such that $s''t(\lambda) = s'st(\lambda)$.

Next, observe that for any $p \in P_\lambda$ and $q \in Q_\lambda$ we have

$$pc_{t(\lambda)}q = (-1)^q c_{t(\lambda)}.$$

LEMMA 1.13. *Let $t(\lambda)$ be a Young tableau and $y \in \mathbb{Q}(S_n)$. Assume that for all $p \in P_{t(\lambda)}$ and $q \in Q_{t(\lambda)}$ we have*

$$pyq = (-1)^q y.$$

Then $y = ac_{t(\lambda)}$ for some $a \in \mathbb{Q}$.

PROOF. Let T be a set of representatives of the double cosets $P_{t(\lambda)} \backslash S_n / Q_{t(\lambda)}$. Then S_n is the disjoint union $\bigsqcup_{s \in T} P_{t(\lambda)} s Q_{t(\lambda)}$ and we can write y in the form

$$\sum_{s \in T} d_s \sum_{p \in P_{t(\lambda)}, q \in Q_{t(\lambda)}} (-1)^q psq = \sum_{s \in T} d_s a_{t(\lambda)} s b_{t(\lambda)}.$$

It suffices to show that if $s \notin P_{t(\lambda)}Q_{t(\lambda)}$ then $a_{t(\lambda)}sb_{t(\lambda)} = 0$. This follows from Exercise 1.12. Indeed, there exists a transposition τ in the intersection $P_{t(\lambda)} \cap Q_{st(\lambda)}$. Therefore

$$a_{t(\lambda)}sb_{t(\lambda)}s^{-1} = a_{t(\lambda)}b_{st(\lambda)} = (a_{t(\lambda)}\tau)(\tau b_{st(\lambda)}) = -a_{t(\lambda)}b_{st(\lambda)} = 0.$$

$\square$

This lemma implies

COROLLARY 1.14. *We have $c_{t(\lambda)}\mathbb{Q}(S_n)c_{t(\lambda)} \subset \mathbb{Q}c_{t(\lambda)}$.*

Now we are ready to prove the first assertion of Theorem 1.7.

LEMMA 1.15. *The ideal $\mathbb{Q}(S_n)c_{t(\lambda)}$ is a minimal left ideal of $\mathbb{Q}(S_n)$.*

PROOF. Consider a left ideal $W \subset \mathbb{Q}(S_n)c_{t(\lambda)}$. Then by Corollary 1.14 either $c_{t(\lambda)}W = \mathbb{Q}c_{t(\lambda)}$ or $c_{t(\lambda)}W = 0$.

If $c_{t(\lambda)}W = \mathbb{Q}c_{t(\lambda)}$, then $\mathbb{Q}(S_n)c_{t(\lambda)}W = \mathbb{Q}(S_n)c_{t(\lambda)}$. Hence $W = \mathbb{Q}(S_n)c_{t(\lambda)}$. If $c_{t(\lambda)}W = 0$, then $W^2 = 0$. But $\mathbb{Q}(S_n)$ is a semisimple ring, hence $W = 0$. $\square$

Note that Corollary 1.14 also implies that V_λ is absolutely irreducible (statement (4) in Theorem 1.7) because

$$\mathrm{End}_{S_n}(\mathbb{Q}(S_n)c_{t(\lambda)}) = c_{t(\lambda)}\mathbb{Q}(S_n)c_{t(\lambda)} \simeq \mathbb{Q}.$$

COROLLARY 1.16. *For every Young tableau $t(\lambda)$ we have $c_{t(\lambda)}^2 = n_\lambda c_{t(\lambda)}$, where $n_\lambda = \frac{n!}{\dim V_\lambda}$.*

PROOF. By Corollary 1.14 we know that $c_{t(\lambda)}^2 = n_\lambda c_{t(\lambda)}$ for some $n_\lambda \in \mathbb{Q}$. Moreover, there exists a primitive idempotent $e \in \mathbb{Q}(S_n)$ such that $c_{t(\lambda)} = n_\lambda e$. To find n_λ note that the trace of e in the regular representation equals $\dim V_\lambda$, and the trace of $c_{t(\lambda)}$ in the regular representation equals $n!$. □

EXERCISE 1.17. Introduce the lexicographical order on partitions by setting $\mu > \lambda$ if there exists i such that $\mu_j = \lambda_j$ for all $j < i$ and $\mu_i > \lambda_i$. Show that if $\mu > \lambda$, then for any two Young tableaux $t(\mu)$ and $t'(\lambda)$ there exist entries i and j which lie in the same row of $t(\mu)$ and in the same column of $t'(\lambda)$.

LEMMA 1.18. *Let $t(\lambda)$ and $t'(\mu)$ be two Young tableaux such that $\lambda < \mu$. Then $c_{t(\lambda)}\mathbb{Q}(S_n)c_{t'(\mu)} = 0$.*

PROOF. We have to check that $c_{t(\lambda)}sc_{t'(\mu)}s^{-1} = 0$ for any $s \in S_n$, which is equivalent to $c_{t(\lambda)}c_{st'(\mu)} = 0$. Therefore it suffices to prove that $c_{t(\lambda)}c_{t'(\mu)} = 0$. By Exercise 1.17 there exists a transposition τ which belongs to the intersection $Q_{t(\lambda)} \cap P_{t'(\mu)}$. Then, repeating the argument from the proof of Lemma 1.13, we obtain

$$c_{t(\lambda)}c_{t'(\mu)} = c_{t(\lambda)}\tau^2 c_{t'(\mu)} = -c_{t(\lambda)}c_{t'(\mu)}.$$

□

Now we show the second statement of Theorem 1.7.

LEMMA 1.19. *Two irreducible representations V_λ and V_μ are isomorphic if and only if $\lambda = \mu$.*

PROOF. It suffices to show that if $\lambda \neq \mu$, then V_λ and V_μ are not isomorphic. Without loss of generality we may assume $\lambda < \mu$ and take some Young tableaux $t(\lambda)$ and $t'(\mu)$. By Lemma 1.18 we obtain that $c_{t(\lambda)}$ acts by zero on V_μ. On the other hand, by Corollary 1.16, $c_{t(\lambda)}$ does not annihilate V_λ. Hence the statement. □

By Remark 1.8 the proof of Theorem 1.7 is complete.

REMARK 1.20. Note that in fact we have proven that if $\lambda \neq \mu$, then $c_{t(\lambda)}\mathbb{Q}(S_n)c_{t'(\mu)} = 0$ for any pair of tableaux $t(\lambda), t'(\mu)$. Indeed, if $c_{t(\lambda)}\mathbb{Q}(S_n)c_{t'(\mu)} \neq 0$, then

$$\mathbb{Q}(S_n)c_{t(\lambda)}\mathbb{Q}(S_n)c_{t'(\mu)} = \mathbb{Q}(S_n)c_{t'(\mu)}.$$

But this is impossible since all the irreducible components of $\mathbb{Q}(S_n)c_{t(\lambda)}\mathbb{Q}(S_n)$ are isomorphic to V_λ.

LEMMA 1.21. *Let $\rho : S_n \to \mathrm{GL}(V)$ be a finite-dimensional representation of S_n. Then the multiplicity of V_λ in V equals the rank of $\rho(c_{t(\lambda)})$.*

PROOF. The rank of $c_{t(\lambda)}$ in V_λ is 1 and $c_{t(\lambda)}V_\mu = 0$ for all $\mu \neq \lambda$. Hence the statement. □

EXERCISE 1.22. Let λ be a partition and χ_λ denote the character of V_λ.

(1) Prove that $\chi_\lambda(s) \in \mathbb{Z}$ for all $s \in S_n$.
(2) Prove that $\chi_\lambda(s) = \chi_\lambda(s^{-1})$ for all $s \in S_n$ and hence V_λ is self-dual.
(3) For a tableau $t(\lambda)$ let $\bar{c}_{t(\lambda)} = b_{t(\lambda)} a_{t(\lambda)}$. Prove that $\mathbb{Q}(S_n) c_{t(\lambda)}$ and $\mathbb{Q}(S_n) \bar{c}_{t(\lambda)}$ are isomorphic $\mathbb{Q}(S_n)$-modules.

EXERCISE 1.23. Let λ be a partition. We define the *conjugate* partition $\lambda^\perp$ by setting $\lambda_i^\perp$ to be equal to the length of the i-th column in the Young diagram λ. For example, if $\lambda =$ ⬚ , then $\lambda^\perp =$ ⬚ .

Prove that for any partition λ, the representation $V_{\lambda^\perp}$ is isomorphic to the tensor product of V_λ with the sign representation.

Since $\mathbb{Q}$ is a splitting field for S_n, Theorem 1.7 provides classification of irreducible representations of S_n over any field of characteristic zero.

2. Schur–Weyl duality.

2.1. Dual pairs. We will start with the following general statement.

THEOREM 2.1. *Let G and H be two groups and $\rho : G \times H \to \mathrm{GL}(V)$ be a representation in a vector space V. Assume that V has a decomposition*

$$V = \bigoplus_{i=1}^{m} V_i \otimes \mathrm{Hom}_G(V_i, V)$$

for some absolutely irreducible representations $V_1, \ldots, V_m$ of G, and that the sub-algebra generated by $\rho(H)$ equals $\mathrm{End}_G(V)$. Then every $W_i := \mathrm{Hom}_G(V_i, V)$ is an absolutely irreducible representation of H and W_i is not isomorphic to W_j if $i \neq j$.

PROOF. Since every V_i is an absolutely irreducible representation of G, we have

$$\mathrm{End}_G(V) = \prod_{i=1}^{m} \mathrm{End}_k(W_i).$$

By our assumption the homomorphism

$$\rho : k(H) \to \prod_{i=1}^{m} \mathrm{End}_k(W_i)$$

is surjective. Hence the statement. □

REMARK 2.2. In general, we say that G and H satisfying the conditions of Theorem 2.1 form a *dual pair*.

EXAMPLE 2.3. Let k be an algebraically closed field of characteristic zero, G be a finite group. Let ρ be the regular representation of G in $k(G)$ and σ be the representation of G in $k(G)$ defined by

$$\sigma_g(h) = hg^{-1}$$

for all $g, h \in G$. Then $k(G)$ has the structure of a $G \times G$-module and we have a decomposition

$$k(G) = \bigoplus_{i=1}^{r} V_i \boxtimes V_i^*,$$

where $V_1, \ldots, V_r$ are all irreducible representations of G (up to isomorphism).

2.2. Duality between $GL(V)$ and S_n. Let V be a vector space over a field k of characteristic zero. Then V is an irreducible representation of the group $GL(V)$. We would like to understand $V^{\otimes n}$ as a $GL(V)$-module. Is it semisimple? If so, what are its simple components?

Let us define the representation $\rho : S_n \to \mathrm{GL}\,(V^{\otimes n})$ by setting

$$s\,(v_1 \otimes \cdots \otimes v_n) := v_{s(1)} \otimes \cdots \otimes v_{s(n)},$$

for all $v_1, \ldots, v_n \in V$ and $s \in S_n$. One can easily check that the actions of $GL(V)$ and S_n in the space $V^{\otimes n}$ commute. We will show that $GL(V)$ and S_n form a dual pair.

THEOREM 2.4. *(Schur–Weyl duality)* Let $m = \dim V$ and $\Gamma_{n,m}$ denote the set of all Young diagrams λ with n boxes such that the number of rows of λ is not bigger than m. Then

$$V^{\otimes n} = \bigoplus_{\lambda \in \Gamma_{n,m}} V_\lambda \otimes S_\lambda(V),$$

where V_λ is the irreducible representation of S_n associated to λ and

$$S_\lambda(V) := \mathrm{Hom}_{S_n}(V_\lambda, V^{\otimes n})$$

is an irreducible representation of $\mathrm{GL}(V)$. Moreover, $S_\lambda(V)$ and $S_\mu(V)$ are isomorphic if and only if $\lambda = \mu$.

REMARK 2.5. If $\lambda \in \Gamma_{n,m}$ and $t(\lambda)$ is an arbitrary Young tableau of shape λ, then the image of the Young symmetrizer $c_{t(\lambda)}$ in $V^{\otimes n}$ is a simple $GL(V)$-module isomorphic to $S_\lambda(V)$.

EXAMPLE 2.6. Let $n = 2$. Then we have a decomposition $V \otimes V = S^2(V) \oplus \Lambda^2(V)$. Theorem 2.4 implies that $S^2(V) = S_{(2)}(V)$ and $\Lambda^2(V) = S_{(1,1)}(V)$ are irreducible representations of $GL(V)$. More generally, $S_{(n)}(V)$ is isomorphic to $S^n(V)$ and $S_{(1,\ldots,1)}(V)$ is isomorphic to $\Lambda^n(V)$.

Let us prove Theorem 2.4.

LEMMA 2.7. *Let* $\sigma : k(GL(V)) \to \mathrm{End}_k(V^{\otimes n})$ *be the homomorphism of algebras induced by the action of* $GL(V)$ *on* $V^{\otimes n}$. *Then*

$$\mathrm{End}_{S_n}\left(V^{\otimes n}\right) = \sigma(k(GL(V))).$$

PROOF. Let $E = \mathrm{End}_k(V)$. Then we have an isomorphism of algebras

$$\mathrm{End}_k\left(V^{\otimes n}\right) \simeq E^{\otimes n}.$$

We define the action of S_n on $E^{\otimes n}$ by setting

$$s(X_1 \otimes \cdots \otimes X_n) := X_{s(1)} \otimes \cdots \otimes X_{s(n)}$$

for all $s \in S$ and $X_1, \ldots, X_n \in E$. Then $\mathrm{End}_{S_n}\left(V^{\otimes n}\right)$ coincides with the subalgebra of S_n-invariants in $E^{\otimes n}$, that is with the n-th symmetric power $S^n(E)$ of E. Therefore, it suffices to show that $S^n(E)$ is the linear span of $\sigma(g)$ for all $g \in GL(V)$.

We will need the following

EXERCISE 2.8. Let W be a vector space. Prove that for all $n \geq 2$, the following identity holds in the symmetric algebra $S\left(W\right)$:

$$2^{n-1}n!x_1 \ldots x_n = \sum_{i_2=0,1;\ldots;i_n=0,1} (-1)^{i_2+\cdots+i_n}\left(x_1 + (-1)^{i_2}\, x_2 + \cdots + (-1)^{i_n}\, x_n\right)^n.$$

Let us choose a basis $e_1, \ldots, e_{m^2}$ of E such that all non-zero linear combinations $a_1 e_1 + \cdots + a_{m^2} e_{m^2}$ with coefficients $a_i \in \{-n, \ldots, n\}$ belong to $GL(V)$. (The existence of such a basis follows from the density of $GL(V)$ in E for the Zariski topology.) By Exercise 2.8, the set

$$\{(a_1 e_{i_1} + \cdots + a_n e_{i_n})^{\otimes n} \mid a_i = \pm 1,\ i_1, \ldots i_n \leq m^2\}$$

spans $S^n(E)$. On the other hand, every non-zero $(a_1 e_{i_1} + \cdots + a_n e_{i_n})$ belongs to $GL(V)$. By construction, we have

$$(a_1 e_{i_1} + \cdots + a_n e_{i_n})^{\otimes n} = \sigma(a_1 e_{i_1} + \cdots + a_n e_{i_n}),$$

hence $S^n(E)$ is the linear span of $\sigma(g)$ for $g \in GL(V)$.   □

LEMMA 2.9. *Let* $\lambda = (\lambda_1, \ldots, \lambda_p)$ *be a partition of* n. *Then* $S_\lambda(V) \neq 0$ *if and only if* $\lambda \in \Gamma_{n,m}$.

PROOF. Consider the tableau $t(\lambda)$ with entries $1, \ldots, n$ placed in increasing order from top to bottom of the Young diagram λ starting from the first column. For instance, for $\lambda =$ we consider the tableau $t(\lambda) =$. By Remark 2.5 $S_\lambda(V) \neq 0$ if and only if $c_{t(\lambda)}(V^{\otimes n}) \neq 0$.

If $\lambda^\perp = \mu = (\mu_1, \ldots, \mu_r)$, then

$$b_{t(\lambda)}(V^{\otimes n}) = \otimes_{i=1}^r \Lambda^{\mu_i}(V).$$

If λ is not in $\Gamma_{n,m}$, then $\mu_1 > m$ and $b_{t(\lambda)}(V^{\otimes n}) = 0$. Hence $c_{t(\lambda)}(V^{\otimes n}) = 0$.

Let $\lambda \in \Gamma_{n,m}$. Choose a basis $v_1, \ldots, v_m$ in V, then

$$B := \{v_{i_1} \otimes \cdots \otimes v_{i_n} \mid 1 \leq i_1, \ldots, i_n \leq m\}$$

is a basis of $V^{\otimes n}$. Consider the particular basis vector

$$u := v_1 \otimes \cdots \otimes v_{\mu_1} \otimes \cdots \otimes v_1 \otimes \cdots \otimes v_{\mu_r} \in B.$$

One can easily see that, in the decomposition of $c_{t(\lambda)}(u)$ in the basis B, u occurs with coefficient $\prod_{i=1}^{p} \lambda_i!$. In particular, $c_{t(\lambda)}(u) \neq 0$. Hence the statement. $\square$

Lemma 2.7, Lemma 2.9 and Theorem 2.1 imply Theorem 2.4. Furthermore, Theorem 2.4 together with the Jacobson density theorem (Theorem 2.9 Chapter 5) implies the double centralizer property:

COROLLARY 2.10. *Under the assumptions of Theorem 2.4 we have*

$$\mathrm{End}_{GL(V)}(V^{\otimes n}) = \rho(k(S_n)).$$

DEFINITION 2.11. Let λ be a partition of n. The Schur functor S_λ is the functor from the category of vector spaces to itself defined by

$$V \mapsto S_\lambda(V) = \mathrm{Hom}_{S_n}(V_\lambda, V^{\otimes n}).$$

REMARK 2.12. Note that, if $n \geq 2$, the functor S_λ is not additive, namely if V and W are not zero,

$$S_\lambda(V \oplus W) \neq S_\lambda(V) \oplus S_\lambda(W).$$

The decomposition of $S_\lambda(V \oplus W)$ into simple $GL(V) \times GL(W)$-modules will be discussed later on (see subsection 8.3).

Schur–Weyl duality holds for an infinite-dimensional space in the following form.

PROPOSITION 2.13. *Let V be an infinite-dimensional vector space and Γ_n be the set of all partitions of n. Then we have the decomposition*

$$V^{\otimes n} = \bigoplus_{\lambda \in \Gamma_n} V_\lambda \otimes S_\lambda(V),$$

each $S_\lambda(V)$ is a simple $GL(V)$-module and $S_\lambda(V)$ is not isomorphic to $S_\mu(V)$ if $\lambda \neq \mu$.

PROOF. The existence of the decomposition is straightforward (consider the collection of finite-dimensional subspaces of V). For any embedding of a finite-dimensional subspace W of V, we have the corresponding embedding $S_\lambda(W) \hookrightarrow S_\lambda(V)$. Furthermore, $S_\lambda(W) \neq 0$ if and only if $\dim W \geq \lambda_1^\perp$. Hence $S_\lambda(V) \neq 0$ for all $\lambda \in \Gamma_n$.

Furthermore, $S_\lambda(V)$ is the union of $S_\lambda(W)$ for all finite-dimensional subspaces $W \subset V$. Since $S_\lambda(W)$ is a simple $GL(W)$-module when $\dim W$ is sufficiently large, we obtain that $S_\lambda(V)$ is a simple $GL(V)$-module.

To prove the last assertion we notice that Corollary 2.10 holds by the Jacobson density theorem (Theorem 2.9 Chapter 5), hence $S_\lambda(V)$ is not isomorphic to $S_\mu(V)$ if $\lambda \neq \mu$. $\square$

Schur–Weyl duality provides a link between tensor product of $GL(V)$-modules and induction-restriction of representations of symmetric groups.

DEFINITION 2.14. Let λ be a partition of p and μ a partition of q. Note that $S_\lambda(V) \otimes S_\mu(V)$ is a $GL(V)$-submodule of $V^{\otimes(p+q)}$, hence it is a semisimple $GL(V)$-module and can be written as a direct sum of simple factors $S_\nu(V)$, each occurring with some multiplicities. These multiplicities are called *Littlewood-Richardson coefficients*. More precisely, we define $N^\nu_{\lambda,\mu}$ as the function of three partitions λ, μ and ν given by

$$N^\nu_{\lambda,\mu} := \dim \operatorname{Hom}_{GL(V)}(S_\nu(V), S_\lambda(V) \otimes S_\mu(V)).$$

Clearly, $N^\nu_{\lambda,\mu} \neq 0$ implies that ν is a partition of $p + q$.

PROPOSITION 2.15. *Let λ be a partition of p and μ a partition of q, $n = p + q$ and $\dim V \geq n$. Consider the injective homomorphism $S_p \times S_q \hookrightarrow S_n$ which sends S_p to the permutations of $1, \ldots, p$ and S_q to the permutations of $p + 1, \ldots, n$. Then for any partition ν of n we have*

$$N^\nu_{\lambda,\mu} = \dim \operatorname{Hom}_{S_n}(V_\nu, \operatorname{Ind}_{S_p \times S_q}^{S_n}(V_\lambda \boxtimes V_\mu)) = \dim \operatorname{Hom}_{S_p \times S_q}(V_\nu, V_\lambda \boxtimes V_\mu).$$

PROOF. Let us choose three tableaux $t(\nu)$, $t'(\lambda)$ and $t''(\mu)$. We use the identifications

$$S_\nu(V) \simeq c_{t(\nu)}(V^{\otimes n}), \ S_\lambda(V) \simeq c_{t'(\lambda)}(V^{\otimes p}), \ S_\mu(V) \simeq c_{t''(\mu)}(V^{\otimes q}).$$

Since $V^{\otimes n}$ is a semisimple $GL(V)$-module, we have

$$\operatorname{Hom}_{GL(V)}(c_{t'(\lambda)}(V^{\otimes p}) \otimes c_{t''(\mu)}(V^{\otimes q}), c_{t(\nu)}(V^{\otimes n})) = c_{t(\nu)}k(S_n)c_{t'(\lambda)}c_{t''(\mu)}.$$

Now we use the isomorphism of S_n-modules

$$\operatorname{Ind}_{S_p \times S_q}^{S_n}(V_\lambda \boxtimes V_\mu) \simeq k(S_n)c_{t'(\lambda)}c_{t''(\mu)}.$$

Then, by Lemma 1.21, we obtain

$$N^\nu_{\lambda,\mu} = \dim \operatorname{Hom}_{S_n}(V_\nu, \operatorname{Ind}_{S_p \times S_q}^{S_n}(V_\lambda \boxtimes V_\mu)) = \dim c_{t(\nu)}k(S_n)c_{t'(\lambda)}c_{t''(\mu)}.$$

The second equality follows by Frobenius reciprocity (Chapter 2, Theorem 5.3). □

3. General facts on Hopf algebras

Let Z be a commutative unital ring.

Let A be a unital Z-algebra, we denote by $m : A \otimes A \to A$ the Z-linear multiplication. Since A is unital, there is an Z-linear map $e : Z \to A$. Moreover, we assume we are given two Z-linear maps $m^* : A \to A \otimes A$ (called the comultiplication) and $e^* : A \to Z$ (called the counit) such that the following axioms hold:

- (A): the multiplication m is associative, meaning the following diagram is commutative

$$
\begin{array}{ccc}
& m \otimes id_A & \\
A \otimes A \otimes A & \longrightarrow & A \otimes A \\
id_A \otimes m \downarrow & & \downarrow \; m \\
A \otimes A & \longrightarrow & A \\
& m &
\end{array}
$$

- (A^*): the comultiplication is coassociative, namely the following diagram commutes:

$$
\begin{array}{ccc}
& m^* & \\
A & \longrightarrow & A \otimes A \\
m^* \downarrow & & \downarrow \quad id_A \otimes m^* \\
A \otimes A & \longrightarrow & A \otimes A \otimes A \\
& m^* \otimes id_A &
\end{array}
$$

Note that this is the transpose of the diagram of (A).

- (U): The fact that $e(1) = 1$ can be expressed by the commutativity of the following diagrams:

$$
\begin{array}{ccccc}
Z \otimes A & \simeq & A & & \\
e \otimes 1 \downarrow & & \downarrow \; Id, & & \\
A \otimes A & \longrightarrow & A & & \\
& m & & &
\end{array}
\qquad
\begin{array}{ccc}
A \otimes Z & \simeq & A \\
1 \otimes e \downarrow & & \downarrow \; Id \\
A \otimes A & \longrightarrow & A \\
& m &
\end{array}
$$

- (U^*): similarly, the following diagrams commute

$$
\begin{array}{ccc}
A & \simeq & A \otimes Z \\
Id \uparrow & & \uparrow \quad 1 \otimes e^*, \\
A & \longrightarrow & A \otimes A \\
& m^* &
\end{array}
\qquad
\begin{array}{ccc}
A & \simeq & Z \otimes A \\
Id \uparrow & & \uparrow \quad e^* \otimes 1, \\
A & \longrightarrow & A \otimes A \\
& m^* &
\end{array}
$$

- (Antipode): there exists a Z-linear isomorphism $S : A \to A$ such that the following diagrams commute:

$$
\begin{array}{ccc}
& S \otimes Id_A & \\
A \otimes A & \longrightarrow & A \otimes A \\
m^* \uparrow & & \downarrow \; m, \\
A & \longrightarrow & A \\
& e \circ e^* &
\end{array}
\qquad
\begin{array}{ccc}
& Id_A \otimes S & \\
A \otimes A & \longrightarrow & A \otimes A \\
m^* \uparrow & & \downarrow \; m, \\
A & \longrightarrow & A \\
& e \circ e^* &
\end{array}
$$

DEFINITION 3.1. This set of data is called a *Hopf algebra* if the following property holds:

(H): the map $m^* : A \to A \otimes A$ is a homomorphism of Z-algebras.

Moreover, if the antipode axiom is missing, then we call it a *bialgebra*.

EXERCISE 3.2. Show that if an antipode exists, then it is unique. Moreover, if S is a left antipode and S' is a right antipode, then $S = S'$. Furthermore $S : A \to A^{op}$ is a homomorphism of algebras.

REMARK 3.3. Assume that A is a commutative Hopf algebra, for any commutative algebra B, set $X_B := \mathrm{Hom}_{Z-alg}(A, B)$, then the composition with m^* induces a map $X_B \times X_B = \mathrm{Hom}_{Z-alg}(A \otimes A, B) \to X_B$ which defines a group law on X_B. This property characterizes commutative Hopf algebras.

EXAMPLE 3.4. If M is a Z-module, then the symmetric algebra $S^\bullet(M)$ has a Hopf algebra structure, for the comultiplication m^* defined by: if Δ denotes the diagonal map $M \to M \oplus M$, then $m^* : S^\bullet(M \oplus M) = S^\bullet(M) \otimes S^\bullet(M)$ is the canonical morphism of Z-algebras induced by Δ.

EXERCISE 3.5. Find m^* when Z is a field and M is finite dimensional.

DEFINITION 3.6. Let A be a bialgebra, an element $x \in A$ is called *primitive* if $m^*(x) = x \otimes 1 + 1 \otimes x$.

EXERCISE 3.7. Show that if k is a field of characteristic zero and if V is a finite dimensional k-vector space, then the primitive elements in $S^\bullet(V)$ are exactly the elements of V.

We say that a bialgebra A is *connected graded* if

(1) $A = \bigoplus_{n \in \mathbb{N}} A_n$ is a graded algebra;
(2) $m^* : A \to A \otimes A$ is a homomorphism of graded algebras, where the grading on $A \otimes A$ is given by the sum of gradings;
(3) $A_0 = Z$;
(4) the counit $e^* : A \to Z$ is a homomorphism of graded rings.

LEMMA 3.8. *Let A be a graded connected bialgebra, $I = \bigoplus_{n>0} A_n$. Then for any $x \in I$, $m^*(x) = x \otimes 1 + 1 \otimes x + m_+^*(x)$ for some $m_+^*(x) \in I \otimes I$. In particular every element of A of degree 1 is primitive.*

PROOF. From the properties (3) and (4) we have that $I = \mathrm{Ker}\, e^*$. Write

$$m^*(x) = y \otimes 1 + 1 \otimes z + m_+^*(x).$$

We have to check that $y = z = x$. But this immediately follows from the counit axiom. □

PROPOSITION 3.9. *Let A be a connected graded bialgebra and $\mathcal{P}$ be the set of primitive elements of A. Assume that $I^2 \cap \mathcal{P} = 0$. Then A is commutative and admits an antipode.*

PROOF. Let us prove first that A is commutative. Assume the opposite. Let $x \in A_k, y \in A_l$ be some homogeneous element of A such that $[x, y] := xy - yx \neq 0$ and $k + l$ as small as possible. Then $m^*([x, y]) = [m^*(x), m^*(y)]$. By minimality of

$k+l$, we have that $[m_+^*(x), m_+^*(y)] = 0$, hence $[x, y]$ is primitive. On the other hand, $m^*([x, y]) \in I \otimes I$, hence $[x, y] = 0$. Contradiction.

Next, let us prove the existence of an antipode. For every $x \in A_n$ we construct $S(x) \in A_n$ recursively. We set

$$S(x) := -x \text{ for } n = 1, \quad S(x) = -x - m \circ (\mathrm{Id} \otimes S) \circ m_+^*(x) \text{ for } n > 1.$$

EXERCISE 3.10. Check that S satisfies the antipode axiom and is an involution.

$\square$

4. The Hopf algebra associated to the representations of symmetric groups

Let us consider the free $\mathbb{Z}$-module $\mathcal{A} = \oplus_{n \in \mathbb{N}} \mathcal{A}(S_n)$ where $\mathcal{A}(S_n)$ is freely generated by the characters of the irreducible representations (in $\mathbb{C}$-vector spaces) of the symmetric group S_n. Note that since every S_n-module is semi-simple, $\mathcal{A}(S_n)$ is the Grothendieck group of the category S_n-mod of finite dimensional representations of S_n. It is a $\mathbb{N}$-graded module, where the homogeneous component of degree n is equal to $\mathcal{A}(S_n)$ if $n \geq 1$ and the homogeneous part of degree 0 is $\mathbb{Z}$ by convention. Moreover, we equip it with a $\mathbb{Z}$-valued symmetric bilinear form, denoted $\langle ; \rangle$, for which the given basis of characters is an orthonormal basis, and with the positive cone $\mathcal{A}^+$ generated over the non-negative integers by the orthonormal basis.

In order to define the Hopf algebra structure on $\mathcal{A}$, we use the *induction* and *restriction* functors:

$$I_{p,q} : (S_p \times S_q)\text{-mod} \longrightarrow S_{p+q}\text{-mod},$$

$$R_{p,q} : S_{p+q}\text{-mod} \longrightarrow (S_p \times S_q)\text{-mod}.$$

REMARK 4.1. Frobenius (see Theorem 5.3) observed that the induction functor is left-adjoint to the restriction.

Since the restriction and induction functors are exact, they define maps in the Grothendieck groups. Moreover, the following lemma holds:

EXERCISE 4.2. Show that we have a group isomorphism

$$\mathcal{A}(S_p \times S_q) \simeq \mathcal{A}(S_p) \otimes_{\mathbb{Z}} \mathcal{A}(S_q).$$

We deduce, from the collections of functors $I_{p,q}, R_{p,q}, p, q \in \mathbb{N}$, two maps:

$$m : \mathcal{A} \otimes \mathcal{A} \longrightarrow \mathcal{A},$$

$$m^* : \mathcal{A} \longrightarrow \mathcal{A} \otimes \mathcal{A}.$$

More explicitly, if M (resp. N) is an S_p (resp. S_q)-module and if $[M]$ (resp. $[N]$) denotes its class in the Grothendieck group,

$$m([M] \otimes [N]) = [I_{p,q}(M \otimes N)],$$

and if P is an S_n-module,

$$m^*([P]) = \sum_{p+q=n} [R_{p,q}(P)].$$

EXERCISE 4.3. Show that m is associative and hence m^* is coassociative (use adjunction), m is commutative and m^* is cocommutative (use adjunction).

The tricky point is to show the following lemma:

LEMMA 4.4. *The map m^* is an algebra homomorphism.*

PROOF. (Sketch) We will use Theorem 7.4 Chapter 2 to compute

$$\mathrm{Res}^{S_n}_{S_p \times S_q} \mathrm{Ind}^{S_n}_{S_k \times S_l} M \otimes N,$$

where $p + q = k + l = n$, M and N are representations of S_k and S_l respectively. The double cosets $S_p \times S_q \backslash S_n / S_k \times S_l$ are enumerated by quadruples $(a, b, c, d) \in \mathbb{N}^4$ satisfying $a + b = p, c + d = q, a + c = k, b + d = l$. So we have

$$\mathrm{Res}^{S_n}_{S_p \times S_q} \mathrm{Ind}^{S_n}_{S_k \times S_l} M \otimes N =$$

$$= \bigoplus_{a+b=p,c+d=q,a+c=k,b+d=l} \mathrm{Ind}^{S_p \times S_q}_{S_a \times S_b \times S_c \times S_d} \mathrm{Res}^{S_k \times S_l}_{S_a \times S_b \times S_c \times S_d} M \otimes N.$$

and

$$\mathrm{Res}^{S_k \times S_l}_{S_a \times S_b \times S_c \times S_d} M \otimes N = \mathrm{Res}^{S_k}_{S_a \times S_c} M \otimes \mathrm{Res}^{S_l}_{S_b \times S_d} N.$$

If

$$R_{a,c}(M) \otimes R_{b,d} N = \oplus_i A_i \otimes B_i \otimes C_i \otimes D_i,$$

then

(6.1) $R_{p,q} I_{k,l}(M \otimes N) = \displaystyle\sum_{a+b=p,c+d=q,a+c=k,b+d=l} \sum_i I_{a,b}(A_i \otimes C_i) \otimes I_{c,d}(B_i \otimes D_i).$

The relation (6.1) is the condition

$$m^* m(a, b) = \sum_{i,j} m(a_i, b_j) \otimes m(a^i, b^j),$$

where $m^*(a) = \sum_i a_i \otimes a^i$, $m^*(b) = \sum_j b_j \otimes b^j$, in terms of homogeneous components. ☐

The axiom (U) corresponds to the inclusion $\mathcal{A}_0 \subset \mathcal{A}$ and (U^*) is its adjoint. As we will see in Proposition 6.6 the antipode of the class of a simple S_n-module $[M]$ is the virtual module $(-1)^n[\varepsilon \otimes M]$, where ε is the sign representation of S_n.

Hence we have a structure of Hopf algebra on $\mathcal{A}$, and the following properties are easily checked:

• *positivity*: the cone $\mathcal{A}^+$ is stable under multiplication (for m),

• *self-adjointness*: The maps m and m^* are mutually adjoint with respect to the scalar product on $\mathcal{A}$ and the corresponding scalar product on $\mathcal{A} \otimes \mathcal{A}$.

DEFINITION 4.5. A graded connected bialgebra A over $\mathbb{Z}$ together with a homogeneous basis Ω, equipped with a scalar product $\langle\ ,\ \rangle$ for which Ω is orthonormal, which is positive $m(A^+ \otimes A^+) \subset A^+$, $m^*(A^+) \subset A^+ \otimes A^+$ (for the cone A^+ generated over $\mathbb{N}$ by Ω) and self-adjoint is called a *positive self-adjoint Hopf algebra*, PSH algebra for short. Moreover, the elements of Ω are called *basic elements* of A.

REMARK 4.6. It is easy to see that in a PSH algebra $\mathcal{P} \cap I^2 = 0$. Hence a PSH algebra is automatically a commutative and cocommutative Hopf algebra by Proposition 3.9.

We have just seen that:

PROPOSITION 4.7. *The algebra $\mathcal{A}$ with the basis Ω given by the classes of all irreducible representations is a PSH algebra.*

EXERCISE 4.8. Show that for any $a_1, \ldots, a_n$ in A, the matrix $\mathrm{Gram}(a_1, \ldots, a_n) = (a_{ij})$ such that $a_{ij} := \langle a_i, a_j \rangle$ (called the Gram matrix) is invertible in $M_n(\mathbb{Z})$ (i.e. the determinant is ± 1) if and only if the a_i's form a basis of the sublattice of A generated by some subset of cardinality n of Ω. Note that if the Gram matrix is the identity, then, up to sign, the a_i's belong to Ω.

EXERCISE 4.9. Assume H is a Hopf algebra with a scalar product and assume H is commutative and self-adjoint. Let x be a primitive element in H and consider the map

$$d_x : H \longrightarrow H,\ y \mapsto \sum_i \langle y_i, x \rangle y^i$$

where $m^*(y) = \sum_i y_i \otimes y^i$.

- Show that d_x is a derivation (for all a, b in H, $d_x(ab) = a d_x(b) + d_x(a)b$).
- Show that if x and y are primitive elements in H, then $d_x(y) = \langle y, x \rangle$.
- Show that $\langle xa, b \rangle = \langle a, d_x(b) \rangle$ for all $a, b \in H$.

5. Classification of PSH algebras part 1: decomposition theorem

In this section we classify PSH algebras following Zelevinsky, [38]. Let A be a PSH algebra with the specified basis Ω and positive cone A^+. Let us denote by Π the set of basic primitive elements of A, that is, primitive elements belonging to Ω.

A multi-index α is a finitely supported function from Π to $\mathbb{N}$. For such an α we denote by π^α the monomial $\prod_{p \in \Pi} p^{\alpha(p)}$. We denote by M the set of such monomials.

For $a \in A$ we denote by $Supp(a)$ (and call support of a) the set of basic elements which appear in the decomposition of a.

LEMMA 5.1. *The supports of π^α and π^β are disjoint whenever $\alpha \neq \beta$.*

PROOF. Since the elements of M belong to A^+, we just have to show that the scalar product $\langle \pi^\alpha, \pi^\beta \rangle$ is zero when $\alpha \neq \beta$. We prove this by induction on the total degree of the monomial π^α. Write $\pi^\alpha = \pi_1 \pi^\gamma$ for some π_1 such that $\alpha(\pi_1) \neq 0$. Then (recall Exercise 4.9)

$$\langle \pi^\alpha, \pi^\beta \rangle = \langle \pi^\gamma, d_{\pi_1}(\pi^\beta) \rangle.$$

Since the total degree of π^γ is less than the degree of π^α, if the scalar product is not zero, we obtain by the induction assumption that $d_{\pi_1}(\pi^\beta)$ is a multiple of π^γ. This implies $\pi^\beta = \pi_1 \pi^\gamma = \pi^\alpha$. □

For every monomial $\pi^\alpha \in M$, denote by A^α the $\mathbb{Z}$-span of $Supp(\pi^\alpha)$.

LEMMA 5.2. *For all π^α, π^β in M,*

$$A^\alpha A^\beta \subset A^{\alpha+\beta}.$$

PROOF. We consider the partial ordering $\leq$ in A whose positive cone is A^+ (i.e. $x \leq y$ if and only if $y - x \in A^+$). Note that if $0 \leq x \leq y$ then $Supp(x) \subset Supp(y)$. Therefore if we pick up ω in $Supp(\pi^\alpha)$ and $\eta \in Supp(\pi^\beta)$ then $\omega \eta \leq \pi^{\alpha+\beta}$, hence the result. □

Let I be the ideal spanned by all elements of positive degree.

EXERCISE 5.3. Show that if $x \in A$ is primitive, then $x \in I$.

Moreover, $x \in I$ is primitive if and only if x is orthogonal to I^2. Indeed for y and z in I, $\langle m^*(x) - 1 \otimes x - x \otimes 1, y \otimes z \rangle = \langle x, yz \rangle$, hence the result by Exercise 5.3.

LEMMA 5.4. *One has:*

$$A = \bigoplus_{\pi^\alpha \in M} A^\alpha.$$

PROOF. Assume the equality doesn't hold, then there exists an $\omega \in \Omega$ which does not belong to this sum. We choose such an ω with minimal degree k. Since ω is not primitive, it is not orthogonal to I^2 and therefore belongs to the support of some $\eta \eta'$ with η, η' belonging to Ω. Hence $k = k' + k''$ where k' (resp. k'') is the degree of η (resp. η'). By minimality of k, η and η' lie in the direct sum, thus, by Lemma 5.2, we obtain a contradiction. □

LEMMA 5.5. *Let $\pi^\alpha, \pi^\beta \in M$ be relatively prime monomials. Then the restriction of multiplication induces an isomorphism $A^\alpha \otimes A^\beta \simeq A^{\alpha+\beta}$ given by a bijection between $Supp(\pi^\alpha) \times Supp(\pi^\beta)$ and $Supp(\pi^{\alpha+\beta})$.*

PROOF. We will prove that the Gram matrix (see Exercise 4.8)

$$\mathrm{Gram}((\omega \eta)_{\omega \in Supp(\pi^\alpha), \eta \in Supp(\pi^\beta)})$$

is the identity. This will be enough since it implies that the products $\omega \eta$ are distinct elements of Ω (again, see Exercise 4.8), and they exhaust the support of $\pi^{\alpha+\beta}$ which belongs to their linear span.

Let ω_1, ω_2 (resp. η_1, η_2) be elements of $Supp(\pi^\alpha)$ (resp. $Supp(\pi^\beta)$), then

$$\langle \omega_1 \eta_1, \omega_2 \eta_2 \rangle = \langle m^*(\omega_1 \eta_1), \omega_2 \otimes \eta_2 \rangle = \langle m^*(\omega_1) m^*(\eta_1), \omega_2 \otimes \eta_2 \rangle.$$

One has $m^*(\omega_1) \in \bigoplus_{\alpha'+\alpha''=\alpha} A^{\alpha'} \otimes A^{\alpha''}$ and $m^*(\eta_1) \in \bigoplus_{\beta'+\beta''=\beta} A^{\beta'} \otimes A^{\beta''}$ (this is just a transposed version of Lemma 5.2), hence

$$m^*(\omega_1)m^*(\eta_1) \in \sum_{\alpha'+\alpha''=\alpha, \beta'+\beta''=\beta} A^{\alpha'+\beta'} \otimes A^{\alpha''+\beta''}.$$

On the other hand, $\omega_2 \otimes \eta_2$ belongs to $A^\alpha \otimes A^\beta$. We must understand in which cases $A^{\alpha'+\beta'} \otimes A^{\alpha''+\beta''} = A^\alpha \otimes A^\beta$ and this occurs if and only if $\alpha' + \beta' = \alpha, \alpha'' + \beta'' = \beta$. Since π^α and π^β are relatively prime, this occurs if and only if $\beta' = 0 = \alpha''$.

The component of $m^*(\omega_1)$ in $A^\alpha \otimes A^0$ is $\omega_1 \otimes 1$ and the component of $m^*(\eta_1)$ in $A^0 \otimes A^\beta$ is $1 \otimes \eta_1$ (see Exercise 5.3), therefore

$$\langle \omega_1\eta_1, \omega_2\eta_2 \rangle = \langle (\omega_1 \otimes 1)(1 \otimes \eta_1), \omega_2 \otimes \eta_2 \rangle = \langle \omega_1 \otimes \eta_1, \omega_2 \otimes \eta_2 \rangle = \langle \omega_1\omega_2, \eta_1\eta_2 \rangle = \delta_{\omega_1,\omega_2}\delta_{\eta_1,\eta_2}.$$

Hence the result. □

The following Theorem is a direct consequence of Lemmas 5.1, 5.2, 5.4, 5.5.

THEOREM 5.6. (Zelevinsky's decomposition theorem). Let A be a PSH algebra with basis Ω, and let Π be the set of basic primitive elements of A. For every $\pi \in \Pi$ we set $A_\pi := \bigoplus_{n \in \mathbb{N}} A^{\pi^n}$. Then

(1) A_π is a PSH algebra and its unique basic primitive element is π,
(2) $A = \bigotimes_{\pi \in \Pi} A_\pi$.

REMARK 5.7. In the second statement, the tensor product might be infinite: it is defined as the span of tensor monomials with a finite number of entries non-equal to 1.

DEFINITION 5.8. The *rank* of the PSH algebra A is the cardinality of the set Π of basic primitive elements in A.

6. Classification of PSH algebras part 2: unicity for the rank 1 case

By the previous section, understanding a PSH algebra is equivalent to understanding its rank one components. Therefore, we want to classify the rank one cases.

Let A be a PSH algebra of rank one with marked basis Ω, and denote by π its unique basic primitive element. We assume that we have chosen the grading of A so that π is of degree 1. We will construct a sequence $(e_i)_{i \in \mathbb{N}}$ of elements of Ω such that:

(1) $e_0 = 1, e_1 = \pi$ and e_n is of degree n (it is automatically homogeneous since it belongs to Ω),
(2) $A \simeq \mathbb{Z}[e_1, e_2, \ldots, e_n, \ldots]$ as graded $\mathbb{Z}$-algebras,
(3) $m^*(e_n) = \sum_{i+j=n} e_i \otimes e_j$.

Actually, we will find exactly two such sequences and we will denote the second one $(h_i)_{i \in \mathbb{N}}$. The antipode map exchanges those two sequences.

We denote by d the derivation of A which is adjoint to the multiplication by π (see Exercise 4.9).

LEMMA 6.1. *There are exactly two elements in Ω of degree 2 and their sum is equal to π^2.*

PROOF. First note that

$$\langle \pi^2, \pi^2 \rangle = \langle \pi, d(\pi^2) \rangle = 2\langle \pi, \pi \rangle = 2.$$

On the other hand, if we write π^2 in the basis Ω, $\pi^2 = \sum_{\omega \in \Omega} \langle \pi^2, \omega \rangle \omega$, we get

$$\langle \pi^2, \pi^2 \rangle = \sum_{\omega \in \Omega} \langle \pi^2, \omega \rangle^2,$$

but $\langle \pi^2, \omega \rangle$ is a non negative integer, hence the result. □

We will denote by e_2 one of those two basic elements and h_2 the other one. Furthermore, we set e_2^* (resp. h_2^*) to be the linear operator on A of degree -2 which is adjoint to the multiplication by e_2 (resp. h_2).

EXERCISE 6.2. Show that the operator e_2^* satisfies the identities

(6.2) $$m^*(e_2) = e_2 \otimes 1 + \pi \otimes \pi + 1 \otimes e_2,$$

(6.3) $$e_2^*(ab) = e_2^*(a)b + ae_2^*(b) + d(a)d(b).$$

LEMMA 6.3. *There is exactly one element e_n (resp. h_n) of degree n in Ω such that $h_2^*(e_n) = 0$ (resp. $e_2^*(h_n) = 0$). This element satisfies $d(e_n) = e_{n-1}$ (resp. $d(h_n) = h_{n-1}$).*

PROOF. We prove this for the sequence e_n by induction on n. The argument for h_n follows by symmetry. For $n = 2$, $h_2^*(e_2)$ is the scalar product $\langle h_2, e_2 \rangle$ which is zero because e_2 and h_2 are two distinct elements of Ω. We assume that the statement of the lemma holds for all $i < n$. If x is of degree n and satisfies $h_2^*(x) = 0$, then $d(x)$ (which is of degree $n - 1$) is proportional to e_{n-1} by the induction hypothesis, since h_2^* and d commute. The scalar is equal to $\langle d(x), e_{n-1} \rangle$ $= \langle x, \pi e_{n-1} \rangle$ since d is the adjoint of the multiplication by π. As in Lemma 6.1, we prove next that $\langle \pi e_{n-1}, \pi e_{n-1} \rangle = 2$: indeed $\langle \pi e_{n-1}, \pi e_{n-1} \rangle = \langle d(\pi e_{n-1}), e_{n-1} \rangle = \langle e_{n-1} + \pi e_{n-2}, e_{n-1} \rangle = 1 + \langle \pi e_{n-2}, e_{n-1} \rangle = 1 + \langle e_{n-2}, d(e_{n-1}) \rangle = 2$.

Therefore πe_{n-1} decomposes as the sum of two distinct basic elements $\omega_1 + \omega_2$. Besides, using Exercise 6.2 equation (6.3), we have $h_2^*(\pi e_{n-1}) = e_{n-2}$. Since h_2^* is a positive operator (i.e. preserves A^+), $h_2^*(\omega_1) + h_2^*(\omega_2) = e_{n-2}$ implies that one of the factors $h_2^*(\omega_i)$ ($i = 1$ or 2) is zero, so that ω_i can be choosen for $x = e_n$. □

PROPOSITION 6.4. *For every $n \geq 1$,*

$$m^*(e_n) = \sum_{k=0}^{n} e_k \otimes e_{n-k}.$$

PROOF. If ω belongs to Ω, we denote by ω^* the adjoint of the multiplication by ω. We just need to show that $\omega^*(e_n) = 0$ except if $\omega = e_k$ for some $0 \leq k \leq n$, in which case $\omega^*(e_n) = e_{n-k}$.

Indeed, we can write

$$\pi^k = \sum_{\omega \in \Omega, deg(\omega)=k} C_\omega \omega$$

where the coefficients C_ω are positive integers, hence

$$d^k = \sum_{\omega \in \Omega, deg(\omega)=k} C_\omega \omega^*.$$

Since

$$\sum_{\omega \in \Omega, deg(\omega)=k} C_\omega \omega^*(e_n) = d^k(e_n) = e_{n-k},$$

all the terms in the sum are zero except one (by integrity of the coefficients). It remains to show that the non-zero term comes from the element e_k of Ω. But this is clear since

$$d^{n-k} e_k^*(e_n) = e_k^* d^{n-k}(e_n) = e_k^*(e_k) = 1.$$

EXERCISE 6.5. (1) Show that for every $n \geq 0$, $e_n^*(ab) = \sum_{0 \leq k \leq n} e_k^*(a) e_{n-k}^*(b)$.

(2) We make the convention that $h_{-1} = 0$. Prove the following equality for any positive integer r and $i_1, \ldots, i_r$ non negative integers:

(6.4) $$e_r^*(h_{i_1} \ldots h_{i_r}) = h_{i_1-1} \ldots h_{i_r-1}.$$

PROPOSITION 6.6. Let t be an indeterminate, the two formal series $\sum_{i \geq 0} e_i t^i$ and $\sum_{i \geq 0} (-1)^i h_i t^i$ are mutually inverse.

PROOF. Since A is a graded bialgebra over $\mathbb{Z}$, we know (Proposition 3.9) that it is equipped with a unique antipode S. Let us show that it exchanges e_n and $(-1)^n h_n$:

First, let us show that S is an isometry for the scalar product of A. Indeed, we have the following commutative diagram

(6.5)
$$\begin{array}{ccc} & Id_A \otimes S & \\ A \otimes A & \longrightarrow & A \otimes A \\ m^* \uparrow & & \downarrow m, \\ A & \longrightarrow & A \\ & e \circ e^* & \end{array}$$

where $e : \mathbb{Z} \to A$ is the unit of A (see section 3). By considering the adjoint of this diagram, we understand that S^* is also an antipode, and by uniqueness of the antipode, $S = S^{-1} = S^*$ hence S is an isometry and so for $\omega \in \Omega$, $S(\omega) = \pm \eta$, for some $\eta \in \Omega$ (ω and η have the same degree). Applying the diagram (6.5) to π, who is primitive, we check that $S(\pi) = -\pi$. In the same way, we obtain $S(\pi^2) = \pi^2$ and $S(e_2) = h_2$. Since e_n is the unique basic element of degree n satisfying the relation $h_2^*(e_n) = 0$, we have $S(e_n) = \pm h_n$ and the sign is $(-1)^n$ since $S(\pi^n) = (-1)^n \pi^n$.

The diagram (6.5) implies that $(m \circ Id_A \otimes S \circ m^*)(e_n) = 0$ for all $n \geq 1$. By Proposition 6.4, we know that $m^*(e_n) = \sum_{0 \leq k \leq n} e_k \otimes e_{n-k}$ and so we have $\sum_{0 \leq k \leq n} (-1)^{n-k} e_k h_{n-k} = 0$. The result follows. □

We will now use the definitions and notations for partitions introduced in Section 1. For a partition $\lambda = (\lambda_1, \ldots, \lambda_n)$, we denote by e_λ the product $e_\lambda = e_{\lambda_1} \ldots e_{\lambda_n}$ and use a similar definition for h_λ. Note that in general, the elements e_λ, h_λ do not belong to Ω.

DEFINITION 6.7. Let $\lambda = (\lambda_1, \ldots, \lambda_r)$ and $\mu = (\mu_1, \ldots, \mu_s)$ be two partitions of the same integer n. We say that λ *is greater or equal than* μ *for the dominance order* and denote it $\lambda \succeq \mu$ if, for every $k \leq \inf(r, s)$, $\lambda_1 + \ldots + \lambda_k \geq \mu_1 + \ldots + \mu_k$.

LEMMA 6.8. *Let* λ *and* μ *be partitions of a given integer* n, *define* $M_{\lambda,\mu}$ *as the number of square matrices with entries belonging to* $\{0,1\}$ *such that the sum of the entries in the* i-*th row (resp. column) is* λ_i *(resp.* μ_i*)* [1]. *Then:*

(1) $\langle e_\lambda, h_\mu \rangle = M_{\lambda,\mu}$,
(2) $M_{\lambda,\lambda^\perp} = 1$,
(3) $M_{\lambda,\mu} \neq 0$ *implies* $\lambda \preceq \mu^\perp$.

PROOF. (Sketch) We write $\lambda = (\lambda_1, \ldots, \lambda_r)$ and $\mu = (\mu_1 \ldots \mu_s)$. By Exercise 6.5, we have

$$e_{\lambda_1}^*(h_{\mu_1} \ldots h_{\mu_s}) = \sum_{\nu_i = 0,1, \sum \nu_i = \lambda_1} h_{\mu_1 - \nu_1} \ldots h_{\mu_s - \nu_s}.$$

Next, we apply $e_{\lambda_2}^*$ to this sum, $e_{\lambda_3}^*$ to the result, and so on. We obtain:

$$\langle e_\lambda, h_\mu \rangle = e_\lambda^*(h_\mu) = \sum_{\nu_{ij} \in \{0,1\}, \sum_j \nu_{ij} = \lambda_i} h_{\mu_1 - \sum_i \nu_{i1}} \ldots h_{\mu_s - \sum_i \nu_{is}}.$$

The terms in the sum of the right-hand side are equal to 0 except when $\mu_j = \sum_i \nu_{ij}$ for all i, in which case the value is 1. The statement (1) follows.

For statement (2), we see easily that the only matrix $N = (\nu_{ij})$ with entries in $\{0,1\}$ such that $\sum_j \nu_{ij} = \lambda_i$ and $\sum_i \nu_{ij} = \lambda_j^\perp$ is the one such that the entries decrease along both the rows and the columns, hence the result.

Finally, consider a matrix $N = (\nu_{ij})$ with entries in $\{0,1\}$ such that $\sum_j \nu_{ij} = \lambda_i$ and $\sum_i \nu_{ij} = \mu_j$. Then $\mu_i^\perp$ is the number of columns with at least i non-zero entries. From this the desired inequality can be seen. □

COROLLARY 6.9. *The matrix* $(\langle e_\lambda, h_{\mu^\perp} \rangle)_{\lambda,\mu \vdash n}$ *is upper triangular with* 1*'s on the diagonal. In particular, its determinant is equal to* 1.

PROPOSITION 6.10. *When* λ *varies along the partitions of* n, *the collection of* e_λ*'s is a basis of the homogeneous component of degree* n, A_n, *of* A.

PROOF. First we notice that every h_i is a polynomial with integral coefficients in the e_j's. This follows immediately from Proposition 6.6. Therefore the base change matrix P from $(h_\lambda)_{\lambda \vdash n}$ to $(e_\lambda)_{\lambda \vdash n}$ has integral entries. Then the Gram matrix G_e of $(e_\lambda)_{\lambda \vdash n}$ satisfies the equality

$$(\langle e_\lambda, h_\mu \rangle)_{\lambda,\mu \vdash n} = P^t G_e,$$

---

[1]It is easy to see that $M_{\lambda,\mu}$ stabilizes as a function of the size of matrices.

where P^t denotes the transposed P. The corollary 6.9 ensures that the left-hand side has determinant ± 1 (the corollary is stated for $\mu^\perp$ and $\mu \mapsto \mu^\perp$ is an involution which can produce a sign). Hence G_e has determinant ± 1: we refer to Exercise 4.8 to ensure that the $\mathbb{Z}$-module generated by $(e_\lambda)_{\lambda \vdash n}$ has a basis contained in Ω. Since the support of e_1^n is the set of all $\omega \in \Omega$ of degree n, we conclude that $(e_\lambda)_{\lambda \vdash n}$ is a basis of A_n. $\qquad\square$

We deduce from the results of this section:

THEOREM 6.11. *(Zelevinsky) Up to isomorphism, there is only one rank one PSH algebra. It has only one non-trivial automorphism ι, preserving the positive cone. This ι takes any homogeneous element x of degree n to $(-1)^n S(x)$ where S is the antipode.*

REMARK 6.12. The sets of algebraically independent generators (e_n) and (h_n) of the $\mathbb{Z}$-algebra A play symmetric roles, and they are exchanged by the automorphism ι of Theorem 6.11.

7. Bases of PSH algebras of rank one

Let A be a PSH algebra of rank one, with basis Ω and scalar product $\langle\,,\rangle$, we will use the sets of generators (e_n) and (h_n). We keep all the notations of the preceding section.

We will first describe the primitive elements of A. We denote $A_\mathbb{Q} := A \otimes \mathbb{Q}$.

EXERCISE 7.1. Consider the algebra $A[[t]]$ of formal power series with coefficients in A. Let $f \in A[[t]]$ such that $m^*(f) = f \otimes f$ and the constant term of f is 1. Show that the logarithmic derivative $g := \frac{f'}{f}$ satisfies $m^*(g) = g \otimes 1 + 1 \otimes g$.

PROPOSITION 7.2.   (1) *For every $n \geq 1$, there is exactly one primitive element p_n of degree n, such that $\langle p_n, h_n \rangle = 1$. Moreover, every primitive element of degree n is a integral multiple of p_n.*
    (2) *In the formal power series ring $A_\mathbb{Q}[[t]]$, we have the following equality:*

$$(6.6) \qquad exp\left(\sum_{n \geq 1} \frac{p_n}{n} t^n\right) = \sum_{n \geq 0} h_n t^n.$$

PROOF. We first show that the set of primitive elements of degree n is a subgroup of rank 1. Indeed, we recall that the primitive elements form the orthogonal complement of I^2 in I (see just below Exercise 5.3). Since all the elements $(h_\lambda)_{\lambda \vdash n}$ except h_n are in I^2, the conclusion follows. Moreover, A_n is its own dual with respect to the scalar product. Let us denote by $(h^\lambda)_{\lambda \vdash n}$ the dual basis of $(h_\lambda)_{\lambda \vdash n}$. Clearly, h^n can be chosen as p_n. Hence statement (1).

Consider the formal series $H(t) := \sum_{n \geq 0} h_n t^n \in A[[t]]$, it satisfies the relation $m^*(H) = H \otimes H$ by Proposition 6.4, re-written in terms of h's instead of e's. Hence, using Exercise 7.1, we get $P(t) := \frac{H'(t)}{H(t)}$ which satisfies $m^*(P) = P \otimes 1 + 1 \otimes P$. Hence all the coefficients of P are primitive elements in A. Write $P(t) = \sum_{i \geq 0} \varpi_{i+1} t^i$. To prove statement (2), it remains to check that $\langle \varpi_n, h_n \rangle = 1$. We have $P(t)H(t) = H'(t)$, so when we compare the terms on both sides we get

(6.7) $$\varpi_n + h_1 \varpi_{n-1} + \ldots + h_{n-1} \varpi_1 - n h_n = 0.$$

By induction on n, this implies

$$\varpi_n = (-1)^n h_1^n + \sum_{\lambda \vdash n, \lambda \neq (1,\ldots,1)} c_\lambda h_\lambda,$$

where the c_λ's are integers. Now, we compute the scalar product with e_n: we apply Lemma 6.8 and find that there is no contribution from the terms indexed by λ if $\lambda \neq (1,\ldots,1)$. Therefore, $\langle \varpi_n, e_n \rangle = (-1)^n$. Finally, we use the automorphism ι of A to get the conclusion that $p_n = \varpi_n$ since $\iota(e_n) = h_n$ and $\iota(\varpi_n) = (-1)^n \varpi_n$ by Proposition 6.6. □

For every partition $\lambda = (\lambda_1, \ldots, \lambda_r)$, we set $p_\lambda = p_{\lambda_1} \ldots p_{\lambda_r}$. Let us compute their Gram matrix:

PROPOSITION 7.3. *The family* (p_λ) *is an orthogonal basis of* $A_\mathbb{Q}$ *and*

$$\langle p_\lambda, p_\lambda \rangle = \prod_{j=1}^{|\lambda|} (\lambda_j^\perp - \lambda_{j+1}^\perp)! \prod_i \lambda_i.$$

PROOF. Since p_i is primitive, the operator p_i^* is a derivation of A. Moreover, since p_i is of degree i, p_i and p_j are orthogonal when $i \neq j$. We compute $\langle p_i, p_i \rangle$: we use the formula (6.7) (recall that we proved that $p_n = \varpi_n \forall n$) and since $p_i^*(h_r p_{i-r}) = p_i^*(h_r)p_{i-r} + h_r p_i^*(p_{i-r}) = 0$ if $1 \leq r \leq i-1$, we obtain $p_i^*(p_i) = \langle p_i, p_i \rangle = \langle p_i, i h_i \rangle = i$ by Proposition 7.2.

To show that p_λ is orthogonal to p_μ if $\lambda \neq \mu$, we repeat the argument of the proof of Lemma 5.1.

Finally, we compute $\langle p_i^s, p_i^s \rangle$: we use the fact that p_i^* is a derivation such that $p_i^*(p_i) = i$, hence $\langle p_i^s, p_i^s \rangle = s! i^s$. This implies the formula giving $\langle p_\lambda, p_\lambda \rangle$ for any λ. □

Now we want to compute the transfer matrices between the bases (h_λ) (or (e_λ)) and Ω.

LEMMA 7.4. *Let* λ *be a partition, then the intersection of the supports* $Supp(e_{\lambda^\perp})$ *and* $Supp(h_\lambda)$ *is of cardinality one. We will denote this element* ω_λ.

PROOF. By Lemma 6.8, $\langle e_{\lambda^\perp}, h_\lambda \rangle = 1$ and, by the positivity of those elements, this implies the statement. □

Our first goal is to express h_λ's in terms of ω_μ's. First, we compute $h_i^*(\omega_\lambda)$, and for this, we need to introduce some notations.

Let λ be a partition, or equivalently a Young diagram. We denote by $r(\lambda)$ (resp. $c(\lambda)$) the number of rows (resp. columns) of λ.

We denote by $\mathbf{R}_i^\lambda$ the set of all μ's such that μ is obtained from λ by removing exactly i boxes, at most one in every row of λ. Similarly, $\mathbf{C}_i^\lambda$ is the set of all μ's such that μ is obtained from λ by removing exactly i boxes, at most one in every column of λ. In the specific case where $i = r(\lambda)$, there is only one element in the set $\mathbf{R}_i^\lambda$ and this element will be denoted by $\lambda^\leftarrow$, it is the diagram obtained by removing the first column of λ, similarly, if $i = c(\lambda)$ the unique element of $\mathbf{C}_i^\lambda$ will be denoted by $\lambda^\downarrow$ and is the diagram obtained by suppressing the first row of λ.

REMARK 7.5. Note that if $\mu \in \mathbf{C}_i^\lambda$, then $\mu^\perp \in \mathbf{R}_i(\lambda^\perp)$.

THEOREM 7.6. (Pieri's rule) One has:

$$h_i^*(\omega_\lambda) = \sum_{\mu \in \mathbf{C}_i^\lambda} \omega_\mu,$$

and

$$e_j^*(\omega_\lambda) = \sum_{\mu \in \mathbf{R}_j^\lambda} \omega_\mu.$$

We need several lemmas to show this statement.

LEMMA 7.7. For all i, j in $\mathbb{N}$,

$$e_i^* \circ h_j = h_j \circ e_i^* + h_{j-1} \circ e_{i-1}^*$$

PROOF. From Exercise 6.5 statement (1), we obtain that

$$e_i^*(h_j x) = e_1^*(h_j)e_{i-1}^*(x) + h_j e_i^*(x) \ \forall x \in A,$$

hence the Lemma. □

LEMMA 7.8. Let p, q be two integers, let $a \in A$. Let us assume that $h_i^*(a) = 0$ for $i > p$ and $e_j^*(a) = 0$ for $j > q$, then

$$h_p^* \circ e_q^*(a) = 0$$

and

$$h_{p-1}^* \circ e_q^*(a) = h_p^* \circ e_{q-1}^*(a).$$

PROOF. Using a transposed version of Proposition 6.6, we get:

$$\sum_{i+j=n} (-1)^j h_i^* \circ e_j^* = 0.$$

The lemma follows. □

LEMMA 7.9. One has:

$$e_{r(\lambda)}^*(\omega_\lambda) = \omega_{\lambda^\leftarrow}.$$

PROOF. Applying Equation (6.4), we get

$$e^*_{r(\lambda)}(h_\lambda) = h_{\lambda\leftarrow}.$$

Since $e^*_{r(\lambda)}$ is a positive operator and $\omega_\lambda < h_\lambda$ by definition, we have $Supp(e^*_{r(\lambda)}(\omega_\lambda)) \subset Supp(h_{\lambda\leftarrow})$.

By definition of ω_λ, we know that $\langle e_{\lambda\perp}, \omega_\lambda \rangle = 1$, so we have $\langle e_{(\lambda\leftarrow)\perp}, e^*_{r(\lambda)}(\omega_\lambda)\rangle = 1$. Therefore, $\omega_{\lambda\leftarrow} \in Supp(e^*_{r(\lambda)}(\omega_\lambda))$.

It is sufficient to show now that $\langle e^*_{r(\lambda)}(\omega_\lambda), h_{\lambda\leftarrow}\rangle = 1$: let us compute:

$$\langle e^*_{r(\lambda)}(\omega_\lambda), h_{\lambda\leftarrow}\rangle = h^*_{\lambda\leftarrow} e^*_{r(\lambda)}(\omega_\lambda),$$

assume $\lambda = (\lambda_1, \ldots, \lambda_r)$,

$$\langle e^*_r(\omega_\lambda), h_{\lambda\leftarrow}\rangle = h^*_{\lambda_r-1} \circ \ldots \circ h^*_{\lambda_1-1} \circ e^*_r(\omega_\lambda)$$

$$= h^*_{\lambda_r-1} \circ \ldots \circ h^*_{\lambda_2-1} \circ e^*_{r-1} \circ h^*_{\lambda_1}(\omega_\lambda)$$

by Lemma 7.8 (the hypothesis is satisfied because if $i > \lambda_1$ then $h^*_i(e_{\lambda\perp}) = 0$ and for all $j > r$, $e^*_j(h_\lambda) = 0$). We use the same trick repeatedly, the enthusiastic reader is encouraged to check that the hypothesis of Lemma 7.8 is satisfied at each step by induction. We finally obtain

$$h^*_{\lambda\leftarrow} \circ e^*_{r(\lambda)}(\omega_\lambda) = h^*_\lambda(\omega_\lambda) = 1.$$

$\square$

For every i in $\mathbb{N}$ and for every partition λ, we set:

$$h^*_i(\omega_\lambda) = \sum_\mu a^i_{\lambda,\mu} \omega_\mu$$

which can also be written as

(6.8)
$$h_i \omega_\mu = \sum_\lambda a^i_{\lambda,\mu} \omega_\lambda$$

the Theorem 7.6 amounts to computing the coefficients $a^i_{\lambda,\mu}$.

LEMMA 7.10. *If $i > 0$, for every partitions λ and μ, then*

$$a^i_{\lambda,\mu} = \begin{cases} a^i_{\lambda\leftarrow,\mu\leftarrow} & if & r(\lambda) = r(\mu) \\ a^{i-1}_{\lambda\leftarrow,\mu\leftarrow} & if & r(\lambda) = r(\mu) + 1 \\ 0 & & otherwise \end{cases}$$

REMARK 7.11. The first equality of Theorem 7.6 is obtained from Lemma 7.10 by induction on $c(\lambda)$, the second one follows via the automorphism ι.

PROOF. First, let us prove that if $a^i_{\lambda,\mu} \neq 0$, then $r(\lambda) = r(\mu)$ or $r(\lambda) = r(\mu)+1$.

Assume $r(\mu) > r(\lambda)$ and $a^i_{\lambda,\mu} \neq 0$: then we have $\omega_\mu \leq h^*_i(\omega_\lambda)$, therefore, applying Lemma 7.9, we get

$$\omega_{\mu\leftarrow} = e^*_{r(\mu)}(\omega_\mu) \leq e^*_{r(\mu)} \circ h^*_i(\omega_\lambda) = h^*_i \circ e^*_{r(\mu)}(\omega_\lambda) = 0,$$

which gives a contradiction.

Assume $r(\mu) < r(\lambda) - 1$ and $a^i_{\lambda,\mu} \neq 0$: then applying the equation (6.8), we have $\omega_\lambda \leq h_i\omega_\mu$, therefore applying Lemma 7.9 and Lemma 7.7, we get

$$\omega_{\lambda\leftarrow} = e^*_{r(\lambda)}(\omega_\lambda) \leq e^*_{r(\lambda)} \circ h_i(\omega_\mu) = h_i \circ e^*_{r(\lambda)}(\omega_\mu) + h_{i-1} \circ e^*_{r(\lambda)-1}(\omega_\mu) = 0,$$

which again gives a contradiction.

Next, we look at the case $r(\lambda) = r(\mu)$. We do a direct computation:

$$h^*_i(\omega_{\lambda\leftarrow}) = e^*_{r(\lambda)} \circ h^*_i(\omega_\lambda) = \sum_\mu a^i_{\lambda,\mu} e^*_{r(\lambda)}(\omega_\mu) = \sum_{r(\lambda)=r(\mu)} a^i_{\lambda,\mu} \omega_{\mu\leftarrow}.$$

Finally, we assume $r(\lambda) = r(\mu) + 1$. We apply Lemma 7.7,

$$e^*_{r(\mu)+1}(h_i\omega_\mu) = h_{i-1}e^*_{r(\mu)}(\omega_\mu) = h_{i-1}(\omega_{\mu\leftarrow}) = \sum_\nu a^{i-1}_{\nu,\mu\leftarrow} \omega_\nu.$$

On the other hand,

$$e^*_{r(\mu)+1}(h_i\omega_\mu) = \sum_\lambda a^i_{\lambda,\mu} e^*_{r(\mu)+1}(\omega_\lambda) = \sum_{r(\lambda)=r(\mu)+1} a^i_{\lambda,\mu} \omega_{\lambda\leftarrow}.$$

Now we compare the coefficients and obtain that

$$a^i_{\lambda,\mu} = a^{i-1}_{\lambda\leftarrow,\mu\leftarrow}.$$

$\square$

For a partition λ, we introduce the notion of *semistandard tableau of shape λ*: this is a Young diagram of shape λ filled with entries which are no longer distinct, with the condition that the entries are non decreasing along the rows and increasing along the columns of λ. For instance,

1	1	1
2	3	
3		
4		

is a semistandard tableau.

To such a semistandard tableau, we associate its *weight*, which is the sequence m_i consisting of the numbers of occurences of the integer i in the tableau: in our example, $m_1 = 3$, $m_2 = 1$, $m_3 = 2$, $m_4 = 1$ and all the other m_i's are zero.

PROPOSITION 7.12. *Let λ be a partition of n. Let $m_1, \ldots, m_r$ be a sequence of non negative integers such that $m_1 + \ldots + m_r = n$, then $\langle h_{m_1} \ldots h_{m_r}, \omega_\lambda \rangle$ is the number of semistandard tableaux of shape λ and weight $m_1, \ldots, m_r$.*

PROOF. We iterate Pieri's rule (see Theorem 7.6):

$$h^*_{m_r}(\omega_\lambda) = \sum_{\mu \in \mathbf{C}^\lambda_{m_r}} \omega_\mu,$$

$$(h_{m_{r-1}} h_{m_r})^*(\omega_\lambda) = \sum_{\mu_1 \in \mathbf{C}^\lambda_{m_r}} \sum_{\mu_2 \in \mathbf{C}^{\mu_1}_{m_{r-1}}} \omega_{\mu_2},$$

and eventually

$$\langle h_{m_1} \ldots h_{m_r}, \omega_\lambda \rangle = (h_{m_1} \ldots h_{m_{r-1}} h_{m_r})^*(\omega_\lambda) = \sum_{\mu_1 \in \mathbf{C}^\lambda_{m_1}} \sum_{\mu_2 \in \mathbf{C}^{\mu_1}_{m_2}} \cdots \sum_{\mu_r \in \mathbf{C}^{\mu_{r-1}}_{m_r}} 1,$$

because $\omega_{\mu_r} = 1$ due to the fact that $m_1 + \ldots + m_r = n$.

The sequences $\mu_1, \ldots, \mu_r$ indexing the sum in the right-hand side are in bijective correspondence with the semistandard tableaux of shape λ and weight $m_1, \ldots, m_r$, indeed given such a semistandard tableau, we set μ_i to be the union of the boxes filled with numbers $\leq i$: μ_i is a semistandard tableau. Hence the result. $\square$

REMARK 7.13. Since the product is commutative, $\langle h_{m_1} \ldots h_{m_r}, \omega_\lambda \rangle$ depends only on the non-increasing rearrangement μ of the sequence $m_1, \ldots, m_r$. Note that the partition μ we just obtained satisfies $\lambda \succeq \mu$.

DEFINITION 7.14. Let λ, μ be two partitions of n, we define the *Kostka number* $K_{\lambda\mu}$ to be the number of semistandard tableaux of shape λ and weight μ.

THEOREM 7.15. *(Jacobi-Trudi) For any partition λ of n,*

$$\omega_\lambda = \det\left((h_{\lambda_i - i + j})_{1 \leq i, j \leq r(\lambda)} \right).$$

PROOF. The theorem is proven by induction on n. For $n = 1$ the statement is clear and we assume that the equality holds for all partitions μ of m with $m < n$.

We will use the automorphism H of the PSH algebra A defined by

$$H(h_i) = \sum_{j \leq i} h_j,$$

(the automorphism H is the formal sum $\sum_{k \in \mathbb{N}} h^*_k$).

First, we notice that the linear map $H - Id : I \to A$ is injective: indeed this amounts to saying that its adjoint restricts to the surjective linear map

$$\sum_{0 \leq j \leq n} A_j \longrightarrow A_n, \quad (a_0, \ldots, a_{n-1}) \mapsto a_0 h_n + \ldots + a_{n-1} h_1,$$

and this assertion is clear since A is the polynomial algebra $\mathbb{Z}[(h_i)_{i \in \mathbb{N}}]$.

Let us explain the induction step: we denote by ϖ_λ the determinant of the theorem. We know (Pieri's rule, Theorem 7.6) that $H(\omega_\lambda) = \sum_{i\geq 0, \mu \in C_i^\lambda} \omega_\mu$, and we will show that $H(\varpi_\lambda)$, when λ varies, satisfies the same equality (with obvious changes of notations). This will conclude the proof since $H - Id : I \to A$ is injective.

Since H is an algebra homomorphism, we have

$$H(\varpi_\lambda) = H\left(\det\left((h_{\lambda_i-i+j})_{1\leq i,j\leq r(\lambda)}\right)\right) = \det\left((H(h_{\lambda_i-i+j}))_{1\leq i,j\leq r(\lambda)}\right).$$

We know that $H(h_{\lambda_i-i+j}) = \sum_{k\leq \lambda_i} h_{k-i+j}$ by definition of H. Hence

$$H(\varpi_\lambda) = \det\left(\left(\sum_{k_i\leq \lambda_i} h_{k_i-i+j}\right)_{1\leq i,j\leq r}\right).$$

We notice that, in this determinant, every entry is a partial sum of the entry which is just above it, we are led to subtract the $i + 1$-th row from the i-th row for all i. This doesn't affect the value of the determinant, therefore we obtain the equality

$$H(\varpi_\lambda) = \det\left(\left(\sum_{\lambda_2\leq k_1\leq \lambda_1,\ldots,\lambda_r\leq k_{r-1}\leq \lambda_{r-1},k_r\leq \lambda_r} h_{k_i-i+j}\right)_{1\leq i,j\leq r}\right).$$

Since the determinant is a multilinear function of its rows, we deduce

$$H(\varpi_\lambda) = \sum_{\lambda_2\leq k_1\leq \lambda_1,\ldots,\lambda_r\leq k_{r-1}\leq \lambda_{r-1},k_r\leq \lambda_r} \det\left((h_{k_i-i+j})_{1\leq i,j\leq r}\right).$$

Now each family of indices $k_1,\ldots,k_r$ gives rise to a partition μ belonging to C_m^λ for $m = n - k_1 - \ldots - k_r$. For the terms which do not give rise to a partition the determinant vanishes. This implies the result. $\square$

8. Harvest

In the previous four sections, we defined and classified PSH algebras and we obtained precise results in the rank one case. Now it is time to see why this was useful. In this section, we will meet two avatars of the rank one PSH algebra, namely the algebra $\mathcal{A}$ of section 4, and the Grothendieck group of polynomial representations of the group GL_∞: this interpretation will give us precious information concerning the representation theory in both cases. The final section of this chapter will be devoted to another very important application of PSH algebras, in the infinite rank case, associated to linear groups over finite fields. We will only state the main results without proof and refer the reader to Zelevinsky's seminal book.

8.1. Representations of symmetric groups revisited. We use the notations of Section 4. We know by Proposition 4.7 that $\mathcal{A}$ is a PSH algebra.

PROPOSITION 8.1. *The PSH algebra $\mathcal{A}$ is of rank one with basic primitive element π, the class (in the Grothendieck group) of the trivial representation of the trivial group S_1.*

PROOF. Our goal is to show that every irreducible representation of S_n ($n \in \mathbb{N}\backslash\{0\}$) appears in π^n. It is clear that π^n is the regular representation of S_n, hence the result. $\square$

COROLLARY 8.2. *For any PSH algebra A of rank 1, together with a choice of indexation of the basis Ω by the partitions (there are two such choices conjugate by ι) the multiplication table of the ω_λs is given by the Littlewood-Richardson coefficients (see Definition 2.14):*

$$\omega_\mu \omega_\nu = \sum_\lambda N_{\mu,\nu}^\lambda \omega_\lambda.$$

By adjunction,

$$m^*(\omega_\lambda) = \sum_{\mu,\nu} N_{\mu,\nu}^\lambda \omega_\mu \otimes \omega_\nu.$$

The choice of π gives us two isomorphisms between $\mathcal{A}$ and A (one is obtained from the other by application of the automorphism ι). We choose the isomorphism which send h_2 to the trivial representation of S_2 (hence it sends e_2 to the sign representation of S_2).

Let us give an interpretation of the different bases (e_λ), (h_λ), (ω_λ), (p_λ) in this setting.

EXERCISE 8.3.   (1) Check that e_i corresponds to the sign representation of S_i and that h_i corresponds to the trivial representation of S_i.

(2) Show that ω_λ corresponds to the class of the irreducible representation V_λ defined in Section 1.

REMARK 8.4. In the case of symmetric groups, the Grothendieck group is also the direct sum of $\mathbb{Z}$-valued class functions on S_n when n varies. See Chapter 1 together with the fact that the characters of the symmetric groups take their values in $\mathbb{Z}$.

EXERCISE 8.5.   (1) Show that the primitive element $\frac{p_i}{i}$ is the characteristic function of the circular permutation of S_i.

(2) Interpreting the induction functors involved, show that, for every partition λ, p_λ is a multiple of the characteristic function θ_λ of the conjugacy class c_λ corresponding to λ. More precisely $p_\lambda = \frac{|\lambda|!}{|c_\lambda|}\theta_\lambda$. *Hint*: check that $\langle p_\lambda, p_\lambda \rangle = \frac{|\lambda|!}{|c_\lambda|}$.

The following Proposition is now clear:

PROPOSITION 8.6. *The character table of S_n is just the transfer matrix express-ing the ω_μ's in terms of $\frac{|c_\lambda|}{|\lambda|!}p_\lambda$'s when μ and λ vary along the partitions of n.*

Our goal now is to prove the Hook formula:
Let λ be a partition, let $a = (i,j)$ be any box in the Young diagram λ, we denote $h(a)$ the number of boxes (i',j') of the Young diagram such that $i' = i$ and $j' \geq j$ or $i' \geq i$ and $j' = j$: $h(a)$ is called the *hook length* of a.

THEOREM 8.7. *(Hook formula) For every partition λ of n, the dimension of the S_n-module V_λ is equal to*

$$\dim V_\lambda = \frac{n!}{\prod_{a \in \lambda} h(a)}$$

PROOF. For any S_n-module V, let us denote by $rdimV$ the reduced dimension of V, that is the quotient $\frac{\dim V}{n!}$: this defines a ring homomorphism from $\mathcal{A}$ to $\mathbb{Q}$, as one can easily see computing the dimension of an induced module.

We write $\lambda = (\lambda_1, \ldots, \lambda_r)$. Set $L_i = \lambda_i + r - i$ and consider the new partition consisting of $(L_1, \ldots, L_r) := L$. We apply Theorem 7.15 and notice that $rdim(h_p) = \frac{1}{p!}$: therefore

$$rdim(\omega_\lambda) = \det\left(\left(\frac{1}{(L_i - r + j)!}\right)_{1 \leq i,j \leq r}\right).$$

Since $L_i! = (L_i - r + j)! P_{r-j}(L_i)$ where $P_k(X)$ is the polynomial $X(X-1)\ldots(X - k+1)$, the right-hand side becomes

$$\frac{1}{L_1! \ldots L_r!} \det\left((P_{r-j}(L_i))_{1 \leq i,j \leq r}\right).$$

Now P_k is a polynomial of degree k with leading coefficient 1, hence this determinant is a Vandermonde determinant and is equal to $\prod_{1 \leq i < j \leq r}(L_i - L_j)$ and we get

$$\dim V_\lambda = \frac{n!}{L_1! \ldots L_r!} \prod_{1 \leq i < j \leq r}(L_i - L_j).$$

Noting that $\frac{L_i!}{\prod_{i<j}(L_i - L_j)}$ is the product of the hook lengths of boxes of the i-th row of λ, we obtain the wanted Hook formula. $\square$

THEOREM 8.8. *For every partition λ of n, the restriction of V_λ to S_{n-1} is the direct sum $\oplus_\mu V_\mu$ where the Young diagram of μ is obtained from the Young diagram of λ by deleting exactly one box.*

PROOF. This restriction is $h_1^*(\omega_\lambda)$. Hence the result. $\square$

EXERCISE 8.9. Compute the dimension of the S_6-module V_λ for $\lambda = (3,2,1)$. Calculate the restriction of V_λ to S_5.

8.2. Symmetric polynomials in infinitely many variables over $\mathbb{Z}$. Let R be a unital commutative ring, let us define the ring S_R of symmetric polynomials in a fixed infinite sequence $(X_i)_{i \in \mathbb{N}_{>0}}$ of variables with coefficients in R. Recall that the symmetric group S_n acts on the polynomial ring $R[X_1, \ldots, X_n]$ by $\sigma(X_i) := X_{\sigma(i)}$ and the ring of invariants consists of the symmetric polynomials in n variables. There is a surjective algebra homomorphism which preserves the degree

$$\psi_n : R[X_1, \ldots, X_{n+1}]^{S_{n+1}} \to R[X_1, \ldots, X_n]^{S_n}$$
$$P(X_1, \ldots, X_{n+1}) \mapsto P(X_1, \ldots, X_n, 0).$$

By definition, S_R is the projective limit of the maps $(\psi_n)_{n \in \mathbb{N}_{>0}}$ in the category of graded rings.

In order to be more explicit, we need to introduce the ring of formal power series $R[[X_1, \ldots, X_n, \ldots]]$ consisting of (possibly infinite) formal linear combinations, with coefficients in R, $\sum_\alpha a_\alpha X^\alpha$, where α runs along multi-indices $(\alpha_i)_{i \geq 1}$ of integers with finite support and the degrees of X^α with $a_\alpha \neq 0$ are bounded. There is no difficulty in defining the product since, for any multi-index α, there are only finitely many ways of expressing α as a sum $\alpha_1 + \alpha_2$. We set S_∞ to be the groups of permutations of all positive integers generated by the transpositions. Then S_R is the subring of $R[[X_1, \ldots, X_n, \ldots]]$ whose elements are invariant under S_∞ and such that the degrees of the monomials are bounded.

Let A be the PSH algebra of rank one.

THEOREM 8.10. *The map $\psi : A \to S_{\mathbb{Z}}$, given by, for all $a \in A$,*

(6.9)
$$\psi(a) = \sum_\alpha \langle a, \prod_i h_{\alpha_i} \rangle X^\alpha,$$

is an algebra isomorphism.

REMARK 8.11. We immediately deduce from the formula for ψ the following statements:

(1) $\psi(h_n) = \sum_{|\alpha|=n} X^\alpha$, where $|\alpha| := \sum_i \alpha_i$ if $\alpha = (\alpha_1, \ldots, \alpha_i \ldots)$,
(2) $\psi(e_n) = \sum_{\alpha=(\alpha_1, \ldots)} X^\alpha$, where every α_i is either 0 or 1 and $|\alpha| = n$,
(3) $\psi(p_n) = \sum_{i \geq 1} X_i^n$.

Finally, if we denote by $h_\lambda^\diamond$ the dual basis of h_λ with respect to the scalar product on A, then

(4) $\psi(h_\lambda^\diamond) = \sum_\alpha X^\alpha$ where α runs along the multi-indices whose non-increasing rearrangement is λ.

PROOF. We follow the proof given in Zelevinsky's book, attributed to Bernstein. Let us first define the homomorphism ψ: we iterate the comultiplication $A \to A \otimes A$ and obtain an algebra homomorphism $\mu_n : A \to A^{\otimes n}$ for any n ($\mu_2 = m^*$).

Furthermore, the counit ε induces a map $\varepsilon_n : A^{\otimes n+1} \to A^{\otimes n}$ such that the following diagram is commutative:

(6.10)

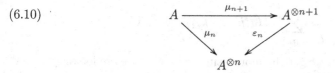

If B is a $\mathbb{N}$-graded commutative ring and t is an indeterminate, we can define a canonical homomorphism $\beta_B : B \to B[t]$ by setting, for any $b \in B$ of degree k, $\beta_B(b) := bt^k$: thus we obtain a homomorphism $\beta_A^{\otimes n} : A^{\otimes n} \to A[X_1, \ldots, X_n]$. Note that, in order to obtain a homogeneous homomorphism, we have to forget the grading of A for the definition of the degree in $A[X_1, \ldots, X_n]$: in this algebra, the elements of A have degree 0. Note also that the image of μ_n is always contained in $(A^{\otimes n})^{S_n}$.

EXERCISE 8.12. Show that the following diagram is commutative (the symmetric group acts on both the set of variables and the factors of $A^{\otimes k}$):

(6.11)
$$
\begin{array}{ccc}
(A^{\otimes n+1})^{S_{n+1}} & \xrightarrow{\ \beta_A^{\otimes n+1}\ } & (A^{\otimes n+1}[X_1, \ldots, X_{n+1}])^{S_{n+1}} \\
\Big\downarrow{\varepsilon_n} & & \Big\downarrow \\
(A^{\otimes n})^{S_n} & \xrightarrow[\ \beta_A^{\otimes n}\ ]{} & (A^{\otimes n}[X_1, \ldots, X_n])^{S_n}
\end{array}
$$

Let $a \in A$, then, by definition of μ_n,

$$
\mu_n(a) = \sum_{\lambda_1, \ldots, \lambda_n} \langle a, \omega_{\lambda_1} \ldots \omega_{\lambda_n} \rangle \omega_{\lambda_1} \otimes \ldots \otimes \omega_{\lambda_n}
$$

where $\lambda_1, \ldots, \lambda_n$ are partitions. Thus,

(6.12) $$\beta_A^{\otimes n}(\mu_n(a)) = \sum_{\lambda_1, \ldots, \lambda_n} \langle a, \omega_{\lambda_1} \ldots \omega_{\lambda_n} \rangle \omega_{\lambda_1} \otimes \ldots \otimes \omega_{\lambda_n} X_1^{|\lambda_1|} \ldots X_n^{|\lambda_n|}.$$

LEMMA 8.13. *There are exactly two positive algebra homomorphisms from A to $\mathbb{Z}$, conjugate up to ι (see Theorem 6.11) which transform the basic primitive element π into 1. One of them, denoted by δ, is such that $\delta(h_i) = 1$ for all i and $\delta(\omega_\lambda) = 0$ whenever ω_λ is not one of the h_is.*

PROOF. Such a homomorphism maps π^2 onto 1, but $\pi^2 = e_2 + h_2$ and since it is positive, either e_2 or h_2 is sent to one 1 (and the other to 0). Since ι exchanges e_2 and h_2, we can assume that h_2 is sent to 1 (and e_2 to 0). We denote this homomorphism by δ. Let ω be a basic element of degree n in A, distinct from h_n. By Lemma 6.3, $e_2^*(\omega) \neq 0$, hence $\omega \leq e_2 \pi^{n-2}$ and since $\delta(e_2) = 0$ and δ is positive, $\delta(\omega) = 0$. Then, since $\delta(\pi^n) = 1$, we obtain $\delta(h_n) = 1$, hence the Lemma. $\qquad\square$

Set $\psi_n = \delta^{\otimes n} \circ \beta_A^{\otimes n} \circ \mu_n : A \to \mathbb{Z}[X_1, \ldots, X_n]$. Applying Lemma 8.13 and (6.12), we obtain

$$(6.13) \qquad \psi_n(a) = \sum_{(i_1, \ldots, i_n) \in \mathbb{N}^n} \langle a, h_{i_1} \ldots h_{i_n} \rangle X_1^{i_1} \ldots X_n^{i_n}.$$

Taking the projective limit, we get the morphism $\psi : A \to S_{\mathbb{Z}}$ we are looking for and the item (4) of Remark 8.11 ensures that ψ is an isomorphism. $\square$

We now compute $\psi(\omega_\lambda)$ for any partition λ, and more precisely $\psi_n(\omega_\lambda)$ for any $n \geq |\lambda|$.

We first introduce the following notation for the generalized Vandermonde determinant: let $\mu = (\mu_1, \ldots, \mu_n)$ be a decreasing sequence of non-negative integers, we set $V_\mu(X_1, \ldots, X_n) = \det((X_i^{\mu_j})_{1 \leq i,j \leq n})$. Notice that $V_{(n-1, n-2, \ldots, 1, 0)}(X_1, \ldots, X_n)$ is the usual Vandermonde determinant.

PROPOSITION 8.14. *One has*

$$(6.14) \qquad \psi_n(\omega_\lambda) = \frac{V_{(\lambda_1+n-1, \lambda_2+n-2, \ldots \lambda_n)}(X_1, \ldots, X_n)}{V_{(n-1, n-2, \ldots, 1, 0)}(X_1, \ldots, X_n)}.$$

EXERCISE 8.15. Prove Proposition 8.14. Hints: use the Jacobi-Trudi formula (see Theorem 7.15), the duality in the exterior algebra of $\mathbb{C}[X_1, \ldots, X_n]^{\lambda_1+n-1}$ (seen as a free $\mathbb{C}[X_1, \ldots, X_n]$-module), the linear relations

$$\psi_n(e_0)X_i^{n+j} - \psi_n(e_1)X_i^{n+j-1} + \ldots + (-1)^n \psi_n(e_n)X_i^j = 0$$

(Vieta's formula) and

$$e_0 h_{n+j} - e_1 h_{n+j-1} + \ldots + (-1)^n e_n h_j = 0$$

(see the end of the proof of Proposition 6.6).

8.3. Complex general linear group for an infinite countable dimensional vector space.
Let V be an infinite countable dimensional complex vector space, we consider the group $G = GL(V)$. Denote by $\mathcal{T}$ the full subcategory of the category of G-modules whose objects are submodules of direct sums of tensor powers of V. We saw in section 2 that $\mathcal{T}$ is a semisimple category. The simple modules are indexed by partitions and we denote by $S_\lambda(V)$ the simple module associated to the partition λ. We denote by $K(\mathcal{T})$ the Grothendieck group of $\mathcal{T}$.

Our aim is to equip $K(\mathcal{T})$ with a structure of PSH algebra of rank one.

We define the *multiplication*: $m([M], [N]) = [M \otimes N]$ for M and N in $\mathcal{T}$ (recall that if $M \in \mathcal{T}$, we denote by $[M]$ its class in the Grothendieck group).

We define the *scalar product*: $\langle [M], [N] \rangle = \dim \mathrm{Hom}_G(M, N)$, and the *grading*: by convention, the degree of $V^{\otimes n}$ is n.

Finally we proceed to define the *comultiplication* m^*, and it is a trifle more tricky. Since V is infinite dimensional, we can choose an isomorphism $\varphi : V \to V \oplus V$. By composition with φ, we obtain a group morphism $\Phi : G \times G$ to G,

$$\Phi \begin{pmatrix} A & 0 \\ 0 & B \end{pmatrix} = \varphi^{-1} \circ \begin{pmatrix} A & 0 \\ 0 & B \end{pmatrix} \circ \varphi.$$

We have two canonical projectors of $V \oplus V$ and we denote by V_1 (resp. V_2) the image of the first (resp. second) one.

EXERCISE 8.16. Show that $V_1^{\otimes p} \otimes V_2^{\otimes q}$ is a semisimple $G \times G$-module and that its irreducible components are of the form $S_\lambda(V_1) \otimes S_\mu(V_2)$, where λ is a partition of p and μ is a partition of q.

Therefore, if we denote $\tilde{\mathcal{T}}$ the full sucategory of the category of $G \times G$-modules whose objects are submodules of direct sums of $V_1^{\otimes p} \otimes V_2^{\otimes q}$, then its Grothendieck group is isomorphic to $K(\mathcal{T}) \otimes K(\mathcal{T})$.

Hence, the restriction functor Res (with respect to the inclusion of $G \times G$ in G) maps the category $\mathcal{T}$ to the category $\tilde{\mathcal{T}}$. Therefore it induces a linear map $m^* : K(\mathcal{T}) \to K(\mathcal{T}) \otimes K(\mathcal{T})$.

THEOREM 8.17. *The Grothendieck group $K(\mathcal{T})$, equipped the operations described above and the basis given by the classes of simple modules, is a PSH algebra of rank one, and the basic primitive element is the class of V, $[V]$.*

PROOF. The only axiom of the definition of PSH algebras which is not straightforward and needs to be checked is the self-adjointness, namely the fact that m and m^* are mutually adjoint with respect to the scalar product. For this, we have to find a functorial bijective map $\mathrm{Hom}_G(M \otimes N, P) \to \mathrm{Hom}_{G \times G}(M \otimes N, \mathrm{Res}(P))$ (where M, N, P are objects of $\mathcal{T}$). Since any G-module is the direct sum of its homogeneous components, we may asume that M, N, P are homogeneous of degree p, q, n respectively, with $n = p + q$.

For any object $W \in \mathcal{T}$ homogeneous of degree r, set $\Pi_W := \mathrm{Hom}_G(V^{\otimes r}, W)$ which is an S_r-module; Schur-Weyl duality (see Proposition 2.13) can be reformulated in saying that there is a canonical isomorphism of G-modules $W \simeq \Pi_W \otimes_{\mathbb{C}(S_r)} V^{\otimes r}$. We set $M_1 := \Pi_M \otimes_{\mathbb{C}(S_p)} V_1^{\otimes p} \hookrightarrow M$, $N_2 := \Pi_N \otimes_{\mathbb{C}(S_q)} V_2^{\otimes q} \hookrightarrow N$.

Then we have an inclusion $M_1 \otimes N_2 \subset M \otimes N$, and the restriction defines a map $\mathrm{Hom}_G(M \otimes N, P) \to \mathrm{Hom}_{G \times G}(M_1 \otimes N_2, \mathrm{Res}(P))$. This is the functorial map we were looking for.

In order to show this map is bijective, it is enough (by the semisimplicity of the categories $\mathcal{T}$ and $\tilde{\mathcal{T}}$) to check it for $M = V^{\otimes p}$, $N = V^{\otimes q}$ and $P = V^{\otimes n}$ with $p + q = n$: indeed, on one hand,

$$\dim \mathrm{Hom}_G(V^{\otimes p} \otimes V^{\otimes q}, V^{\otimes n}) = \dim \mathrm{Hom}_{G \times G}(V_1^{\otimes p} \otimes V_2^{\otimes q}, V^{\otimes n}) = n!$$

the first equality coming from the Schur-Weyl duality and the second equality comes from the formula

$$V^{\otimes n} \simeq \oplus_{r=0}^n \left(V_1^{\otimes r} \otimes V_2^{\otimes (n-r)} \right)^{\oplus \frac{n!}{r!(n-r)!}}.$$

On the other hand, the map is injective because $V_1^{\otimes p} \otimes V_2^{\otimes q}$ spans the G-module $V^{\otimes n}$.

$\square$

9. General linear groups over a finite field

This section is a summary without proofs of Chapter III, sections 9–11 of Zelevinsky's book [38].

Let p be a prime and q a power of p. We denote by $\mathbb{F}_q$ the finite field with q elements and by G_n the linear group $GL_n(\mathbb{F}_q)$. We consider, for every integer n, the category $\mathcal{C}_n$ of G_n-modules over the complex numbers and we denote by $\mathcal{K}_n$ the Grothendieck group of $\mathcal{C}_n$. We want to consider the collection of all G_n at the same time, so we set $A_q := \oplus_{n \geq 0} \mathcal{K}_n$, where by convention $\mathcal{K}_0 = \mathbb{Z}$.

As a $\mathbb{Z}$-module, A_q has a basis Ω consisting of the classes of all the irreducible modules over G_n, $n \geq 1$, and $1 \in \mathcal{K}_0$. There is a non-degenerate bilinear form on A_q which is defined by, for $[V], [W] \in \mathcal{K}_n$, $\langle V, W \rangle = \dim \mathrm{Hom}_{G_n}(V, W)$ and $\langle \mathcal{K}_m, \mathcal{K}_n \rangle = 0$ if $m \neq n$.

We equip A_q with a multiplication $\mu_{m,n} : \mathcal{K}_m \otimes \mathcal{K}_n \to \mathcal{K}_{m+n}$, which comes from the *parabolic induction*: let $P_{m,n}$ denote the subgroup of G_{m+n}:

$$P_{m,n} = \left\{ \begin{pmatrix} A & B \\ 0 & C \end{pmatrix}, A \in G_m, C \in G_n \right\}.$$

If $V \in \mathcal{C}_m$ and $W \in \mathcal{C}_n$, we extend the $G_m \times G_n$-module $V \otimes W$ into a $P_{m,n}$-module with trivial action of the normal subgroup of $P_{m,n}$:

$$U_{m,n} = \left\{ \begin{pmatrix} \mathrm{Id}_m & B \\ 0 & \mathrm{Id}_n \end{pmatrix} \right\}.$$

We then define a G_{m+n}-module by induction from $P_{m,n}$ to G_{m+n}, and this functor $I_{m,n}$ is called the parabolic induction. By definition, the class of this module is $\mu_{m,n}([V] \otimes [W])$.

Now we equip A_q with a comultiplication μ^*, defined over $\mathcal{K}_r$ by the direct sum (for $m + n = r$) of the maps $\mu_{m,n}^*$ coming from the functor (called parabolic restriction) $R_{m,n}$ which is right-adjoint to $I_{m,n}$. The functor $R_{m,n}$ transforms the G_r-module M into the $G_m \times G_n$-module $\mathrm{Hom}_{U_{m,n}}(\mathbb{C}, M)$.

THEOREM 9.1. *(Zelevinsky) When equipped with μ, μ^*, the basis Ω and the scalar product defined above, A_q is a PSH algebra.*

REMARK 9.2. (1) The proof of this theorem essentially follows the same lines as the proof of Proposition 4.7. Nevertheless, contrary to the symmetric case, the commutativity of the multiplication is not obvious (but is automatic, see Proposition 3.9).

(2) In order to prove that the comultiplication is a ring homomorphism, Zelevinsky uses the identification (which follows from the Bruhat decomposition) between $P_{m,n} \backslash G_{m+n} / P_{m,n}$ and $(S_m \times S_n) \backslash S_{m+n} / (S_m \times S_n)$.

Contrary to the previous cases we studied, this PSH algebra is not of rank one. In order to decompose it, we need the following definition:

DEFINITION 9.3. Let V be an irreducible module over G_n. If V is not a direct summand in any G_n-module obtained by (strict) parabolic induction, then V is called a *cuspidal representation* of G_n.

REMARK 9.4. (1) This amounts to saying that $[V] \in \mathcal{K}_n$ is a primitive element of A_q.

(2) We will call the integer n the cuspidal degree of V and denote it by $d(V)$.

We denote by Γ the set of classes in A_q of all cuspidal representations for all G_n. The following proposition is a direct consequence of Zelevinsky's decomposition theorem (Theorem 5.6):

PROPOSITION 9.5. *One has:*

$$A_q = \bigotimes_{\gamma \in \Gamma} A_\gamma,$$

where A_γ *is the rank one PSH algebra with basic primitive element* γ, *which, as a* $\mathbb{Z}$-*module, is spanned by the supports of the powers of* γ.

The table of characters of the group $GL_n(\mathbb{F}_q)$ is understood since the work of J. A. Green (J.A. Green, The characters of the finite general linear groups, Trans. AMS, 80 (1955) pp. 402–447). The problem is to understand the transition matrix, in the $\mathbb{C}$-vector space $A_q \otimes_{\mathbb{Z}} \mathbb{C}$, from the set Ω of characters of irreducible representations to the set Υ of characteristic functions of conjugacy classes in the collection of groups G_n. The multiplication of $A_q \otimes_{\mathbb{Z}} \mathbb{C}$ in the basis Υ is exactly the multiplication of the classical *Hall algebra* $\mathcal{H}(\mathbb{F}_q[t, t^{-1}])$:

An element $v \in \Upsilon$ of degree n is the isomorphism class of a $\mathbb{F}_q[t, t^{-1}]$-module of finite length: n is the dimension of the underlying $\mathbb{F}_q$-vector space V and t is seen as an automorphism of V. The multiplication rule is given by $v_1 v_2 = \sum_{v \in \Upsilon} c_{v_1, v_2}^v v$, where, for a given $\mathbb{F}_q[t, t^{-1}]$-module M of isomorphism class v, the constant c_{v_1, v_2}^v is the number of submodules $N \subset M$ of type v_1 such that M/N is of type v_2.

This statement is an interpretation of parabolic induction in terms of conjugacy classes.

The Hall algebra $\mathcal{H}(\mathbb{F}_q[t, t^{-1}])$ is the infinite tensor product of local Hall algebras $\mathcal{H}(\mathbb{F}_q[t, t^{-1}]_\mathfrak{p})$ where $\mathfrak{p}$ is a maximal ideal of $\mathbb{F}_q[t, t^{-1}]$ (which is automatically generated by a monic irreducible polynomial with non-zero constant term). This decomposition amounts to the equality $M = \oplus_\mathfrak{p} M_\mathfrak{p}$ for any torsion $\mathbb{F}_q[t, t^{-1}]$-module M of finite type.

REMARK 9.6. The definition of local Hall algebras remains meaningful if we replace $\mathbb{F}_q[t, t^{-1}]_\mathfrak{p}$ by any discrete valuation ring $\mathfrak{o}$, with finite residue field k. This algebra $\mathcal{H}(\mathfrak{o})$ has a marked basis, the classes of $\mathfrak{o}$-modules of finite length, which can be indexed by partitions in the obvious way. We abuse notation and denote the element associated to the partition λ in this basis by v_λ. The multiplication table, that is the matrix of the product $\mathcal{H}(\mathfrak{o}) \otimes \mathcal{H}(\mathfrak{o}) \to \mathcal{H}(\mathfrak{o})$, is given by the family $g_{\lambda,\mu}^\nu(|k|)$ (λ, μ, ν vary among the partitions) where $g_{\lambda,\mu}^\nu(x)$ is the so-called Hall polynomial associated to the partitions λ, μ, ν; therefore $\mathcal{H}(\mathfrak{o})$ depends only on the cardinality of the residue field k.

Moreover, there is a scalar product $\langle, \rangle_{|k|}$ on $\mathcal{H}(\mathfrak{o})$ for which the basis $(v_\lambda)_\lambda$ is orthogonal and, if λ is a partition of n, $\langle v_\lambda, v_\lambda \rangle_{|k|}$ is the number of $\mathfrak{o}$-module structures of shape λ on k^n. We do not give a combinatorial formula for these numbers since it is rather complicated to state and we will need it only for small values of n.

This is the beginning of a long story told by I.G. Macdonald, in [25].

Hence we are now given two tensor decompositions of $A_q \otimes_\mathbb{Z} \mathbb{C}$, the one we just obtained and the one coming from Proposition 9.5 after tensoring with $\mathbb{C}$. Understanding the character table is now theoretically "reduced" to understanding the composed maps

$$\varphi_{\gamma,\mathfrak{p}} : A_\gamma \to A_q \to A_q \otimes_\mathbb{Z} \mathbb{C} \to \mathcal{H}(\mathbb{F}_q[t, t^{-1}]_\mathfrak{p})$$

when the cuspidal representation γ and the maximal ideal $\mathfrak{p}$ vary.

Note that $A_q \otimes_\mathbb{Z} \mathbb{C}$ is the algebra of central functions on the collection of groups G_n. Hence it comes with a hermitian scalar product and a comultiplication which is the complexification of μ^*. We carry them into the Hall algebra $\mathcal{H}(\mathbb{F}_q[t, t^{-1}])$. The resulting scalar product is obviously the tensor product of the family of scalar products $\langle, \rangle_{\mathbb{F}_q[t,t^{-1}]/\mathfrak{p}}$ when the prime ideal $\mathfrak{p}$ varies. It is a formal consequence of those facts that the local Hall algebras are equipped with a comultiplication, which is adjoint to the given multiplication. Finally, we note that the projection morphism $\mathcal{H}(\mathbb{F}_q[t, t^{-1}]) \to \mathcal{H}(\mathbb{F}_q[t, t^{-1}]_\mathfrak{p})$ is the orthogonal projection, and that the maps $\varphi_{\gamma,\mathfrak{p}}$ are morphisms of Hopf algebras.

REMARK 9.7. This last fact is important, since we know that a graded cocommutative Hopf algebra over $\mathbb{C}$ is the symmetric algebra $S(P)$ of the graded vector space P consisting of its primitive elements, and a morphism of graded Hopf algebras $F : S(P_1) \to S(P_2)$ always restricts to a linear map $f : P_1 \to P_2$, therefore F is obtained from f after taking symmetric algebras.

In his book, Zelevinsky computes the maps $\varphi_{\gamma,\mathfrak{p}}$ for all γ and for $\mathfrak{p} = (t - 1)$, which will be denoted by $\mathfrak{q}$ in what follows. In other words, he computes the values of each character on every unipotent conjugacy class. Later on, Springer and Zelevinsky ([31]) were able to extend this description to every map $\varphi_{\gamma,\mathfrak{p}}$ (i.e. for every maximal ideal $\mathfrak{p}$).

Let us give an idea of Zelevinsky's argument.

We denote by $c \in \Gamma$ the trivial representation of G_1: it is a cuspidal representation. Since A_c is of rank one, it is isomorphic to A, and there a unique way of sending

$h_n \in A$ to the trivial representation of G_n. We compute the map

$$\varphi_{c,\mathfrak{q}} : A_c \to \mathcal{H} := \mathcal{H}(\mathbb{F}_q[t, t^{-1}]_{\mathfrak{q}}),$$

starting from

(6.15)
$$\varphi_{c,\mathfrak{q}}(h_n) = \sum_{v \text{ unipotent}, \deg(v)=n} v.$$

We use the multiplication table of the elements v in order to compute $\varphi_{c,\mathfrak{q}}(p_n)$ where p_n is the primitive element in A defined in Proposition 7.2. Zelevinsky gives the formula

$$\varphi_{c,\mathfrak{q}}(p_n) = \sum_{\lambda \vdash n} (1-q)(1-q^2) \cdots (1-q^{l(\lambda)-1}) v_\lambda^{\mathfrak{q}}$$

where $l(\lambda)$ is the number of parts of λ and $v_\lambda^{\mathfrak{q}}$ is the unipotent basic element of shape λ (see [38], Theorem 10.3.(b)). This formula implies that $\varphi_{c,\mathfrak{q}}$ is injective and induces an isomorphism $A_c \otimes_{\mathbb{Z}} \mathbb{C} \simeq \mathcal{H}$ since the dimension of the spaces of homogeneous elements coincide (see Remark 9.7).

REMARK 9.8. In other words, $\mathcal{H}$ can be identified (as a Hopf algebra) with the algebra of symmetric polynomials in infinitely many variables over $\mathbb{C}$, but this identification is not compatible with the scalar products. We do not wish to make this identification in the long run, but let us proceed with it for a little while. Denote by $\langle , \rangle_{\mathcal{H}}$ the scalar product induced by the one of $A_q \otimes \mathbb{C}$. In the book [25], Chap. 3, Macdonald expresses the symmetric function $\frac{v_\lambda}{\langle v_\lambda, v_\lambda \rangle_{\mathcal{H}}}$, which corresponds to the unipotent conjugacy class of shape λ divided by the order of its stabilizer, as reciprocal *Hall-Littlewood polynomials* in the variable t specialized at q, $t^{n(\lambda)} P_\lambda(x; t^{-1})$. Here $n(\lambda) = \sum_{i \geq 0} \binom{\lambda'_i}{2}$, where λ' is the conjugate partition of λ and x is the set of variables. Hall-Littlewood polynomials are given by an explicit formula, see [25] Chap. 3, formulae (2.1), (2.2); the degree in t of $P_\lambda(x; t)$ is $\leq n(\lambda)$, and the transition matrix from the Schur polynomials (ω_λ) to (v_λ) is upper unitriangular with respect to the dominant order on partitions. Tables are given in Macdonald's book, pp. 239–241.

REMARK 9.9. If we replace $\mathcal{H}$ by any other Hall algebra over a discrete valuation ring $\mathfrak{o}$ with the same residue field, then an equivalent of the equation (6.15) still holds when we replace $\sum_{v \text{ unipotent}, \deg(v)=n} v$ by the sum of classes of $\mathfrak{o}$-modules of length n.

Fix $\gamma \in \Gamma$ a cuspidal representation, of cuspidal degree d, the map $\varphi_{\gamma,\mathfrak{q}}$ transforms primitive elements of A_γ into primitive elements of $\mathcal{H}$, we can identify A_γ (in two ways due to the automorphism ι) with the PSH algebra A where the degree has been multiplied by d. Thus, we have $\varphi_{\gamma,\mathfrak{q}}(p_n) = \alpha_n \varphi_{c,\mathfrak{q}}(p_{dn})$, for some constant α_n. If we change the identification (by applying ι), α_n is changed into $-\alpha_n$. Our problem amounts to computing the constants α_n.

In order to do that, we need to introduce the *Gel'fand-Graev modules*,[16], $\mathbb{G}_{n,\chi}$: we choose a non-trivial additive character χ of the base field $\mathbb{F}_q$. Take U_n to be the unipotent subgroup of G_n consisting of upper triangular matrices with 1s on the diagonal. We denote by $\mathbb{C}_{n,\chi}$ the 1-dimensional U_n-module such that, if $u = ((u_{i,j})_{i,j}) \in U_n$, u acts on $\mathbb{C}_{n,\chi}$ by $\prod_i \chi(u_{i,i+1})$. Then we set

$$\mathbb{G}_{n,\chi} = \mathrm{Ind}_{U_n}^{G_n} \mathbb{C}_{n,\chi}.$$

REMARK 9.10. The class of the Gel'fand-Graev module in the Grothendieck group does not depend on the choice of the characer χ. Indeed, if χ_1 and χ_2 are two non-trivial additive characters of $\mathbb{F}_q$, there is a unique $a \in \mathbb{F}_q^*$ such that $\chi_2 = a\chi_1$. Therefore the diagonal matrix $Diag(a, 1, \ldots, 1) =: D$ conjugates $\mathbb{C}_{n,\chi_1}$ and $\mathbb{C}_{n,\chi_2}$ as U_n-modules and thus the two induced modules are isomorphic.

For any fixed n, consider the linear form $\delta_n : \mathcal{K}_n \to \mathbb{Z}$ given by the scalar product with $[\mathbb{G}_{n,\chi}]$. Let $\delta : A_q \to \mathbb{Z}$ be the linear form which restricts to δ_n on $\mathcal{K}_n$.

By Remark 9.10, this map does not depend on the choice of χ.

THEOREM 9.11. *(Gel'fand-Gel'fand-Graev) The linear form δ is multiplicative and maps every cuspidal representation to 1.*

We do not provide any outline of proof but we apply it to the problem of computing the constants α_n.

The first thing is that we can fix an identification between A_γ and the PSH algebra A in such a way that $\delta(e_n) = 1$, for all n, hence $\delta(p_n) = (-1)^n$.

On the other hand, the complexification of δ factors through a linear form on $\mathcal{H}$ since, by definition, δ vanishes on all conjugacy classes which are not unipotent. Therefore we get $\delta(\varphi_{c,\mathfrak{q}}(p_{dn})) = (-1)^{dn}$ and $\delta(\varphi_{\gamma,\mathfrak{q}}(p_n)) = (-1)^n$, thus

COROLLARY 9.12. *One has $\varphi_{\gamma,\mathfrak{q}}(p_n) = (-1)^{(d-1)n}\varphi_{c,\mathfrak{q}}(p_{dn})$.*

The reader should feel that this corollary allows to compute the values of the irreducible characters of G_n on unipotent conjugacy classes. Let us make this more precise.

NOTATION 9.13. For any family of partitions $(\lambda_\gamma)_{\gamma \in \Gamma}$, with $\sum_{\gamma \in \Gamma} d(\gamma)|\lambda_\gamma| = n$, we set

$$[V_{(\lambda_\gamma)}] = \bigotimes_{\gamma \in \Gamma} \omega_{\lambda_\gamma}^\gamma.$$

This notation is parallel to the following parametrization of conjugacy classes in the union of all groups $GL_n(\mathbb{F}_q)$: we denote by $\mathcal{X}$ the set of maximal ideals in $\mathbb{F}_q[t, t^{-1}]$, if $\mathfrak{p} \in \mathcal{X}$ we set $\mathsf{d}(\mathfrak{p})$ to be the degree of the corresponding residue field over $\mathbb{F}_q$, and we index Υ_n by the families of partitions $(\lambda_\mathfrak{p})_{\mathfrak{p} \in \mathcal{X}}$ such that $\sum_\mathfrak{p} \mathsf{d}(\mathfrak{p})|\lambda_\mathfrak{p}| = n$.

We thus obtain a parametrization of all the irreducible representations of G_n for every n. We apologise for the intricate notations.

Let $\omega = \otimes_{\gamma \in \Gamma} \omega_{\lambda_\gamma}^\gamma$ be the class in A_q of an irreducible G_n-module (we have $\sum |\lambda_\gamma| = n$), let v_μ be the characteristic function of the unipotent class in G_n of shape μ, we want to compute the scalar product $\langle \omega, v_\mu \rangle$. In this computation, we may replace ω by its orthogonal projection in the Hall algebra $\mathcal{H}$, which is by definition $\Pi_{\gamma \in \Gamma} \varphi_{\gamma, q}(\omega_{\lambda_\gamma}^\gamma)$. For each γ, we express $\omega_{\lambda_\gamma}^\gamma$ as a linear combination with rational coefficients of the Newton functions p_ν, where $|\nu| = |\lambda_\gamma|$; by Corollary 9.12, $\varphi_{\gamma, q}(\omega_{\lambda_\gamma}^\gamma)$ is expressed as a linear combination of $\varphi_{c, q}(p_{\nu'})$ with $|\nu'| = d(\gamma)|\lambda_\gamma|$, and their product, when γ varies, is given as an explicit expression in terms of $\varphi_{c, q}(p_i)$, $i \in \mathbb{N}$.

For instance, we obtain a dimension formula for a cuspidal representation γ of cuspidal degree d:

$$\dim \gamma = (q^{d-1} - 1)(q^{d-2} - 1) \ldots (q - 1).$$

REMARK 9.14. The basis (ω_λ) of $\mathcal{H}$ is not orthogonal with respect to $\langle , \rangle_{\mathcal{H}}$, but the Newton polynomials p_λ are (see the proof of Lemma 5.5, and use the fact that the elements p_n are primitive and their degrees are all distinct).

Furthermore, Theorem 9.11 leads to a branching rule for the representations of $GL_n(\mathbb{F}_q)$.

COROLLARY 9.15. We fix the integer n and the character χ of $\mathbb{F}_q$. The Gel'fand-Graev representation $\mathbb{G}_{n, \chi}$ is multiplicity-free.

An irreducible representation M satisfies $\delta([M]) = 0$ or 1. Gel'fand-Graev call the representations such that $\delta([M]) = 1$ non-degenerate. Therefore the non-degenerate modules are the ones appearing in the decomposition of Gel'fand-Graev representations.

COROLLARY 9.16. Consider a family of non-negative integers (k_γ) for $\gamma \in \Gamma$, with finite support. Set $[V_{(k_\gamma)}] := \otimes_{\gamma \in \Gamma} \omega_{1^{k_\gamma}}^\gamma$. Then every non-degenerate irreducible representation is isomorphic to a unique $V_{(k_\gamma)}$.

Our next goal is to state an analogue of Theorem 8.8 for the present situation.

We denote by Aff_k the group of affine transformations of the space $\mathbb{F}_q^k$ and by $\mathcal{K}(\mathrm{Aff}_k)$ the Grothendieck group of the category of finite dimensional Aff_k-modules. We notice that Aff_{n-1} can be seen as the subgroup of G_n consisting of matrices with last row $(0, \ldots, 0, 1)$.

By definition, Aff_k is the semi-direct product $\mathbb{F}_q^k \rtimes G_k$, and $\mathcal{K}(\mathrm{Aff}_k)$ can be determined using [30] 8.2, Prop. 25. One has

$$\mathcal{K}(\mathrm{Aff}_k) = \mathcal{K}_k \oplus \mathcal{K}(\mathrm{Aff}_{k-1}),$$

where $\mathcal{K}_k$ is seen as a subgroup of $\mathcal{K}(\mathrm{Aff}_k)$ through the quotient map $\mathrm{Aff}_k \to G_k$. In order to define the map $\mathcal{K}(\mathrm{Aff}_{k-1}) \to \mathcal{K}(\mathrm{Aff}_k)$, we view Aff_{k-1} as the stabilizer in G_k of the character $(1, \ldots, 1, \chi)$ of the additive group $\mathbb{F}_q^k$ and we use Serre's construction 8.2 Proposition 25 in [30]. This decomposition does not depend on the choice of χ.

Using induction, we immediately obtain that

$$\mathcal{K}(\mathrm{Aff}_{n-1}) = \bigoplus_{0 \leq i \leq n-1} \mathcal{K}_i.$$

We construct a ring homomorphism $D : A_q \to A_q$ as the composition map:

$$A_q \xrightarrow{m^*} A_q \otimes A_q \xrightarrow{\mathrm{Id} \otimes \delta} A_q \otimes \mathbb{Z} = A_q.$$

PROPOSITION 9.17. (1) *The restriction functor from G_n to Aff_{n-1} induces the map $D - \mathrm{Id} : \mathcal{K}_n \to \mathcal{K}_{n-1} \oplus \ldots \oplus \mathcal{K}_1 \simeq \mathcal{K}(\mathrm{Aff}_{n-1})$.*

(2) *The induction functor from G_{n-1} to Aff_{n-1} induces the map $D : \mathcal{K}_{n-1} \to \mathcal{K}_{n-1} \oplus \ldots \oplus \mathcal{K}_1 \simeq \mathcal{K}(\mathrm{Aff}_{n-1})$.*

We now apply Pieri's rule (Theorem 7.6) and Theorem 9.11. Recall the notations introduced just before Pieri's rule, and set $\mathbf{R}^\lambda$ the union of all the $\mathbf{R}^\lambda_i$. We obtain the following result:

THEOREM 9.18. (1) *Let V be the irreducible G_n-module indexed by the family of partitions λ_γ (with $\sum_\gamma d(\gamma)|\lambda_\gamma| = n$), then the restriction of V to Aff_{n-1} can be expressed as the direct sum $\bigoplus V_{(\mu_\gamma)}$, with $\mu_\gamma \in \mathbf{R}^{\lambda_\gamma}$ and $(\mu_\gamma) \neq (\lambda_\gamma)$.*

(2) *Let W be an irreducible Aff_{n-1}-module. Then there exists a family of partitions $(\mu_\gamma)_{\gamma \in \Gamma}$, with $\sum_\gamma d(\gamma)|\mu_\gamma| < n$, representing $[W]$ in $\mathcal{K}(\mathrm{Aff}_{n-1})$. Then the irreducible components of the image of W by the induction to G_n are labelled by the families of partitions (λ_γ) such that $\mu_\gamma \in \mathbf{R}_{\lambda_\gamma}$ and $\sum_\gamma d(\gamma)|\lambda_\gamma| = n$, each with multiplicity one.*

REMARK 9.19. In the first statement, each $V_{(\mu_\gamma)}$ is an irreducible representation of G_i for some $i < n$ since $\sum_\gamma d(\gamma)|\mu_\gamma| < n$, hence it is a representation of Aff_{n-1}.

Our next goal is to state the main result of [31], which computes the maps $\varphi_{\gamma, \mathfrak{p}}$ for all maximal ideals $\mathfrak{p}$. The main difficulty resides in the construction of a bijection between the set Γ_d of cuspidal representations of cuspidal degree d and the set of conjugacy classes of primitive multiplicative characters of $\mathbb{F}_{q^d}$ under the action of the Frobenius automorphism $x \mapsto x^q$ of $\mathbb{F}_{q^d}$ (a multiplicative character of $\mathbb{F}_{q^d}$ is primitive if it doesn't factorize through any norm map $\mathbb{F}_{q^d} \to \mathbb{F}_{q^{d'}}$ where d' is a strict divisor of d). One of the tools is Brauer's theory of modular characters which we do not intend to explain in this book (nevertheless, this theory is developed in [30]).

Let $\gamma \in \Gamma_d$ be a cuspidal representation, choose a multiplicative character ξ of $\mathbb{F}_{q^d}$ representing γ. Let $\mathfrak{p}$ be a maximal ideal of $\mathbb{F}_q[t, t^{-1}]$, generated by a polynomial P of degree $\mathbf{d}$. The graded group of primitive elements in A_γ has the natural basis $(p_{n,\gamma})_{n \in \mathbb{N}}$, which is the image of the family $(p_n) \subset A$ under the identification explained after Theorem 9.11; the degree of $p_{n,\gamma}$ is nd. Similarly, the graded vector space of primitive elements in the Hall algebra $\mathcal{H}(\mathbb{F}_q[t, t^{-1}]_\mathfrak{p})$ has the natural basis $(p_{n,\mathfrak{p}})_{n\in\mathbb{N}}$, which is the image of the family $(p_n) \subset A$ by the isomorphism $A \otimes_{\mathbb{Z}} \mathbb{C} \to \mathcal{H}(\mathbb{F}_q[t, t^{-1}]_{\mathfrak{p}})$ (see remark 9.9); the degree of $p_{n,\mathfrak{p}}$ is nd.

THEOREM 9.20. *(Springer-Zelevinsky) One has:*

$$\varphi_{\gamma,\mathfrak{p}}(p_{n,\gamma}) = (-1)^{n(d-1)} \left(\sum_{x,P(x)=0} \frac{1}{\alpha(x)} \xi(N_{\mathbb{F}_{q^d}(x)/\mathbb{F}_{q^d}}(x)) \right) p_{\frac{nd}{d},\mathfrak{p}}$$

where $\alpha(x) = [\mathbb{F}_{q^d}(x) : \mathbb{F}_{q^d}]$ and N is the norm of field extensions. Note that the expression vanishes if d does not divide nd.

We want to test this compact formula with two examples of character tables: the cases of $GL_3(\mathbb{F}_2)$ and $GL_2(\mathbb{F}_q)$. The second example is developed in [12] with classical methods. For both cases we also refer to Steinberg's thesis ([34]).

EXAMPLE 9.21. Character table of $GL_3(\mathbb{F}_2)$.

In this case, there are lots of tremendous simplifications in the computations due to the small size of the base field $\mathbb{F}_2$ and the uniqueness of the character of its multiplicative group. Thus we are able to give details which would necessitate more notations with a bigger base field.

The list of irreducible polynomials with non-zero constant term of degree ≤ 3 is $X + 1, X^2 + X + 1, X^3 + X + 1, X^3 + X^2 + 1$. We denote by j the chosen root of $X^2 + X + 1$ in $\mathbb{F}_4$ and by y (resp. y^{-1}) the chosen root of $X^3 + X + 1$ (resp. $X^3 + X^2 + 1$) in $\mathbb{F}_8$. Therefore, the list of conjugacy classes of cuspidal characters of degree ≤ 3 is parametrized by $\chi_0 = c$, the trivial character of $\mathbb{F}_2^*$, χ_j (conjugate to χ_{j^2}) character of $\mathbb{F}_4^*$ and two non-conjugate characters of $\mathbb{F}_8^*$, η and $\bar{\eta}$.

The 6 conjugacy classes in $GL_3(\mathbb{F}_2)$ are Id, the Jordan block $J_{(2,1)}$, the maximal Jordan block $J_{(3)}$, $Diag(j,1)$ (j is a 2×2 matrix, and we abuse notation), y and y^{-1}.

The northwest 3×3 block of the table is obtained with the table for $n = 3$ in [25] p. 239, after taking the reciprocal polynomials evaluated at $t = 2$ (see Remark 9.8).

The southwest 3×3 block is deduced from the northwest block after multiplication on the left by the matrix $M = \begin{pmatrix} -1 & 0 & 1 \\ 1 & -1 & 1 \\ 1 & -1 & 1 \end{pmatrix}$: indeed, ^tM is the matrix of the character table of the group S_3, taking the columns indexed by the partitions (21) and (twice) (3) because the image of $(1)_{\chi_j}(1)_c$ (resp. $(1)_\eta$ and $(1)_{\bar{\eta}}$) in $\mathcal{H}$ is $p_{(21)}$ (resp. $p_{(3)}$).

The northeast 3×3 block comes from the composed homomorphism

$$\begin{array}{ccccc} & m^* & & \varphi_{c,X^2+X+1} \otimes \varphi_{c,X+1} & \\ A_c & \to & A_c \otimes A_c & \longrightarrow & \mathcal{H}_{X^2+X+1} \otimes \mathcal{H}_{X+1} \end{array}.$$

Finally, let us find the southeast 3×3 block: the northwest coefficient is trivially 1, and since $\mathbb{F}_4$ is not included in $\mathbb{F}_8$ the other entries of the first line and the first column are 0. The remaining 2×2 matrix is $\begin{pmatrix} \beta & \overline{\beta} \\ \overline{\beta} & \beta \end{pmatrix}$ where $\beta = \xi(x) + \xi(x^2) + \xi(x^4)$ in the notations of Theorem 9.20. We get $\beta = \frac{-1 \pm \sqrt{7}}{2}$, since $\xi(x)$ is a primitive 7-th root of 1.

	Id	$J_{(2,1)}$	$J_{(3)}$	$Diag(j,1)$	y	y^{-1}
$(3)_c$	1	1	1	1	1	1
$(2,1)_c$	6	2	0	0	-1	-1
$(1,1,1)_c$	8	0	0	-1	1	1
$(1)_{\chi_j}(1)_c$	7	-1	-1	1	0	0
$(1)_\eta$	3	-1	1	0	$\frac{-1+i\sqrt{7}}{2}$	$\frac{-1-i\sqrt{7}}{2}$
$(1)_{\overline{\eta}}$	3	-1	1	0	$\frac{-1-i\sqrt{7}}{2}$	$\frac{-1+i\sqrt{7}}{2}$

EXAMPLE 9.22. Character table of $GL_2(\mathbb{F}_q)$.

The irreducible representations of $GL_2(\mathbb{F}_q)$ give the basis Ω_2 of the homogeneous component of degree 2 of the PSH algebra A_q, we use the notation 9.13 and check that they come in four families:

- if χ_1, χ_2 are two distinct characters of $\mathbb{F}_q^*$, they belong to Ω_1 and we denote by $(1)_{\chi_1}(1)_{\chi_2}$ their (commutative) product in Ω_2.
- if χ is a character of $\mathbb{F}_q^*$, one sees it as a cuspidal representation of $GL_1(\mathbb{F}_q)$ and the square of χ in A_q is the sum $e_{2,\chi} + h_{2,\chi}$ (of elements of Ω_2) according to the isomorphism $A \to A_\chi$. This clearly gives two families indexed by the partitions $(1,1)_\chi$ for $e_{2,\chi}$ and $(2)_\chi$ for $h_{2,\chi}$ (this one corresponds to the trivial representation if $\chi = 1$).
- if η is a primitive character of $\mathbb{F}_{q^2}^*$, it corresponds by Theorem 9.20 to a cuspidal representation of $GL_2(\mathbb{F}_q)$, which we denote by $(1)_\eta$.

The conjugacy classes of $GL_2(\mathbb{F}_q)$ are indexed by Υ_2, they also come in four families, we indicate the cardinality of the corresponding conjugacy classes:

- if x_1, x_2 are two distinct elements of $\mathbb{F}_q^*$, we denote by $Diag(x_1, x_2)$ the class of the corresponding diagonal element in Υ_2, the cardinality of the conjugacy class is $q^2 + q$.
- if $x \in \mathbb{F}_q^*$, we denote by $Diag(x, x)$ and $Jord(x)$ the classes of the diagonal element and the Jordan block element in Υ_2, the cardinality of the conjugacy classes is 1 for $Diag(x, x)$ and $q^2 - 1$ for $Jord(x)$.
- if y is a primitive element in $\mathbb{F}_{q^2}^*$, we denote by $P_{min,y}$ its minimal polynomial on $\mathbb{F}_q$ and get an element of Υ_2, the cardinality of the conjugacy class is $q^2 - q$.

The character table is the following (we used a different ordering in enumeration in order to be consistent with usual notations in character tables).

	$Diag(x,x)$	$Diag(x_1,x_2)$	$Jord(x)$	$P_{min,y}$
$(2)_\chi$	$\chi(x^2)$	$\chi(x_1 x_2)$	$\chi(x^2)$	$\chi(N(y))$
$(1)_{\chi_1}(1)_{\chi_2}$	$(q+1)\chi_1(x)\chi_2(x)$	$\chi_1(x_1)\chi_2(x_2)+$ $+\chi_1(x_2)\chi_2(x_1)$	$\chi_1(x)\chi_2(x)$	0
$(1,1)_\chi$	$q\chi(x^2)$	$\chi(x_1 x_2)$	0	$-\chi(N(y))$
$(1)_\eta$	$(q-1)\eta(x)$	0	$-\eta(x)$	$-(\eta(x)+\eta(x^q))$

CHAPTER 7

Introduction to representation theory of quivers

Who killed Cock Robin? I, said the Sparrow, with my bow and arrow, I killed Cock Robin. (Nursery Rhyme)

In which we enter the "non-semisimple" woods: armed with quivers, we encounter tame and wild problems.

1. Representations of quivers

A *quiver* is an oriented graph. For example

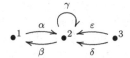

is a quiver.

In this chapter we consider only finite quivers, namely quivers with finitely many vertices and arrows.

The *underlying graph* of a quiver Q is the graph obtained from Q by forgetting the orientation of the arrows.

If Q is a quiver, we denote by Q_0 the set of vertices of Q and by Q_1 the set of arrows of Q. In the example above, $Q_0 = \{1, 2, 3\}$ and $Q_1 = \{\alpha, \beta, \gamma, \delta, \varepsilon\}$.

A quiver Q' is a *subquiver* of a quiver Q if $Q'_0 \subset Q_0$ and $Q'_1 \subset Q_1$.

For every arrow $\gamma \in Q_1 \colon i \xrightarrow{\gamma} j$ we define $s(\gamma) = i$ as the *source* or *tail* of γ and $t(\gamma) = j$ as the *target* or *head* of γ. In the example the vertex 1 is the source of α and the target of β.

An *oriented cycle* is a subgraph with vertices $C_0 := \{s_1, \ldots, s_r\} \subset Q_0$ and arrows $C_1 = \{\gamma_1, \ldots, \gamma_r\} \subset Q_1$ such that γ_i goes from s_i to s_{i+1} if $i < r$ and γ_r goes from s_r to s_1. In our example $\bullet^1 \underset{\beta}{\overset{\alpha}{\rightleftarrows}} \bullet^2$ is a oriented cycle. A *loop* is an arrow whose head and tail coincide. In our example, there is only one loop $\bullet^2$.

© Springer Nature Switzerland AG 2018
C. Gruson and V. Serganova, *A Journey Through Representation Theory*,
Universitext, https://doi.org/10.1007/978-3-319-98271-7_7

DEFINITION 1.1. Fix a field k. Let Q be a quiver. Consider a k-vector space

$$V = \bigoplus_{i \in Q_0} V_i$$

and a collection of k-linear maps

$$\rho = \{\rho_\gamma : V_i \to V_j \mid \gamma \in Q_1, \ s(\gamma) = i, \ t(\gamma) = j\}.$$

Then (V, ρ) is called a *representation* of Q. The *dimension* of the representation (V, ρ) is the vector $d \in \mathbb{Z}^{Q_0}$ such that $d_i = \dim V_i$. In this text we always assume that V is finite-dimensional.

We sometimes use a diagram to visualize a representation of a quiver. For example, if Q is of shape

$$\bullet^1 \xrightarrow{\alpha} \bullet^2 \xleftarrow{\beta} \bullet^3,$$

and if $V_1 = k^2, V_2 = k, V_3 = k$ and $\rho_\alpha = 0$, $\rho_\beta = \mathrm{id}$, we present it as the following diagram:

$$k^2 \xrightarrow{0} k \xleftarrow{\mathrm{id}} k.$$

DEFINITION 1.2. Let (V, ρ) and (W, σ) be two representations of Q. A *morphism* of representations $\varphi \colon (V, \rho) \to (W, \sigma)$ is a set of linear maps $\{\varphi_i \colon V_i \to W_i \mid i \in Q_0\}$ such that the diagram

$$
\begin{array}{ccc}
V_i & \xrightarrow{\rho_\gamma} & V_j \\
\varphi_i \downarrow & & \downarrow \varphi_j \\
W_i & \xrightarrow{\sigma_\gamma} & W_j
\end{array}
$$

is commutative for every $\gamma \in Q_1$, where $i = s(\gamma), j = t(\gamma)$.

We say that two representations (V, ρ) and (W, σ) of Q are *isomorphic* if there exists a morphism $\varphi \colon (V, \rho) \to (W, \sigma)$ such that φ_i is an isomorphism for every $i \in Q_0$.

The *direct sum* $(V \oplus W, \rho \oplus \sigma)$ of two representations (V, ρ) and (W, σ) of a quiver Q is defined in the obvious way.

A representation (W, σ) is a *subrepresentation* of (V, ρ) if, for every $i \in Q_0$, we are given an inclusion $W_i \subset V_i$ such that, for every $\gamma \in Q_1$, with $s(\gamma) = i$, the restriction of ρ_γ to W_i coincides with σ_γ.

A non-zero representation (V, ρ) is called *irreducible* if it does not contain any non-trivial proper subrepresentation; it is called *indecomposable* if it cannot be written as the direct sum of two non-trivial subrepresentations.

EXAMPLE 1.3. Consider the quiver $1 \xrightarrow{\gamma} 2$ and the representation (V, ρ) which corresponds to the diagram $k \xrightarrow{\mathrm{id}} k$. Then (V, ρ) has only one non-trivial proper subrepresentation, given by the diagram $0 \xrightarrow{0} k$. Therefore (V, ρ) is indecomposable but not irreducible.

In most cases it is not difficult to classify irreducible representations of a given quiver. On the other hand, classifying all indecomposable representations up to isomorphism is very hard. Many classical problems of linear algebra can be viewed as particular cases of this general problem. Let us see a few examples.

EXAMPLE 1.4. Let Q be the quiver of Example 1.3. A representation of Q can be seen as a pair of vector spaces V_1 and V_2 together with a linear map $\rho_\gamma : V_1 \to V_2$. Let us fix the dimension (d_1, d_2) and identify V_1 with k^{d_1}, V_2 with k^{d_2}. Classifying the representations of Q of dimension (d_1, d_2) is equivalent to the following problem of linear algebra. Consider the space of matrices of size $d_2 \times d_1$. Then the linear groups $GL(d_1)$ and $GL(d_2)$ act on this space by multiplication on the left and on the right respectively. We would like to describe all the orbits for this action.

Consider a representation (V, ρ) of Q. Choose subspaces $W_1 \subset V_1$ and $W_2 \subset V_2$ such that $V_1 = \operatorname{Ker} \rho_\gamma \oplus W_1$ and $V_2 = \rho_\gamma(W_1) \oplus W_2$. Note that ρ_γ induces an isomorphism $\alpha : W_1 \to \rho_\gamma(W_1)$. Then (V, ρ) is the direct sum of the subrepresentations $\operatorname{Ker} \rho_\gamma \xrightarrow{0} 0$, $0 \xrightarrow{0} W_2$ and $W_1 \xrightarrow{\alpha} \rho_\gamma(W_1)$. It is clear that the first representation can be written as a direct sum of several copies of $k \xrightarrow{0} 0$, the second one is a direct sum of several copies of $0 \xrightarrow{0} k$. These decompositions are not unique, they depend on the choice of basis in $\operatorname{Ker} \rho_\gamma$ and W_2. Finally the representation $W_1 \xrightarrow{\alpha} \rho_\gamma(W_1)$ can be written as a direct sum of several copies of $k \xrightarrow{\mathrm{id}} k$.

Therefore there are three indecomposable representations of Q (up to isomorphism). Their dimensions are $(1, 0)$, $(0, 1)$ and $(1, 1)$. Furthermore, in every dimension there are finitely many non-isomorphic representations. Quivers with the latter property are called *quivers of finite type*.

EXAMPLE 1.5. Consider the quiver Q with one vertex and one loop. Then a finite-dimensional representation of Q is a pair (V, T), where V is a finite-dimensional vector space and T is a linear operator in V. Isomorphism classes of representations of this quiver are the same as conjugacy classes of $n \times n$ matrices where n is the dimension of V. If k is algebraically closed, this classification problem amounts to describing Jordan canonical forms of $n \times n$ matrices. In particular, indecomposable representations correspond to matrices with one Jordan block.

If k is not algebraically closed, the problem of classifying the conjugacy classes of matrices is more tricky. This example shows that representation theory of quivers depends very much on the base field.

EXAMPLE 1.6. Consider the Kronecker quiver $\bullet \rightrightarrows \bullet$. The classification of finite-dimensional representations of this quiver is also a classical problem of linear algebra. It amounts to the classification of pairs of linear operators $S, T : V_1 \to V_2$ up to multiplication by some $X \in GL(V_1)$ on the left and by some $Y \in GL(V_2)$ on the right. It is still possible to obtain this classification by brute force. We will solve this problem using the general theory of quivers in the next chapter.

EXAMPLE 1.7. Now let Q be the quiver with one vertex and two loops. The representation theory of Q is equivalent to classifying pairs of linear operators (T, S) in a vector space V up to conjugation. In contrast with all previous examples, in this case the number of variables parametrizing indecomposable representations of dimension n grows as n^2. We call a pair (T, S) generic if T is diagonal in some basis $e_1, \ldots, e_n$ with distinct eigenvalues and the matrix of S in this basis does not have any zero entry.

EXERCISE 1.8. Check that if (T, S) is generic and $W \subset V$ is both T-stable and S-stable, then $W = 0$ or $W = V$. Thus, the corresponding representation of Q is irreducible.

Therefore every generic pair of operators (T, S) gives rise to an irreducible representation of Q. The eigenvalues of T give n distinct parameters. If T is diagonalized, we can conjugate S by linear operators diagonal in the eigenbasis of T. Thus, we have $n^2 - n$ parameters for the choice of S.

The situation which appears in this example is referred to as *wild*. There is a precise definition of wild quivers and we refer the reader to Chapter 9 for further reading on this subject.

2. Path algebra

As in the case of groups, we can reduce the representation theory of a quiver to the representation theory of some associative ring. In the case of groups, this ring is the group algebra, while in the case of quivers it is the path algebra.

DEFINITION 2.1. Let Q be a quiver. A *path* p is a sequence $\gamma_1, \ldots, \gamma_k$ of arrows such that $s(\gamma_i) = t(\gamma_{i+1})$. Set $s(p) = s(\gamma_k)$, $t(p) = t(\gamma_1)$. The number k of arrows is called the length of p.

DEFINITION 2.2. Let $p_1 = \gamma_1, \ldots, \gamma_k$ and $p_2 = \delta_1, \ldots, \delta_l$ be two paths of Q. We define the product of p_1 and p_2 to be the path $\delta_1, \ldots, \delta_l, \gamma_1, \ldots \gamma_k$ if $t(\gamma_1) = s(\delta_l)$ and zero otherwise.

Next, we introduce an element e_i for each vertex $i \in Q_0$ and define the product of e_i and e_j by the formula

$$e_i e_j = \delta_{ij} e_i.$$

For a path p, we set

$$e_i p = \begin{cases} p, & \text{if } i = t(p) \\ 0 & \text{otherwise} \end{cases},$$

$$p e_i = \begin{cases} p, & \text{if } i = s(p) \\ 0 & \text{otherwise} \end{cases}.$$

The *path algebra* $k(Q)$ of Q is the vector space of k-linear combinations of all paths of Q and elements $\{e_i\}_{i \in Q_0}$, with the multiplication law obtained by extending the product defined above by bilinearity.

Note that every e_i, $i \in Q_0$, is an idempotent in $k(Q)$ and that $\sum_{i \in Q_0} e_i = 1$.

EXAMPLE 2.3. Let Q be the quiver with one vertex and n loops then $k(Q)$ is the free associative algebra with n generators.

EXERCISE 2.4. Let Q be a quiver such that the underlying graph of Q does not contain any cycle or loop. Let $Q_0 = \{1, \ldots, n\}$. Show that the path algebra $k(Q)$ is isomorphic to the subalgebra of the matrix algebra $\mathrm{Mat}_n(k)$ generated by the subset of elementary matrices $\{E_{ii} \mid i \in Q_0\}$, $\{E_{ji} \mid \gamma \in Q_1, s(\gamma) = i, t(\gamma) = j\}$.

In particular, show that the path algebra of the quiver

$$\bullet \leftarrow \bullet \leftarrow \cdots \leftarrow \bullet$$

is isomorphic to the algebra B_n of upper triangular matrices, see Example 7.19 in Chapter 5.

LEMMA 2.5. *Let Q be a quiver.*

(1) *The path algebra $k(Q)$ is generated by the idempotents $\{e_i \mid i \in Q_0\}$ and the paths $\{\gamma \mid \gamma \in Q_1\}$ of length 1.*
(2) *The algebra $k(Q)$ is finite-dimensional if and only if Q does not contain any oriented cycle.*
(3) *If Q is the disjoint union of two quivers Q' and Q'', then $k(Q)$ is isomorphic to the direct product $k(Q') \times k(Q'')$.*
(4) *The path algebra has a natural $\mathbb{Z}$-grading*

$$k(Q) = \bigoplus_{n=0}^{\infty} k(Q)_{(n)},$$

where $k(Q)_{(0)}$ is the span of the idempotents e_i for all $i \in Q_0$ and $k(Q)_{(n)}$ is the span of all paths of length n.
(5) *For every vertex $i \in Q_0$ the element e_i is a primitive idempotent of $k(Q)$, and hence $k(Q)e_i$ is an indecomposable projective $k(Q)$-module.*

PROOF. The first four assertions are straightforward and we leave them to the reader as an exercise. Let us prove (5).

Let $i \in Q_0$. By Exercise 7.14 Chapter 5, proving (5) amounts to checking that if $\varepsilon \in k(Q)e_i$ is an idempotent such that $e_i \varepsilon = \varepsilon e_i = \varepsilon$, then $\varepsilon = e_i$ or $\varepsilon = 0$. We use the grading of $k(Q)$ defined in (4). By definition, the left ideal $k(Q)e_i$ inherits this grading. Hence we can write

$$k(Q)e_i = \bigoplus_{n=0}^{\infty} k(Q)_{(n)}e_i,$$

where $k(Q)_{(0)}e_i = ke_i$ and, for $n > 0$, the graded component $k(Q)_{(n)}e_i$ is spanned by the paths of length n with source at i. We can write $\varepsilon = \varepsilon_0 + \cdots + \varepsilon_l$ with $\varepsilon_n \in k(Q)_{(n)}e_i$. Since ε is an idempotent, we have $\varepsilon_0^2 = \varepsilon_0$, which implies $\varepsilon_0 = e_i$ or $\varepsilon_0 = 0$. In the latter case let ε_p be the first non-zero term in the decomposition of ε. Then the first non-zero term in the decomposition of ε^2 has degree no less than $2p$. This implies $\varepsilon = 0$. If $\varepsilon_0 = e_i$, consider the idempotent $e_i - \varepsilon$ and apply the above argument again. $\qquad\square$

Given a representation (V, ρ) of a quiver Q, $V = \bigoplus_{i \in Q_0} V_i$ one can equip V with a structure of $k(Q)$-module in the following way

(1) The idempotent e_i acts on V_j by $\delta_{ij} \operatorname{Id}_{V_j}$.
(2) For $\gamma \in Q_1$ and $v \in V_i$ we set $\gamma v = \rho_\gamma(v)$ if $i = s(\gamma)$ and zero otherwise.
(3) We extend this action for the whole $k(Q)$ using Lemma 2.5 (1).

Conversely, every $k(Q)$-module V gives rise to a representation ρ of Q when one sets $V_i = e_i V$.

This implies the following Theorem. [1]

THEOREM 2.6. *The category of representations of Q over a field k is equivalent to the category of $k(Q)$-modules.*

EXERCISE 2.7. Let Q be a quiver and $J(Q)$ be the ideal of $k(Q)$ generated by all arrows $\gamma \in Q_1$. Then the quotient $k(Q)/J$ is a semisimple commutative ring isomorphic to k^{Q_0}.

EXERCISE 2.8. Let Q' be a subquiver of a quiver Q. Let $I(Q')$ be the ideal of $k(Q)$ generated by e_i for all $i \notin Q'_0$ and by all $\gamma \notin Q'_1$. Prove that $k(Q')$ is isomorphic to the quotient ring $k(Q)/I(Q')$.

LEMMA 2.9. *Let $A = \bigoplus_{i=0}^{\infty} A_{(i)}$ be a graded algebra and R be the Jacobson radical of A. Then*

(1) *R is a graded ideal, i.e. $R = \bigoplus_{i=0}^{\infty} R_{(i)}$, where $R_{(i)} = R \cap A_{(i)}$;*
(2) *If $u \in R_{(p)}$ for some $p > 0$, then u is nilpotent.*

PROOF. Assume first that the ground field k is infinite. Let $t \in k^*$. Consider the automorphism φ_t of A such that $\varphi_t(u) = t^p u$ for all $u \in A_{(p)}$. Observe that $\varphi_t(R) = R$. Suppose that u belongs to R and write it as the sum of homogeneous components $u = u_0 + \cdots + u_n$ with $u_j \in A_{(j)}$. We have to show that $u_i \in R$ for all $i = 1, \ldots, n$. Indeed,

$$\varphi_t(u) = u_0 + tu_1 + \cdots + t^n u_n \in R$$

---

[1]Compare with the analogous result for groups in Chapter 2.

for all $t \in k^*$. Since k is infinite, this implies $u_i \in R$ for all i. If k is finite, consider the algebra $A \otimes_k \bar{k}$ where $\bar{k}$ is the algebraic closure of k, and use the fact that $R \otimes_k \bar{k}$ is included in the radical of $A \otimes_k \bar{k}$.

Let $u \in R_{(p)}$. Then $1 - u$ is invertible. Hence there exists $a_i \in A_{(i)}$, for some $i = 0, \ldots, n$ such that

$$(a_0 + a_1 + \cdots + a_n)(1 - u) = 1.$$

This relation implies $a_0 = 1$ and $a_{pj} = u^j$ for all $j > 0$. Thus $u^j = 0$ for sufficiently large j. □

Let us call a path p of a quiver Q a *one-way path* if there is no path from $t(p)$ to $s(p)$.

EXERCISE 2.10. The span of all one-way paths of Q is a two-sided nilpotent ideal in $k(Q)$.

LEMMA 2.11. *The Jacobson radical of the path algebra $k(Q)$ is the span of all one-way paths of Q.*

PROOF. Let N be the span of all one-way paths. By Exercise 2.10 N is contained in the radical of $k(Q)$.

Assume now that y belongs to the radical of $k(Q)$. Exercise 2.7 implies that $y \in J(Q)$ and moreover by Lemma 2.9(2) we may assume that y is a linear combination of paths of the same length. We want to prove that $y \in N$. Note that $e_i y e_j$ belongs to the radical for all $i, j \in Q_0$. Assume that the statement is false. Then there exist i and j such that $z := e_i y e_j$ is not in N, in other words there exists a path u with source j and target i. Furthermore zu is a linear combination of oriented cycles $u_1, \ldots, u_l$ of the same length. By Lemma 2.9, zu must be nilpotent. But it is clearly not nilpotent, because $(zu)^k = 0$ would give a linear relation between oriented cycles. Contradiction. □

Lemma 2.11 implies the following

PROPOSITION 2.12. *Let Q be a quiver which does not contain oriented cycles. Then $k(Q)/\operatorname{rad} k(Q) \simeq k^n$, where n is the number of vertices. In particular, every simple $k(Q)$-module is one dimensional.*

PROOF. The assumption on Q implies that every path is a one-way path. Hence the radical of $k(Q)$ is equal to $J(Q)$. □

3. Standard resolution and consequences

3.1. Construction of the standard resolution. A remarkable property of path algebras is the fact that every module has a projective resolution of length at most 2:

THEOREM 3.1. *Let Q be a quiver, A denote the path algebra $k(Q)$ and V be an A-module. Recall that $V = \bigoplus_{i \in Q_0} V_i$. Then the following sequence of A-modules is exact:*

$$0 \to \bigoplus_{\gamma \in Q_1} Ae_{t(\gamma)} \otimes V_{s(\gamma)} \xrightarrow{f} \bigoplus_{i \in Q_0} Ae_i \otimes V_i \xrightarrow{g} V \to 0,$$

where

$$f\left(ae_{t(\gamma)} \otimes v\right) = ae_{t(\gamma)}\gamma \otimes v - ae_{t(\gamma)} \otimes \gamma v$$

for all $\gamma \in Q_1, v \in V_{s(\gamma)}$, and

$$g\left(ae_i \otimes v\right) = av$$

for any $i \in Q_0, v \in V_i$. Hence it is a projective resolution of V.

REMARK 3.2. The structures of A-module considered in the statement are defined by the action of A on the left-hand side of every tensor product.

PROOF. (of Theorem 3.1) The fact that f and g are morphisms of A-modules is left to the reader. First, let us check that $g \circ f = 0$. Indeed,

$$g\left(f\left(ae_j \otimes v\right)\right) = g\left(ae_j\gamma \otimes v - ae_j \otimes \gamma v\right) = ae_j\gamma v - ae_j\gamma v = 0.$$

Since $V = \oplus_{i \in Q_0} V_i$ and $V_i = e_i V$, g is surjective.

Now let us check that f is injective. To simplify notations we set

$$X = \bigoplus_{\gamma \in Q_1} Ae_{t(\gamma)} \otimes V_{s(\gamma)}, \quad Y = \bigoplus_{i \in Q_0} Ae_i \otimes V_i.$$

Consider the $\mathbb{Z}$-grading

$$A \otimes V = \bigoplus_{p=0}^{\infty} A_{(p)} \otimes V.$$

Since all Ae_i for $i \in Q_0$ are homogeneous left ideals of A, there are induced gradings $X = \oplus_{p \geq 0} X_{(p)}$ and $Y = \oplus_{p \geq 0} Y_{(p)}$. Define $f_0 : X \to Y$ and $f_1 : X \to Y$ by

$$f_1\left(ae_{t(\gamma)} \otimes v\right) = ae_{t(\gamma)}\gamma \otimes v, \quad f_0\left(ae_{t(\gamma)} \otimes v\right) = ae_{t(\gamma)} \otimes \gamma v.$$

Note that, for any $p \geq 0$, we have $f_1(X_{(p)}) \subset Y_{(p+1)}$ and $f_0(X_{(p)}) \subset Y_{(p)}$. Moreover, it is clear from the definition that f_1 is injective. Since $f = f_1 - f_0$, we obtain that f is injective by a simple argument on gradings.

It remains to prove that $\operatorname{Im} f = \operatorname{Ker} g$. The exercise just below (Exercise 3.3) implies that for any $y \in Y$ there exists $y_0 \in Y_{(0)}$ such that $y \equiv y_0 \mod \operatorname{Im} f$. Let $y \in \operatorname{Ker} g$, then $y_0 \in \operatorname{Ker} g$. But g restricted to $Y_{(0)}$ is injective. Thus, $y_0 = 0$ and $y \in \operatorname{Im} f$. $\square$

EXERCISE 3.3. Show that for any $p > 0$ and $y \in Y_{(p)}$ there exists $y' \in Y_{(p-1)}$ such that $y' \equiv y \mod \operatorname{Im} f$. (Hint: it suffices to check the statement for $x = u \otimes v$ where $v \in V_i$ and u is a path of length p with source i).

3.2. Extension groups. Let X and Y be two $k(Q)$-modules. We will use the notations $\operatorname{End}_Q(X)$ for $\operatorname{End}_{k(Q)}(X)$, $\operatorname{Hom}_Q(X, Y)$ for $\operatorname{Hom}_{k(Q)}(X, Y)$ and more generally $\operatorname{Ext}_Q^\bullet(X, Y)$ for $\operatorname{Ext}_{k(Q)}^\bullet(X, Y)$.

We define a linear map

(7.1) $$d : \bigoplus_{i \in Q_0} \operatorname{Hom}_k (X_i, Y_i) \to \bigoplus_{\gamma \in Q_1} \operatorname{Hom}_k \left(X_{s(\gamma)}, Y_{t(\gamma)} \right)$$

by the formula

(7.2) $$d\varphi (x) = \varphi (\gamma x) - \gamma \varphi (x)$$

for any $\gamma \in Q_1$, $x \in X_{s(\gamma)}$ and $\varphi \in \operatorname{Hom}_k \left(X_{t(\gamma)}, Y_{t(\gamma)} \right)$. The kernel of d is equal to $\operatorname{Hom}_Q(X, Y)$, and Theorem 3.1 implies that the cokernel of d is isomorphic to $\operatorname{Ext}^1 (X, Y)$.

According to Section 6.4 Chapter 5, every non-zero $\psi \in \operatorname{Ext}^1 (X, Y)$ induces a non-split exact sequence

$$0 \to Y \to Z \to X \to 0.$$

In our situation we can describe the $k(Q)$-module structure on Z precisely. Indeed, consider $\psi \in \bigoplus_{\gamma \in Q_1} \operatorname{Hom}_k \left(X_{s(\gamma)}, Y_{t(\gamma)} \right)$ and denote by ψ_γ the component of $\psi \in \operatorname{Hom}_k \left(X_{s(\gamma)}, Y_{t(\gamma)} \right)$. We set $Z_i = X_i \oplus Y_i$ for every $i \in Q_0$. Furthermore, for every $\gamma \in Q_1$ with source i and target j, we set

$$\gamma (x, y) = (\gamma x, \gamma y + \psi_\gamma x).$$

Obviously, we obtain an exact sequence of $k(Q)$-modules

$$0 \to Y \xrightarrow{i} Z \xrightarrow{\pi} X \to 0,$$

where $i(y) = (0, y)$ and $\pi(x, y) = x$. This exact sequence splits if and only if there exists $\eta \in \operatorname{Hom}_Q (X, Z)$ such that $\pi \circ \eta = \operatorname{Id}$. Note that $\eta = \oplus_{i \in Q_0} \eta_i$ with $\eta_i \in \operatorname{Hom}_k(X_i, Z_i)$ and, for every $x \in X_i$, we have

$$\eta_i(x) = (x, \varphi_i x),$$

for some $\varphi_i \in \operatorname{Hom}_k (X_i, Y_i)$. The condition that η is a morphism of $k(Q)$-modules implies that, for every arrow $\gamma \in Q_1$ with source i and target j, we have

$$\gamma (x, \varphi_i x) = (\gamma x, \gamma \varphi_i x + \psi_\gamma x) = (\gamma x, \varphi_j \gamma x).$$

Hence we have

$$\psi_\gamma x = \varphi_j \gamma x - \gamma \varphi_i x.$$

If we write $\varphi = \oplus_{i \in Q_0} \varphi_i$, then the latter condition is equivalent to $\psi = d\varphi$.

Note also that Theorem 3.1 implies the following:

PROPOSITION 3.4. *In the category of representations of Q,*

$$\operatorname{Ext}^i (X, Y) = 0 \text{ for all } i \geq 2.$$

COROLLARY 3.5. *Let*

$$0 \to Y \to Z \to X \to 0$$

be a short exact sequence of representations of Q, then the maps

$$\text{Ext}^1(V, Z) \to \text{Ext}^1(V, X) \text{ and } \text{Ext}^1(Z, V) \to \text{Ext}^1(Y, V)$$

are surjective for every representation V of Q.

PROOF. Follows from Proposition 3.4 and the long exact sequence for extension groups, Theorem 5.7 Chapter 5. □

LEMMA 3.6. *If X and Y are indecomposable finite-dimensional $k(Q)$-modules and if $\text{Ext}^1(Y, X) = 0$, then every non-zero $\varphi \in \text{Hom}_Q(X, Y)$ is either surjective or injective.*

PROOF. Consider the exact sequences

(7.3) $$0 \to \text{Ker}\,\varphi \to X \xrightarrow{\beta} \text{Im}\,\varphi \to 0,$$

(7.4) $$0 \to \text{Im}\,\varphi \xrightarrow{\delta} Y \to S \cong Y/\text{Im}\,\varphi \to 0.$$

Note that neither sequence splits. Let $\psi \in \text{Ext}^1(S, \text{Im}\,\varphi)$ be the element associated to the sequence (7.4). By Corollary 3.5 and (7.3) we have a surjective map

$$g : \text{Ext}^1(S, X) \to \text{Ext}^1(S, \text{Im}\,\varphi).$$

Let $\psi' \in g^{-1}(\psi)$. Then ψ' induces a non-split exact sequence

$$0 \to X \xrightarrow{\alpha} Z \to S \to 0.$$

This exact sequence and the sequence (7.4) can be arranged in the following commutative diagram

$$
\begin{array}{ccccccccc}
0 & \to & X & \xrightarrow{\alpha} & Z & \to & S & \to & 0 \\
  &   & \downarrow{\beta} & & \downarrow{\gamma} & & \downarrow{\text{Id}} & & \\
0 & \to & \text{Im}\,\varphi & \xrightarrow{\delta} & Y & \to & S & \to & 0,
\end{array}
$$

here β and γ are surjective. We claim that the sequence

(7.5) $$0 \to X \xrightarrow{\alpha + \beta} Z \oplus \text{Im}\,\varphi \xrightarrow{\gamma - \delta} Y \to 0$$

is exact. Indeed, $\alpha + \beta$ is obviously injective and $\gamma - \delta$ is surjective. Furthermore, $\dim Z = \dim X + \dim S$ and $\dim \text{Im}\,\varphi = \dim Y - \dim S$. Therefore,

$$\dim(Z \oplus \text{Im}\,\varphi) = \dim X + \dim Y,$$

and therefore $\text{Ker}(\gamma - \delta) = \text{Im}(\alpha + \beta)$.

By assumption, $\text{Ext}^1(Y, X) = 0$. Hence the exact sequence (7.5) splits, and we have an isomorphism

$$Z \oplus \text{Im}\,\varphi \cong X \oplus Y.$$

By the Krull–Schmidt theorem (Theorem 4.19 Chapter 5) either $X \cong \mathrm{Im}\,\varphi$ and hence φ is injective or $Y \cong \mathrm{Im}\,\varphi$ and hence φ is surjective.   □

3.3. Canonical bilinear form and Euler characteristic.

DEFINITION 3.7. Let $(C_\bullet, d)$ be a complex of finite-dimensional k-vector spaces, we assume that $C_i = \{0\}$ for almost every i. The *Euler characteristic* of the complex is the number

$$\chi(C_\bullet) = \sum_{i \in \mathbb{Z}} (-1)^i \dim_k C_i = \sum_{i \in \mathbb{Z}} (-1)^i \dim_k H_i(C_\bullet).$$

EXERCISE 3.8. Check the second equality.

Let Q be a quiver and X be a finite-dimensional $k(Q)$-module. We use the notation $x = \dim X \in \mathbb{Z}^{Q_0}$ where $x_i = \dim X_i$ for every $i \in Q_0$.
We define a bilinear form on $\mathbb{Z}^{Q_0}$ by the formula

$$\langle x, y \rangle := \sum_{i \in Q_0} x_i y_i - \sum_{\gamma \in Q_1} x_{s(\gamma)} y_{t(\gamma)} = \dim \mathrm{Hom}_Q(X, Y) - \dim \mathrm{Ext}^1(X, Y),$$

where the second equality follows from calculating Euler characteristic in (7.1). The symmetric form

$$(x, y) := \langle x, y \rangle + \langle y, x \rangle$$

is called the *Tits form* of the quiver Q. We will also consider the corresponding quadratic form

$$q(x) := \langle x, x \rangle.$$

4. Bricks

In the rest of this chapter, we assume that the ground field k is algebraically closed and that the representations are finite-dimensional.

Here we discuss further properties of finite-dimensional representations of the path algebra $k(Q)$ of a quiver Q.

DEFINITION 4.1. A $k(Q)$-module X is called a *brick*, if $\mathrm{End}_Q(X) = k$.

EXERCISE 4.2. Show that:
1) if X is a brick, then X is indecomposable.
2) If X is indecomposable and $\mathrm{Ext}^1(X, X) = 0$, then X is a brick (use Lemma 3.6).

EXAMPLE 4.3. Consider the quiver $\bullet \to \bullet$. Then every indecomposable representation is a brick.
For the Kronecker quiver $\bullet \Rightarrow \bullet$ the representation $k^2 \Rightarrow k^2$ with $\rho_{\gamma_1} = \mathrm{Id}$, $\rho_{\gamma_2} = \begin{pmatrix} 0 & 1 \\ 0 & 0 \end{pmatrix}$ is not a brick because $\varphi = (\varphi_1, \varphi_2)$ with $\varphi_1 = \varphi_2 = \begin{pmatrix} 0 & 1 \\ 0 & 0 \end{pmatrix}$ is a non-scalar element in $\mathrm{End}_Q(X)$.

LEMMA 4.4. *Let X be an indecomposable $k(Q)$-module which is not a brick. Then X contains a submodule W, which is a brick, such that $\mathrm{Ext}^1(W, W) \neq 0$.*

PROOF. We will prove the lemma by induction on the length l of X. The base case $l = 1$ is trivial, since in this case X is irreducible and hence it is a brick by Schur's lemma.

Recall that if X is indecomposable and has finite length, then $\varphi \in \mathrm{End}_Q(X)$ is either an isomorphism or nilpotent. Therefore, since k is algebraically closed and X is not a brick, the algebra $\mathrm{End}_Q(X)$ contains a non-zero nilpotent element. Let $\varphi \in \mathrm{End}_Q(X)$ be a non-zero operator of minimal rank. Then φ is nilpotent and $\mathrm{rk}\,\varphi^2 < \mathrm{rk}\,\varphi$, hence $\varphi^2 = 0$.

Let $Y := \mathrm{Im}\,\varphi$, $Z := \mathrm{Ker}\,\varphi$. Clearly, $Y \subset Z$. Consider a decomposition

$$Z = Z_1 \oplus \cdots \oplus Z_p$$

into a sum of indecomposable submodules. Denote by p_i the projection $Z \to Z_i$. Let i be such that $p_i(Y) \neq 0$. Set $\eta := p_i \circ \varphi$, $Y_i := p_i(Y) = \eta(X)$. Note that by our assumption $\mathrm{rk}\,\eta = \mathrm{rk}\,\varphi$, therefore Y_i is isomorphic to Y. Then $\mathrm{Ker}\,\eta = Z$ and $\mathrm{Im}\,\eta = Y_i$.

Note that the exact sequence

$$0 \to Z \to X \xrightarrow{\eta} Y_i \to 0$$

does not split since X is indecomposable. Let X_i be the quotient of X by the submodule $\oplus_{j \neq i} Z_j$ and $\pi : X \to X_i$ be the canonical projection. Then we have the exact sequence

(7.6) $$0 \to Z_i \to X_i \xrightarrow{\bar{\eta}} Y_i \to 0,$$

where $\bar{\eta} := \eta \circ \pi^{-1}$ is well-defined since $\mathrm{Ker}\,\pi \subset \mathrm{Ker}\,\eta$. We claim that (7.6) does not split. Indeed, if it splits, then X_i decomposes into a direct sum $Z_i \oplus L$ for some submodule $L \subset X_i$ which is isomorphic to Y_i. But then $X = Z_i \oplus \pi^{-1}(L)$, which contradicts the indecomposability of X.

Therefore we have shown that $\mathrm{Ext}^1(Y_i, Z_i) \neq 0$. Recall that Y_i is a submodule of Z_i. By Corollary 3.5 we have the surjection

$$\mathrm{Ext}^1(Z_i, Z_i) \to \mathrm{Ext}^1(Y_i, Z_i).$$

Hence $\mathrm{Ext}^1(Z_i, Z_i) \neq 0$.

The length of Z_i is less than the length of X. If Z_i is not a brick, then it contains a submodule W which is a brick by the induction assumption.   □

COROLLARY 4.5. *Assume that Q is a quiver whose Tits form is positive definite. Then every indecomposable representation X of Q is a brick with trivial $\mathrm{Ext}^1(X, X)$. Moreover, if $x = \dim X$, then $q(x) = 1$.*

PROOF. Assume that X is not a brick, then it contains a brick Y such that $\text{Ext}^1(Y, Y) \neq 0$. Then

$$q(y) = \dim \text{End}_Q(Y) - \dim \text{Ext}^1(Y, Y) = 1 - \dim \text{Ext}^1(Y, Y) \leq 0,$$

but this is impossible. Therefore X is a brick. Then

$$q(x) = \dim \text{End}_Q(X) - \dim \text{Ext}^1(X, X) = 1 - \dim \text{Ext}^1(X, X) \geq 0.$$

By positivity of q, we have $q(x) = 1$ and $\dim \text{Ext}^1(X, X) = 0$. $\qquad\square$

5. Orbits in representation varieties

Fix a quiver Q. For an arbitrary $x \in \mathbb{N}^{Q_0}$ consider the space, called the *representation variety* associated to the dimension x,

$$\text{Rep}(x) := \prod_{\gamma \in Q_1} \text{Hom}_k\left(k^{x_{s(\gamma)}}, k^{x_{t(\gamma)}}\right).$$

It is an affine variety of dimension

$$\dim \text{Rep}(x) = \sum_{\gamma \in Q_1} x_{s(\gamma)} x_{t(\gamma)}.$$

Each point $\rho = (\rho_\gamma)_{\gamma \in Q_1}$ in $\text{Rep}(x)$ is a representation of Q of dimension x, and every representation of Q of dimension x is isomorphic to some $\rho \in \text{Rep}(x)$.

Let us consider the group

$$G = \prod_{i \in Q_0} \text{GL}\left(k^{x_i}\right),$$

of dimension

$$\dim G = \sum_{i \in Q_0} x_i^2,$$

and define an action of G on $\text{Rep}(x)$ by the formula

$$g\rho_\gamma := g_{t(\gamma)} \rho_\gamma g_{s(\gamma)}^{-1} \quad \text{for every} \quad \gamma \in Q_1.$$

Two representations ρ and ρ' of Q are isomorphic if and only if they belong to the same orbit under G. In other words we have a bijection between isomorphism classes of representations of Q of dimension x and G-orbits in $\text{Rep}(x)$. For a representation X we denote by O_X the corresponding G-orbit in $\text{Rep}(x)$.

Note that we have, by definition:

(7.7) $$\dim \text{Rep}(x) - \dim G = -q(x).$$

Let us formulate certain properties of the G-action on $\text{Rep}(x)$, without proof. They follow from the general theory of algebraic groups, see for instance [21]. We work with the Zariski topology.

- Each orbit is open in its closure;

- if O and O' are two distinct orbits and O' is included in the Zariski closure of O, then $\dim O' < \dim O$;
- if (X, ρ) is a representation of Q, then $\dim O_X = \dim G - \dim \mathrm{Stab}_X$, where Stab_X denotes the stabilizer of ρ.

We denote by $\mathrm{Aut}_Q(X)$ the group of invertible elements in $\mathrm{End}_Q(X)$.

LEMMA 5.1. *For any representation (X, ρ) of dimension x, we have $\mathrm{Stab}_X = \mathrm{Aut}_Q(X)$ and*

$$\dim \mathrm{Stab}_X = \dim \mathrm{Aut}_Q(X) = \dim \mathrm{End}_Q(X).$$

PROOF. The condition that $\varphi \in \mathrm{End}_Q(X)$ is not invertible is given by the polynomial equation

$$\prod_{i \in Q_0} \det \varphi_i = 0.$$

Since $\mathrm{Aut}_Q(X)$ is not empty and open in $\mathrm{End}_Q(X)$, $\mathrm{Aut}_Q(X)$ and $\mathrm{End}_Q(X)$ have the same dimension. $\qquad\square$

COROLLARY 5.2. *If (X, ρ) is a representation of Q and $\dim X = x$, then*

$$\mathrm{codim}_{\mathrm{Rep}(x)} O_X = -q(x) + \dim \mathrm{End}_Q(X) = \dim \mathrm{Ext}^1(X, X).$$

LEMMA 5.3. *Let (Z, τ) be a non-trivial extension of (Y, σ) by (X, ρ), i.e. there is a non-split exact sequence*

$$0 \to X \to Z \to Y \to 0.$$

Then $O_{X \oplus Y}$ is included in the closure of O_Z and $O_{X \oplus Y} \neq O_Z$.

PROOF. Following the construction just before Proposition 3.4 in Section 3.2, for every $i \in Q_0$, we consider a decomposition $Z_i = X_i \oplus Y_i$ such that, for every $\gamma \in Q_1$ and $(x, y) \in X_{s(\gamma)} \oplus Y_{s(\gamma)}$,

$$\tau_\gamma(x, y) = (\rho_\gamma(x) + \psi_\gamma(y), \sigma_\gamma(y))$$

for some $\psi_\gamma \in \mathrm{Hom}(Y_{s(\gamma)}, X_{t(\gamma)})$.

Next, for every $\lambda \in k \setminus 0$ define $g^\lambda \in G$ by setting for every $i \in Q_0$

$$g_i^\lambda|_{X_i} = \mathrm{Id}_{X_i}, \quad g_i^\lambda|_{Y_i} = \lambda \, \mathrm{Id}_{Y_j}.$$

Then we have

$$g^\lambda \tau_\gamma(x, y) = (\rho_\gamma(x) + \lambda \psi_\gamma(y), \sigma_\gamma(y)).$$

The latter formula makes sense even for $\lambda = 0$ and $g^0 \tau$ lies in the closure of $\{g^\lambda \tau \mid \lambda \in k \setminus 0\}$. Furthermore $g^0 \tau$ is the direct sum $X \oplus Y$. Hence $O_{X \oplus Y}$ is included in the closure of O_Z.

It remains to check that $X \oplus Y$ is not isomorphic to Z. This follows immediately from the inequality

$$\dim \mathrm{Hom}_Q(Y, Z) < \dim \mathrm{Hom}_Q(Y, X \oplus Y).$$

$\qquad\square$

The following corollary is straightforward.

COROLLARY 5.4. *If the orbit O_X is closed in $\mathrm{Rep}(x)$, then X is semisimple.*

COROLLARY 5.5. *Let (X, ρ) be a representation of Q and $X = \bigoplus_{j=1}^{m} X_j$ be a decomposition into direct sum of indecomposable submodules. If O_X is an orbit of maximal dimension in $\mathrm{Rep}(x)$, then $\mathrm{Ext}^1(X_i, X_j) = 0$ for all $i \neq j$.*

PROOF. If $\mathrm{Ext}^1(X_i, X_j) \neq 0$, then by Lemma 5.3 we can construct a representation (Z, τ) such that O_X is in the closure of O_Z. Then $\dim O_X < \dim O_Z$, which contradicts the maximality of $\dim O_X$. □

6. Coxeter–Dynkin and affine graphs

6.1. Definition and properties. Let Γ be a *connected* non-oriented graph with vertices Γ_0 and edges Γ_1. We define the Tits form $(\cdot, \cdot)$ on $\mathbb{Z}^{\Gamma_0}$ by

$$(x, y) := \sum_{i \in \Gamma_0} (2 - l(i)) x_i y_i - \sum_{(i,j) \in \Gamma_1} x_i y_j,$$

where $l(i)$ is the number of loops at i. If we equip each edge of Γ with an orientation then this symmetric form coincides with the Tits form of the corresponding quiver. We define a quadratic form q on $\mathbb{Z}^{\Gamma_0}$ by

$$q(x) := \frac{(x, x)}{2}.$$

By $\{\varepsilon_i \mid i \in \Gamma_0\}$ we denote the standard basis in $\mathbb{Z}^{\Gamma_0}$. If Γ does not have loops, then $(\varepsilon_i, \varepsilon_i) = 2$ for all $i \in \Gamma_0$. If $i, j \in \Gamma_0$ and $i \neq j$, then $(\varepsilon_i, \varepsilon_j)$ equals minus the number of edges between i and j. The matrix of the form $(\cdot, \cdot)$ in the standard basis is called the *Cartan matrix* of Γ.

EXAMPLE 6.1. The Cartan matrix of the graph $\bullet\!-\!\bullet$ is $\begin{pmatrix} 2 & -1 \\ -1 & 2 \end{pmatrix}$. The Cartan matrix of the loop is (0).

DEFINITION 6.2. A connected graph Γ is called *Coxeter–Dynkin* if its Tits form $(\cdot, \cdot)$ is positive definite and *affine* if $(\cdot, \cdot)$ is positive semidefinite but not positive definite. If Γ is neither Coxeter–Dynkin nor affine, then we say that it is of *indefinite type*.

REMARK 6.3. For an affine graph Γ the form $(\cdot, \cdot)$ is always degenerate. Furthermore

(7.8) $$\mathrm{Ker}(\cdot, \cdot) = \{x \in \mathbb{Z}^{Q_0} \mid (x, x) = 0\}.$$

LEMMA 6.4. (a) *If Γ is affine then the kernel of $(\cdot, \cdot)$ is of the form $\mathbb{Z}\delta$, for some $\delta \in \mathbb{N}^{\Gamma_0}$ with all $\delta_i > 0$.*
(b) *If Γ is of indefinite type, then there exists $x \in \mathbb{N}^{\Gamma_0}$ such that $(x, x) < 0$.*

PROOF. Let $x \in \mathbb{Z}^{Q_0}$. We define the *support* of x, $\operatorname{supp} x$, to be the set of vertices $i \in Q_0$ such that $x_i \neq 0$. Let $|x|$ be defined by the condition $|x|_i = |x_i|$ for all $i \in Q_0$. Note that $\operatorname{supp} x = \operatorname{supp} |x|$ and that, by definition of $(\cdot, \cdot)$, we have

$$(7.9) \qquad\qquad (|x|, |x|) \leq (x, x).$$

To prove (b) we just notice that if Γ is of indefinite type, then there exists $x \in \mathbb{Z}^{Q_0}$ such that $(x, x) < 0$. But then (7.9) implies $(|x|, |x|) < 0$.

Now let us prove (a). Let $\delta \in \operatorname{Ker}(\cdot, \cdot)$ and $\delta \neq 0$. Then (7.9) and (7.8) imply that $|\delta|$ also lies in $\operatorname{Ker}(\cdot, \cdot)$. Next we prove that $\operatorname{supp} \delta = Q_0$. Indeed, otherwise we can choose $i \in Q_0 \setminus \operatorname{supp} \delta$ such that i is connected with at least one vertex in $\operatorname{supp} \delta$. Then $(\varepsilon_i, \delta) < 0$, therefore

$$(\varepsilon_i + 2\delta, \varepsilon_i + 2\delta) \leq 2 + 4(\varepsilon_i, \delta) < 0$$

and Γ is not affine.

Finally let $\delta', \delta \in \operatorname{Ker}(\cdot, \cdot)$. Since $\operatorname{supp} \delta = \operatorname{supp} \delta' = Q_0$, one can find $a, b \in \mathbb{Z}$ such that $\operatorname{supp}(a\delta + b\delta') \neq Q_0$. Then $a\delta + b\delta' = 0$. Hence $\operatorname{Ker}(\cdot, \cdot)$ is one-dimensional and the proof of (a) is complete. $\qquad\square$

Note that (a) implies the following

COROLLARY 6.5. *Let Γ be a Coxeter–Dynkin or affine graph. Any proper connected subgraph of Γ is a Coxeter–Dynkin graph.*

DEFINITION 6.6. A non-zero vector $x \in \mathbb{Z}^{\Gamma_0}$ is called a *root* if $q(x) \leq 1$. Note that for every $i \in \Gamma_0$, ε_i is a root. It is called a *simple root*.

EXERCISE 6.7. Let Γ be a connected graph. Show that the number of roots is finite if and only if Γ is a Coxeter–Dynkin graph.

LEMMA 6.8. *Let Γ be a Coxeter–Dynkin or affine graph. If x is a root, then either all $x_i \geq 0$ or all $x_i \leq 0$.*

PROOF. Assume that the statement is false. Let

$$I^+ := \{i \in \Gamma_0 \mid x_i > 0\}, \quad I^- := \{i \in \Gamma_0 \mid x_i < 0\}, \quad x^{\pm} = \sum_{i \in I^{\pm}} x_i \varepsilon_i.$$

Then $x = x^+ + x^-$ and $(x^+, x^-) \geq 0$. Furthermore, since Γ is a Coxeter–Dynkin or affine graph, we have $q(x^{\pm}) > 0$. Therefore

$$q(x) = q(x^+) + q(x^-) + (x^+, x^-) > 1.$$

$\qquad\square$

We call a root x *positive* (resp. *negative*) if $x_i \geq 0$ (resp. $x_i \leq 0$) for all $i \in \Gamma_0$.

6.2. Classification. The following is a complete list of Coxeter–Dynkin graphs (below n is the number of vertices).

A_n

D_n

E_6

E_7

E_8

The affine graphs, except the non-oriented loop

$\tilde{A}_0$

are obtained from the Coxeter–Dynkin graphs by adding a vertex (see Corollary 6.5). Here they are.

$\tilde{A}_1$

For $n > 1$, $\tilde{A}_n$ is a cycle with $n + 1$ vertices. In this case $\delta = (1, \ldots, 1)$. In what follows the numbers are the coordinates of δ.

$\tilde{D}_n$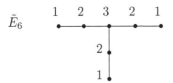

$\tilde{E}_6$

$\tilde{E}_7$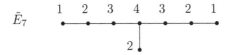

$\tilde{E}_8$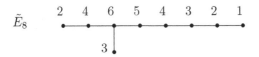

The proof that these graphs form a complete list is presented below in the exercises.

EXERCISE 6.9. Check that A_n, D_n, E_6, E_7, E_8 are Coxeter–Dynkin graphs using the Sylvester criterion and the fact that every subgraph of a Coxeter–Dynkin graph is Coxeter–Dynkin. One can calculate inductively the determinant of the corresponding Cartan matrices. It is $n + 1$ for A_n, 4 for D_n, 3 for E_6, 2 for E_7 and 1 for E_8.

EXERCISE 6.10. Check that the Cartan matrices of $\tilde{A}_n, \tilde{D}_n, \tilde{E}_6, \tilde{E}_7, \tilde{E}_8$ have corank 1 and that every proper connected subgraph is Coxeter–Dynkin. Conclude that these graphs are affine.

EXERCISE 6.11. Let Γ be a Coxeter–Dynkin graph. Using Corollary 6.5, prove that Γ does not have loops, cycles or multiple edges. Prove that Γ has no vertices of degree 4 and at most one vertex of degree 3.

EXERCISE 6.12. Let a Coxeter–Dynkin graph Γ have a vertex of degree 3. Let p, q and r be the lengths of "legs" coming from this vertex. Prove that $\frac{1}{p} + \frac{1}{q} + \frac{1}{r} > 1$. Use this to complete the classification of Coxeter–Dynkin graphs.

EXERCISE 6.13. Complete the classification of affine graphs using Corollary 6.5, Exercise 6.10 and Exercise 6.12.

7. Quivers of finite type and Gabriel's theorem

Recall that a quiver is of finite type if it has finitely many isomorphism classes of indecomposable representations.

EXERCISE 7.1. Prove that a quiver is of finite type if and only if all its connected components are of finite type.

THEOREM 7.2. (Gabriel) Let Q be a connected quiver and Γ be its underlying graph. Then

(1) The quiver Q is of finite type if and only if Γ is a Coxeter–Dynkin graph.
(2) Assume that Γ is a Coxeter–Dynkin graph and (X, ρ) is an indecomposable representation of Q. Then $\dim X$ is a positive root.
(3) If Γ is a Coxeter–Dynkin graph, then for every positive root $x \in \mathbb{Z}^{Q_0}$ there is exactly one indecomposable representation of Q of dimension x.

PROOF. Let us first prove that if Q is of finite type then Γ is a Coxeter–Dynkin graph. Indeed, if Q is of finite type, then for every $x \in \mathbb{N}^{Q_0}$, Rep (x) has finitely many G-orbits. Therefore Rep (x) must contain an open orbit. Assume that Q is not Coxeter–Dynkin. Then there exists a non-zero $x \in \mathbb{Z}^{Q_0}$ such that $q(x) \le 0$. Let $O_X \subset \text{Rep}(x)$ be an open orbit. Then codim $O_X = 0$. But by Corollary 5.2

(7.10) $$\text{codim}\, O_X = \dim \text{End}_Q(X) - q(x) > 0.$$

This is a contradiction.

Now assume that Γ is Coxeter–Dynkin. To show that Q is of finite type it suffices to prove assertions (2) and (3).

Note that (2) follows from Corollary 4.5.

Suppose that x is a positive root. Let (X, ρ) be a representation of Q such that $\dim O_X$ in $\mathrm{Rep}\,(x)$ is maximal. Let us prove that X is indecomposable. Indeed, let $X = X_1 \oplus \cdots \oplus X_s$ be a sum of indecomposable bricks. Then by Corollary 5.5 $\mathrm{Ext}^1\,(X_i, X_j) = 0$. Therefore $q\,(x) = s = 1$ and X is indecomposable.

Finally, if (X, ρ) is an indecomposable representation of Q, then (7.10) implies that O_X is an open orbit in $\mathrm{Rep}\,(x)$. Since $\mathrm{Rep}\,(x)$ is irreducible, it contains at most one open orbit. Hence (3) is proven. $\square$

REMARK 7.3. Gabriel's theorem implies that for a quiver, the property of being of finite type depends only on the underlying graph and does not depend on the orientation.

REMARK 7.4. Theorem 7.2 does not provide an algorithm for finding all indecomposable representations of quivers with Coxeter–Dynkin underlying graphs. We give such an algorithm in the next chapter using the reflection functors.

EXERCISE 7.5. Let Q be a quiver whose underlying graph is A_n. Check that the positive roots are in bijection with the connected subgraphs of A_n. For each positive root x, give a precise construction of an indecomposable representation of dimension x.

CHAPTER 8

Representations of Dynkin and affine quivers

In which we encounter reflection and Coxeter functors and make use of them.

1. Reflection functors

For a quiver Q we denote by mod_Q the category of representations of Q. Let i be a vertex of Q. We denote by $Q_+(i)$ the set of all arrows of Q with target i and by $Q_-(i)$ the set of all arrows of Q with source i. We say that a vertex $i \in Q_0$ is $(+)$-*admissible* if $Q_-(i) = \emptyset$ and $(-)$-*admissible* if $Q_+(i) = \emptyset$. By $\sigma_i(Q)$, we denote the quiver obtained from Q by inverting all arrows belonging to $Q_+(i) \cup Q_-(i)$. Note that we have

$$\sigma_i(Q)_-(i) = Q_+(i), \quad \sigma_i(Q)_+(i) = Q_-(i),$$

if we forget orientation of arrows.

Let $i \in Q_0$ be a $(+)$-admissible vertex and $Q' := \sigma_i(Q)$. Let us introduce the *reflection functor* $F_i^+ : \mathrm{mod}_Q \to \mathrm{mod}_{Q'}$. For every representation (X, ρ) of Q, we define $(X', \rho') := F_i^+(X, \rho)$ as follows: if $j \neq i$, then we set $X_j' := X_j$; if $\gamma \notin Q_+(i)$, then we set $\rho_\gamma' := \rho_\gamma$. Next we consider the map

$$h = \sum_{\gamma \in Q_+(i)} \rho_\gamma : \bigoplus_{\gamma \in Q_+(i)} X_{s(\gamma)} \to X_i$$

and set $X_i' := \mathrm{Ker}\, h$. Finally for every $\gamma \in Q_-'(i)$, we define $\rho_\gamma' : X_i' \to X_{t(\gamma)}$ to be the restriction of the canonical projection $\mathrm{Ker}\, h \to X_{t(\gamma)}$. One defines the action of F_i^+ on morphisms in the natural way.

If $i \in Q_0$ is a $(-)$-admissible vertex, we define the *reflection functor* F_i^- from the category of representations of Q to the category of representations of $Q' := \sigma_i(Q)$ in the following way: we set $(X', \rho') = F_i^-(X, \rho)$, where $X_j' = X_j$ for all $j \neq i$, $\rho_\gamma' = \rho_\gamma$ for all $\gamma \notin Q_-(i)$, $X_i' := \mathrm{Coker}\, \tilde{h}$, where

$$\tilde{h} = \sum_{\gamma \in Q_-(i)} \rho_\gamma : X_i \to \bigoplus_{\gamma \in Q_-(i)} X_{t(\gamma)},$$

and for every $\gamma \in Q_+'(i)$, $\rho_\gamma' : X_{s(\gamma)} \to X_i'$ is the composition

$$X_{s(\gamma)} \hookrightarrow \bigoplus_{\gamma' \in Q_-(i)} X_{s(\gamma')} \to X_i'.$$

C. Gruson and V. Serganova, *A Journey Through Representation Theory*,
Universitext, https://doi.org/10.1007/978-3-319-98271-7_8

EXAMPLE 1.1. Let Q be the quiver $\bullet \to \bullet$. Then $\sigma_1(Q) = \sigma_2(Q) = Q^{op}$. If (X, ρ) is the representation presented by the diagram $k \to 0$, $F_1^-(X, \rho)$ is the zero representation and $F_2^+(X, \rho)$ is presented by the diagram $k \xleftarrow{\text{id}} k$.

EXERCISE 1.2. Let Q be a quiver. Denote by Q^{op} the quiver with the same set of vertices and with all arrows reversed. Define the contravariant *duality functor* $D : \text{mod}_Q \to \text{mod}_{Q^{op}}$ by setting

$$D(X_j) = X_j^* \ (j \in Q_0), \quad D(\rho_\gamma) = \rho_\gamma^* \ (\gamma \in Q_1).$$

Show that if $i \in Q_0$ is $(+)$-admissible, then

$$D \circ F_i^+ = F_i^- \circ D.$$

EXERCISE 1.3. Let $i \in Q_0$ and denote by (L_i, ρ_i) the irreducible representation of Q which has k at the vertex i and zero at all other vertices. Show that if i is $(+)$-admissible, then $F_i^+(L_i) = 0$ and if i is $(-)$-admissible, then $F_i^-(L_i) = 0$.

Let $Q' = \sigma_i(Q)$, $i \in Q_0$ be $(+)$-admissible, X be a representation of Q' and Y be a representation of Q, set $X' = F_i^- X$ and $Y' = F_i^+ Y$. Let $\eta \in \text{Hom}_Q(X, Y')$, we define $\chi \in \text{Hom}_{Q'}(X', Y)$ by setting $\chi_j = \eta_j$ for $j \neq i$ and deducing χ_i from the following commutative diagram

$$
\begin{array}{ccccccc}
X_i & \xrightarrow{\widetilde{h}} & \oplus X_j & \xrightarrow{h} & X_i' & \to & 0 \\
\downarrow{\eta_i} & & \downarrow{\oplus \eta_j} & & \downarrow{\chi_i} & & \\
0 \to \ Y_i' & \xrightarrow{\widetilde{h}} & \oplus Y_j' & \xrightarrow{h} & Y_i & &
\end{array}
$$

Note that χ_i is uniquely determined by η. Similarly, for each $\chi \in \text{Hom}_Q(X', Y)$, one can define $\eta \in \text{Hom}_{Q'}(X, Y')$. A routine check now proves the following:

LEMMA 1.4. *Let $Q' = \sigma_i(Q)$, X be a representation of Q' and Y be a representation of Q, then*

$$\text{Hom}_Q\left(F_i^- X, Y\right) \cong \text{Hom}_{Q'}\left(X, F_i^+ Y\right).$$

This means that the functors F_i^+ and F_i^- are mutually adjoint and it implies:

LEMMA 1.5. *The functor F_i^+ is left-exact and the functor F_i^- is right-exact.*

PROOF. Indeed, let $X' \to X \to X'' \to 0$ be an exact sequence of $k(Q')$-modules and let Y be a $k(Q)$-module. The sequence

$$0 \to \text{Hom}_{Q'}(X'', F_i^+(Y)) \to \text{Hom}_{Q'}(X, F_i^+(Y)) \to \text{Hom}_{Q'}(X', F_i^+(Y))$$

is exact since Hom is left-exact, and is isomorphic to

$$0 \to \text{Hom}_Q(F_i^-(X''), Y) \to \text{Hom}_Q(F_i^-(X), Y) \to \text{Hom}_Q(F_i^-(X'), Y).$$

Since Y is arbitrary, this implies the exactness of the sequence

$$F_i^-(X') \to F_i^-(X) \to F_i^-(X'') \to 0.$$

The left-exactness of F_i^+ can be proven similarly.   □

THEOREM 1.6. *Let (X, ρ) be an indecomposable representation of Q.*

(1) *If i is a $(+)$-admissible vertex of Q then $(X', \rho') := F_i^+(X, \rho)$ is either indecomposable or zero. Furthermore, $F_i^+(X, \rho)$ is zero if and only if X is isomorphic to the irreducible representation (L_i, ρ_i) (notation of Exercise 1.3).*

(2) *If (X, ρ) is not isomorphic to (L_i, ρ_i), then*

$$\dim X_i' = -\dim X_i + \sum_{\gamma \in Q_+(i)} \dim X_{s(\gamma)}$$

and $F_i^-(X', \rho')$ is isomorphic to (X, ρ).

PROOF. Note that if (X, ρ) is not isomorphic to (L_i, ρ_i), then h must be surjective by indecomposability of (X, ρ). This implies the dimension formula. Furthermore, we have the following exact sequence

(8.1) $$0 \to X_i' \xrightarrow{\tilde{h}} \bigoplus_{\gamma \in Q_+(i)} X_{s(\gamma)} \xrightarrow{h} X_i \to 0$$

and thus we have an isomorphism

$$\left(F_i^- X'\right)_i = \operatorname{Coker} \tilde{h} \cong X_i.$$

Lemma 1.4 implies that there is a canonical non-zero homomorphism

$$\varphi : F_i^- F_i^+(X) \to X,$$

If $j \neq i$, then $\varphi : X_j \to X_j$ is the identity. Hence by above we obtain that $F_i^-(X', \rho')$ is isomorphic to (X, ρ).

The analogous statement for F_i^- is easy to obtain using the functor $D : \operatorname{Rep} Q \to \operatorname{Rep} Q^{\operatorname{op}}$, from Exercise 1.2.

Let us show next that (X', ρ') is indecomposable. Indeed, if $X' = \oplus_s Y^s$ is a sum of indecomposable submodules, then $F_i^-(Y^s) = 0$ for all but one $s = s_0$. Hence Y^s is isomorphic to the simple module L_i' concentrated at the vertex i. On the other hand $\tilde{h}$ is injective and therefore X' cannot have a direct summand isomorphic to L_i'.   □

EXERCISE 1.7.   (1) Show that F_i^+ (resp. F_i^-) defines a surjective map

$$\operatorname{Hom}_Q(X, X') \to \operatorname{Hom}_{Q'}(F_i^+(X), F_i^+(X'))$$

(resp. $\operatorname{Hom}_{Q'}(Y, Y') \to \operatorname{Hom}_Q(F_i^-(Y), F_i^-(Y'))$).

(2) Deduce from Lemma 1.4 and Theorem 1.6 that there is a canonical injection

$$\varphi : F_i^- F_i^+(X) \to X,$$

and a canonical surjection

$$\psi : Y \to F_i^+ F_i^-(Y).$$

2. Reflection functors and change of orientation

LEMMA 2.1. *Let Γ be a connected graph without cycles (including no loops or multiple edges), Q and Q' be two quivers on Γ. Then there exists a sequence of reflections $\sigma_{i_1}, \ldots, \sigma_{i_s}$ such that $Q' = \sigma_{i_s} \circ \cdots \circ \sigma_{i_1}(Q)$ and every $j \leq s$ is a $(+)$-admissible vertex for $\sigma_{i_{j-1}} \circ \cdots \circ \sigma_{i_1}(Q)$.*

PROOF. It is sufficient to prove the statement when Q and Q' differ by one arrow $\gamma \in Q_1$. After removing γ, Q splits in two connected components; let Q'' be the component which contains $t(\gamma)$. Since Q'' does not contain any cycle, it is possible to enumerate its vertices in such a way that if $i \to j \in Q_1''$, then $i > j$. Let s be the cardinality of Q_0'', then one can check that $Q' = \sigma_s \circ \cdots \circ \sigma_1(Q)$ and that i is a $(+)$-admissible vertex for $\sigma_{i-1} \circ \cdots \circ \sigma_1(Q)$.   □

THEOREM 2.2. *Let i be a $(+)$-admissible vertex for Q and set $Q' = \sigma_i(Q)$. Then F_i^+ and F_i^- establish a bijection between the indecomposable representations of Q which are not isomorphic to L_i and the indecomposable representations of Q' which are not isomorphic to L_i', where by L_i' we denote the irreducible representation of Q' attached to the vertex i.*

Theorem 2.2 follows from Theorem 1.6. Together with Lemma 2.1 it allows to change the orientation on any quiver whose underlying graph has no cycles. Furthermore, it is sufficient to classify indecomposable representations for one particular orientation to obtain the classification of indecomposables for all possible orientations.

3. Weyl group and reflection functors.

Given any graph Γ without loops, one can associate with it a certain linear group, which is called the *Weyl group* of Γ. We denote by $\varepsilon_1, \ldots, \varepsilon_n$ the vectors in the standard basis of $\mathbb{Z}^{\Gamma_0}$, $\varepsilon_i = \dim L_i$ corresponds to the vertex i. These vectors are called *simple roots*. For each simple root ε_i, set

$$r_i(x) = x - \frac{2(x, \varepsilon_i)}{(\varepsilon_i, \varepsilon_i)} \varepsilon_i = x - (x, \varepsilon_i)\varepsilon_i.$$

The second equality follows from $(\varepsilon_i, \varepsilon_i) = 2$ as Γ has no loops. One can check that r_i preserves the scalar product and $r_i^2 = id$. The linear transformation r_i is called a *simple reflection*. Note that r_i also preserves the lattice generated by simple roots. Hence r_i maps roots to roots. If Γ is Dynkin, the scalar product is positive definite, and r_i is the reflection across the hyperplane orthogonal to ε_i. The *Weyl group* W is the group generated by $r_1, \ldots, r_n$. For a Dynkin diagram, W is finite (since the number of roots is finite).

EXAMPLE 3.1. Let $\Gamma = A_n$. Let $e_1, \ldots, e_{n+1}$ be an orthonormal basis in $\mathbb{R}^{n+1}$. Then one can take the roots of Γ to be $e_i - e_j$, the simple roots to be $\varepsilon_1 := e_1 - e_2$,

$\varepsilon_2 := e_2 - e_3, \ldots, \varepsilon_n := e_n - e_{n+1}$. Note that the action of the simple reflection r_i on the orthonormal basis is the following: $r_i(e_j) = e_j$ if $j \neq i, i+1$, $r_i(e_i) = e_{i+1}$ and $r_i(e_{i+1}) = e_i$. Therefore W is isomorphic to the permutation group S_{n+1}.

One can check by direct calculation, that Theorem 1.6 (2) implies

LEMMA 3.2. *If X is an indecomposable representation of Q with a dimension vector $x \neq \varepsilon_i$ for a given i, then $\dim F_i^{\pm} X = r_i(x)$.*

The element $c = r_n \cdots r_1 \in W$ is called a *Coxeter transformation* or a *Coxeter element*. It depends on the enumeration of simple roots.

EXAMPLE 3.3. In the case $\Gamma = A_n$, every Coxeter element is a cycle of length $n+1$.

LEMMA 3.4. *If $c(x) = x$, then $(x, \varepsilon_i) = 0$ for all i. In particular, for a Dynkin graph, $c(x) = x$ implies $x = 0$.*

PROOF. By definition,

$$c(x) = x + a_1 \varepsilon_1 + \cdots + a_n \varepsilon_n, \ a_i = -\frac{2(\varepsilon_i, x + a_1 \varepsilon_1 + \ldots a_{i-1}\varepsilon_{i-1})}{(\varepsilon_i, \varepsilon_i)}.$$

The condition $c(x) = x$ implies all $a_i = 0$. Hence $(x, \varepsilon_i) = 0$ for all i. □

4. Coxeter functors.

Let Q be a quiver without oriented cycles. An enumeration of its vertices is called *admissible* if $i > j$ for any arrow $i \to j$. Such an enumeration always exists. One can easily see that, in this situation, every vertex i is $(+)$-admissible for $\sigma_{i-1} \circ \cdots \circ \sigma_1(Q)$ and $(-)$-admissible for $\sigma_{i+1} \circ \cdots \circ \sigma_n(Q)$. Furthermore,

$$Q = \sigma_n \circ \sigma_{n-1} \circ \cdots \circ \sigma_1(Q) = \sigma_1 \circ \cdots \circ \sigma_n(Q).$$

Indeed, every arrow $i \to j$ of Q is reversed exactly two times by both sequences of transformations.

We define the *Coxeter functors* $\Phi^{\pm} : \text{mod}_Q \to \text{mod}_Q$ by:

$$\Phi^+ := F_n^+ \circ \cdots \circ F_2^+ \circ F_1^+, \ \Phi^- := F_1^- \circ F_2^- \circ \cdots \circ F_n^-.$$

LEMMA 4.1. (1) *One has: $\text{Hom}_Q(\Phi^- X, Y) \cong \text{Hom}_Q(X, \Phi^+ Y)$;*
(2) *if X is indecomposable and $\Phi^+ X \neq 0$, then $\Phi^- \Phi^+ X \cong X$;*
(3) *if X is indecomposable of dimension x and $\Phi^+ X \neq 0$, then $\dim \Phi^+ X = c(x)$;*
(4) *if Q is Dynkin, then, for any indecomposable X, there exists k such that $(\Phi^+)^k X = 0$.*

PROOF. (1) follows from Lemma 1.4, (2) follows from Theorem 1.6, (3) follows from Lemma 3.2. Let us prove (4). Since the Weyl group W is finite, c has finite order h. It is sufficient to show that for any x there exists k such that $c^k(x)$ is not positive. Assume that this is not true. Then $y = x + c(x) + \cdots + c^{h-1}(x) > 0$ is c-invariant. Contradiction with Lemma 3.4. $\square$

LEMMA 4.2. *The functors $\Phi^{\pm}$ do not depend on the choice of the admissible enumeration.*

PROOF. Note that if i and j are distinct and both $(+)$-admissible (resp. $(-)$-admissible), then $F_i^+ \circ F_j^+ = F_j^+ \circ F_i^+$ (resp. $F_i^- \circ F_j^- = F_j^- \circ F_i^-$). If a sequence $i_1, \ldots, i_n$ gives another admissible enumeration of vertices, and $i_k = 1$, then 1 is distinct from $i_1, \ldots, i_{k-1}$, hence

$$F_1^+ \circ F_{i_{k-1}}^+ \circ \cdots \circ F_{i_1}^+ = F_{i_{k-1}}^+ \circ \cdots \circ F_{i_1}^+ \circ F_1^+.$$

We then proceed by induction. The proof is similar for Φ^-. $\square$

In what follows we usually assume that our enumeration of vertices is admissible.

COROLLARY 4.3. *Let Q be a Dynkin quiver, the dimension function defines a bijection between the set of positive roots of Q and the set of isomorphism classes of indecomposable representations of Q.*

PROOF. Let X be an indecomposable representation of dimension x, and k be the minimal number such that $c^{k+1}(x)$ is not positive. Then there exists a unique vertex i such that

$$x = c^{-k} r_1 \ldots r_{i-1}(\varepsilon_i), \ X \cong \left(\Phi^-\right)^k \circ F_1^- \circ \cdots \circ F_{i-1}^-(L_i'),$$

where L_i' is the simple $k(Q')$-module, attached to the vertex i, for $Q' := \sigma_{i-1} \ldots \sigma_1(Q)$. In particular, x is a positive root and for every positive root x, there is a unique indecomposable representation of dimension x (up to isomorphism). $\square$

5. Further properties of Coxeter functors

Here we assume again that Q is a quiver without oriented cycles or loops and that the enumeration of vertices is admissible. We discuss the properties of the bilinear form $\langle, \rangle$ (see Definition 3.7 in Chapter 7). Since we plan to change the orientation of Q we use a subindex $\langle, \rangle_Q$, whenever it is needed to avoid ambiguity.

LEMMA 5.1. *Let i be a $(+)$-admissible vertex, $Q' = \sigma_i(Q)$, and denote by $\langle, \rangle_Q$ and $\langle, \rangle_{Q'}$ the corresponding bilinear forms. Then*

$$\langle r_i(x), y \rangle_{Q'} = \langle x, r_i(y) \rangle_Q.$$

PROOF. It suffices to check the formula for a subquiver containing i and all its neighbours. Let $x' = r_i(x)$ and $y' = r_i(y)$. Then

$$x_i' = -x_i + \sum_{i \neq j} x_j, \ y_i' = -y_i + \sum_{i \neq j} y_j,$$

$$\langle x', y \rangle_{Q'} = x_i' y_i - x_i' \sum_{i \neq j} y_j + \sum_{i \neq j} x_j y_j = -x_i' y_i' + \sum_{i \neq j} x_j y_j,$$

$$\langle x, y' \rangle_{Q} = x_i y_i' - y_i' \sum_{i \neq j} x_j + \sum_{i \neq j} x_j y_j = -x_i' y_i' + \sum_{i \neq j} x_j y_j.$$

$\square$

COROLLARY 5.2. *Let X be an indecomposable $k(Q)$-module, and Y be an $k(Q')$-module, where $Q' = \sigma_i(Q)$ for some $(-)$-admissible vertex i of Q. Assume that $F_i^- X \neq 0$ and $F_i^+ Y \neq 0$. Then*

$$\operatorname{Ext}_Q^1(X, F_i^+ Y) \simeq \operatorname{Ext}_{Q'}^1(F_i^- X, Y).$$

PROOF. Let x and y be the dimensions of X and Y respectively. Then $r_i(x) = \dim F_i^+(X)$ and $r_i(y) = \dim F_i^-(X)$ and we have $\langle r_i(x), y \rangle_{Q'} = \langle x, r_i(y) \rangle_Q$. Recall that

$$\langle x, r_i(y) \rangle_Q = \dim \operatorname{Hom}_Q\left(X, F_i^+ Y\right) - \dim \operatorname{Ext}_Q^1\left(X, F_i^+ Y\right)$$

and

$$\langle r_i(x), y \rangle_{Q'} = \dim \operatorname{Hom}_{Q'}\left(F_i^- X, Y\right) - \dim \operatorname{Ext}_{Q'}^1\left(F_i^- X, Y\right).$$

Now the statement follows from Lemma 1.4. $\square$

COROLLARY 5.3. *For a Coxeter element c, we have*

$$\left\langle c^{-1}(x), y \right\rangle_Q = \langle x, c(y) \rangle_Q.$$

If X and Y are two indecomposable representations of Q and $\Phi^+ Y \neq 0$, $\Phi^- X \neq 0$ then

$$\operatorname{Ext}_Q^1\left(X, \Phi^+ Y\right) \simeq \operatorname{Ext}_Q^1\left(\Phi^- X, Y\right).$$

Let $A = k(Q)$ be the path algebra of Q (see Chapter 7, Section 2). Recall that any indecomposable projective module is isomorphic to Ae_i for some vertex i of Q.

LEMMA 5.4. *Let j be a $(-)$-admissible vertex of a quiver Q, set $Q' := \sigma_j(Q)$ and $A' := k(Q')$. Then $F_j^-(Ae_i) \simeq A'e_i$ for all $i \neq j$.*

PROOF. First we show that $F_j^-(Ae_i)$ is projective. It suffices to check that

$$\operatorname{Ext}_{Q'}^1(F_j^-(Ae_i), L_k') = 0$$

for every vertex k of Q'. Indeed, if $k \neq j$ we have

$$\operatorname{Ext}_{Q'}^1(F_j^-(Ae_i), L_k') = \operatorname{Ext}_Q^1(Ae_i, F_j^+ L_k') = 0.$$

Now let us consider the case $k = j$. We have

$$\operatorname{Hom}_{Q'}(F_j^-(Ae_i), L_j') = \operatorname{Hom}_Q(Ae_i, F_j^+ L_j') = 0$$

and

$$\operatorname{Ext}_Q^1(Ae_i, L_j) = 0,$$

since Ae_i is a projective $k(Q)$-module. Let $x := \dim Ae_i$. Taking into account that $\dim L_j = \varepsilon_j$ and $r_j(\varepsilon_j) = -\varepsilon_j$ we have

$$\dim \operatorname{Ext}^1_{Q'}(F_j^-(Ae_i), L_j') = -\langle r_j(x), \varepsilon_j \rangle_{Q'} = \langle x, \varepsilon_j \rangle_Q = \dim \operatorname{Hom}_Q(Ae_i, L_j) = 0.$$

To finish the proof we observe that $F_j^+ L_i' = L_i$ if i is not adjacent to j and $F_j^+ L_i'$ is indecomposable with simple subquotients L_j and L_i if i is adjacent to j. Therefore we obtain

$$\dim \operatorname{Hom}_{Q'}(F_j^-(Ae_i), L_i') = \dim \operatorname{Hom}_Q(Ae_i, F_j^+ L_i') = 1.$$

Since $F_j^-(Ae_i)$ is indecomposable, this implies that $F_j^-(Ae_i)$ is a projective cover of L_i'. Therefore $F_j^-(Ae_i) \simeq A'e_i$. □

LEMMA 5.5. *Let Q be a quiver with an admissible enumeration of vertices, i be a vertex of Q and $Q' := \sigma_{i-1} \ldots \sigma_1(Q)$. Then:*

(1) $Ae_i \simeq F_1^- \circ \cdots \circ F_{i-1}^- L_i'$;
(2) $F_{i-1}^+ \circ \cdots \circ F_1^+(Ae_i) \cong L_i'$;
(3) $F_i^+ \circ \cdots \circ F_1^+(Ae_i) = 0$.

PROOF. The first assertion follows from Lemma 5.4 by induction on i. The second one is a consequence of Theorem 1.6 and the last one is an immediate consequence of (2) since $F_i^+ L_i' = 0$. □

COROLLARY 5.6. *For every projective $k(Q)$-module P, $\Phi^+ P = 0$. Similarly, for every injective $k(Q)$-module I, $\Phi^- I = 0$.*

PROOF. The first statement immediately follows from Lemma 5.5 (3). The analogous statement for an injective module follows by duality. □

LEMMA 5.7. *For all $x, y \in \mathbb{Z}^{Q_0}$*

$$\langle y, x \rangle + \langle x, c(y) \rangle = 0.$$

PROOF. Denote by P_j the projective cover of L_j (recall that $P_j = Ae_j$) and let $p_j := \dim P_j$. Moreover, let I_j be the injective hull of L_j and $i_j := \dim I_j$. Using Lemma 5.5 (1) we have

$$p_j = r_1 \ldots r_{j-1}(\varepsilon_j), \quad c(p_j) = r_n \ldots r_j(\varepsilon_j) = -r_n \ldots r_{j+1}(\varepsilon_j).$$

On the other hand, the statement which is dual to Lemma 5.5 (1) implies

$$I_j \simeq F_n^+ \circ \cdots \circ F_{j+1}^+ L_j', \quad i_j = r_n \ldots r_{j+1}(\varepsilon_j).$$

Thus, we obtain $c(p_j) = -i_j$. For any representation X of Q of dimension x, we have

$$\langle p_j, x \rangle_Q = \dim \operatorname{Hom}_Q(P_j, X) = [X : L_j] = \dim \operatorname{Hom}_Q(X, I_j) = \langle x, i_j \rangle_Q.$$

Thus we obtain

$$\langle p_j, x \rangle_Q = -\langle x, c(p_j) \rangle_Q.$$

Since $p_1, \ldots, p_n$ form a basis of $\mathbb{Z}^{Q_0}$, the statement follows. □

6. Affine root systems

We now proceed to describe all the indecomposable representations of affine quivers. From now on we assume that the ground field k is algebraically closed. Let Γ be an affine Dynkin graph. As we have seen in Chapter 7 the kernel of the bilinear symmetric form q (Tits form) in $\mathbb{Z}^{Q_0}$ is one-dimensional. Let δ be the positive generator of this kernel. Then we have

$$\delta = \sum_{i \in Q_0} a_i \varepsilon_i,$$

see Chapter 7, subsection 6.2. Note that $a_i = 1$ at least for one $i \in Q_0$. Removing any vertex i such that $a_i = 1$ from an affine graph give the corresponding Coxeter-Dynkin graph. (Removing other vertices will result in graphs Coxeter–Dynkin of other types.) We fix a vertex $s \in I$ such that $a_s = 1$ and denote by Γ^0 the graph obtained from Γ by removing s. We also set $Q_0^0 = Q_0 \setminus \{s\}$.

A root α of an affine graph Γ is called a *real* root if $q(\alpha) = 1$ and an *imaginary* root if $q(\alpha) = 0$.

LEMMA 6.1. (1) *All the imaginary roots are proportional to δ;*
(2) *Every real root can be written as $\alpha + m\delta$ for some root α of Γ^0;*
(3) *Every positive real root can be obtained from a simple root by the action of the Weyl group W.*

REMARK 6.2. Observe that δ is fixed by any element of the Weyl group.

PROOF. (1) follows from Lemma 6.4 (a) Chapter 7.

To show (2), consider the projection $p : \mathbb{Z}^{n+1} \to \mathbb{Z}^n = \mathrm{span}(\varepsilon_i)_{i \in Q^0}$ with kernel $\mathbb{Z}\delta$. Since $q(\alpha) = q(\alpha + m\delta)$ the projection p maps a real root of Γ to a root of Γ^0 and every vector in the preimage $p^{-1}(\alpha)$ is a root.

Let us prove (3). Let $\alpha = \sum_{i \in I} b_i \varepsilon_i$ be a positive real root. We set $|\alpha| := \sum_{i \in I} b_i$ and prove the statement by induction on $|\alpha|$. The case $|\alpha| = 1$ is clear since then $\alpha = \varepsilon_i$ for some i. Let $|\alpha| > 1$. Since $(\alpha, \alpha) > 0$ there exists j such that $(\alpha, \varepsilon_j) > 0$. Then $\beta := r_j(\alpha) = \alpha - c\varepsilon_j$ for some $c > 0$. Hence β is a positive real root and $|\beta| < |\alpha|$. By the induction assumption $\beta = w(\varepsilon_i)$ for some i and some $w \in W$. Hence $\alpha = r_j w(\varepsilon_i)$. $\qquad\square$

EXAMPLE 6.3. Consider the affine graph $\tilde{A}_1$: $\bullet\!\!=\!\!=\!\!\blacktriangleright$

In this case $\delta = \varepsilon_0 + \varepsilon_1$. All imaginary positive roots are of the form (m, m) where $m \in \mathbb{Z}_{>0}$, the positive real roots are of the form $(m, m + 1)$ or $(m + 1, m)$ for some $m \in \mathbb{N}$. The Weyl group W is the infinite dihedral group, with generators r_0, r_1 and relations $r_0^2 = 1 = r_1^2$. The Coxeter element $c := r_1 r_0$ generates an infinite cyclic normal subgroup in W. One can easily check that

$$c(m, m + 1) = (m + 2, m + 3) = (m, m + 1) + 2\delta,$$

$$c(m + 1, m) = (m - 1, m - 2) = (m + 1, m) - 2\delta.$$

6.1. Representations of the Kronecker quiver. In this subsection we use the Coxeter functors to classify the indecomposable representations of the Kronecker quiver Q:

$$\bullet \rightrightarrows \bullet.$$

The only admissible enumeration of vertices is $1 \rightrightarrows 0$. Let X be an indecomposable representation of Q of dimension $x = (n, m)$. It is given by a pair of linear operators $A, B : k^m \to k^n$.

We will first show that x is a root. Indeed, one can check that $c(x) = x + 2(m - n)\delta$. Therefore if $m > n$, then $c^{-s}(x) < 0$ for $s >> 0$, and if $m < n$ then $c^s(x) < 0$ for $s >> 0$. Since $\dim \Phi^+ X = c(x)$ and $\dim \Phi^- X = c^{-1}(x)$, we know that, in the former case, $(\Phi^-)^s X = 0$ and that, in the latter case, $(\Phi^+)^s X = 0$ for sufficiently large s.

Consider the case $m > n$. Then we have the following two possibilities:

(1) There is exactly one s such that $(\Phi^-)^s X \neq 0$ and $F_1^- \circ (\Phi^-)^s (X) = 0$. Then $(\Phi^-)^s X \simeq L_1$ and hence $X \simeq (\Phi^+)^s L_1$.

(2) There is exactly one s such that $(\Phi^-)^s X = 0$ and $F_0^- \circ (\Phi^-)^{s-1}(X) \neq 0$. Then $F_0^- \circ (\Phi^-)^{s-1}(X) \simeq L_1'$ and hence $X \simeq (\Phi^+)^{s-1} \circ F_0^+ (L_1')$, where L_1' is the irreducible representation of $Q' = \sigma_0(Q) = Q^{op}$ of dimension ε_1.

Similarly in the case $m < n$ we obtain two possibilities:

(1) There is exactly one s such that $(\Phi^+)^s X \neq 0$ and $F_0^+ \circ (\Phi^+)^s (X) = 0$. Then $(\Phi^+)^s X \simeq L_0$ and hence $X \simeq (\Phi^-)^s L_0$.

(2) There is exactly one s such that $(\Phi^+)^s X = 0$ and $F_1^+ \circ (\Phi^+)^{s-1}(X) \neq 0$. Then $(\Phi^+)^{s-1} X \simeq L_0'$ and hence $X \simeq (\Phi^-)^{s-1} \circ F_1^- (L_0')$, where, as in the previous case, L_0' is the irreducible representation of $Q' = \sigma_0(Q) = Q^{op}$ of dimension ε_0.

It follows that either $x = (m, m)$ and x is an imaginary root or $x = (m, m\pm1)$ and x is a real root. Moreover, the above arguments imply that for every real positive root x there exists exactly one indecomposable representation X of dimension x, up to isomorphism. We can describe this X precisely in terms of linear operators $A, B : k^m \to k^n$. If $n = m + 1$ we can set $A = (I_m, 0), B = (0, I_m)$ and if $m = n + 1$ we set $A = (I_m, 0)^t, B = (0, I_m)^t$, where M^t stands for the transposed matrix of M and I_m is the identity $m \times m$ matrix.

It remains to classify indecomposable representations of dimension $x = (m, m)$. We will prove that, for every $m > 0$, there is a one-parameter family of representations of dimension (m, m) enumerated by $t \in \mathbb{P}^1 := \mathbb{P}^1(k)$. In the case $x = (1, 1) = \delta$, the indecomposable representations of dimension δ are in bijection with pairs $(a, b) \in k^2 \setminus (0, 0)$. We denote the corresponding representation by $X_{(a:b)}$. It is easy to see that two representations $X_{(a:b)}$ and $X_{(a':b')}$ are isomorphic if and only if $(a, b) = \lambda(a', b')$ for some invertible $\lambda \in k$. Thus, the indecomposable representations of dimension δ form the family

$$\{X_{(a:b)} \mid (a : b) \in \mathbb{P}^1\}.$$

Note that $\langle \delta, \delta \rangle_Q = 0$ implies

$$\operatorname{Ext}^1_Q(X_{(a:b)}, X_{(c:d)}) = \operatorname{Hom}_Q(X_{(a:b)}, X_{(c:d)}) = 0$$

if $(a : b)$ and $(c : d)$ are two distinct points in $\mathbb{P}^1$.

Let X be an indecomposable representation of dimension (m, m). All representations of dimension (m, m) are in correspondence with the pairs of linear operators $A, B : k^m \to k^m$. Furthermore, $\operatorname{Hom}_Q(X_{(a:b)}, X) \neq 0$ if and only if $bA - aB$ is not invertible. That implies $\operatorname{Hom}_Q(X_{(a:b)}, X) \neq 0$ for at most m points $(a : b) \in \mathbb{P}^1$. In particular, there exists t such that $\operatorname{Hom}_Q(X_{(1:t)}, X) = 0$. This implies the invertibility of the operator $tA - B$. We use this operator to identify the components X_0 and X_1. In other words, we may assume that $tA - B = I_m$. Every A-stable subspace of k^m is also B-stable. Therefore, the indecomposability of X implies that there exists $(a, b) \in k^2 \setminus (0, 0)$ such that $A = aI_m + J_m$ and $B = bI_m + J_m$, where J_m is the nilpotent Jordan block of size $m \times m$. We denote the corresponding representation $X^{(m)}_{(a:b)}$. It is obvious that the isomorphism class of $X^{(m)}_{(a:b)}$ depends only on the point $(a : b) \in \mathbb{P}^1$. Finally we observe that

$$\operatorname{Hom}_Q(X_{(c:d)}, X^{(m)}_{(a:b)}) = \begin{cases} 0 \text{ if } (a : b) \neq (c : d) \\ k \text{ if } (a : b) = (c : d) \end{cases}.$$

This implies that $X^{(m)}_{(a:b)}$ and $X^{(m)}_{(c:d)}$ are not isomorphic when $(a : b)$ and $(c : d)$ are distinct in $\mathbb{P}^1$.

One may get the impression from the example above that any indecomposable representation whose dimension is a real root is annihilated by some power of one of the Coxeter functors. This is not true in general as one can see from the following exercise. We will return to this question in subsection 8.4.

EXERCISE 6.4. Consider the quiver Q:

$$\begin{array}{ccccc} & & 4 & & \\ & & \downarrow & & \\ 1 & \to & 0 & \leftarrow & 3 \\ & & \uparrow & & \\ & & 2 & & \end{array}$$

Let $\alpha := \varepsilon_0 + \varepsilon_1 + \varepsilon_2$. Check that α is a real root, $c(\alpha) = \varepsilon_0 + \varepsilon_3 + \varepsilon_4$ and $c^2(\alpha) = \alpha$. By writing Φ^+ as a composition of reflection functors, check that for any indecomposable X of dimension α the representation $(\Phi^+)^2 X$ is again indecomposable. Hence $(\Phi^+)^m X \neq 0$ for all m.

7. Preprojective and preinjective representations

In this section, we assume that Q is a quiver with a fixed admissible enumeration of vertices.

An indecomposable representation X is called *preprojective*[1] if $(\Phi^+)^s X = 0$ for some s, *preinjective* if $(\Phi^-)^s (X) = 0$ for some s and *regular* if it is neither preprojective nor preinjective.

EXAMPLE 7.1. For the Kronecker quiver, an indecomposable representation is preprojective if its dimension is $(m, m+1)$, preinjective if its dimension is $(m+1, m)$ and regular if its dimension is (m, m).

LEMMA 7.2. *If X is preprojective, then $X = (\Phi^-)^s P$ for some integer s and some projective P. If X is preinjective, then $X = (\Phi^+)^s I$ for some integer s and some injective I.*

PROOF. Suppose $(\Phi^+)^s X \neq 0$, and $(\Phi^+)^{s+1} X = 0$. Then

$$F_{i-1}^+ \ldots F_1^+ (\Phi^+)^s X = L_i', \ X \cong (\Phi^-)^s F_1^- \ldots F_{i-1}^- (L_i')$$

as in Corollary 4.3. Therefore, by Lemma 5.5, we have an isomorphism

$$X \cong (\Phi^-)^s (Ae_i).$$

The statement for preinjective follows by duality.   □

COROLLARY 7.3. *If X is an (indecomposable) preprojective or preinjective representation, then $\dim X$ is a real root.*

PROOF. The proof of Lemma 7.2 shows that there exist $i \in Q_0$ and $w \in W$ such that $\dim X = w(\varepsilon_i)$.   □

LEMMA 7.4. *Let X, Y be indecomposable representations of Q. If X is preprojective and Y is not, then $\operatorname{Hom}_Q (Y, X) = \operatorname{Ext}_Q^1 (X, Y) = 0$.*
If X is preinjective and Y is not, then $\operatorname{Hom}_Q (X, Y) = \operatorname{Ext}_Q^1 (Y, X) = 0$.

PROOF. Assume X is preprojective. Then $X = (\Phi^-)^s P$ for some projective P and some s, thus

$$\operatorname{Ext}_Q^1 \left((\Phi^-)^s P, Y\right) = \operatorname{Ext}_Q^1 \left(P, (\Phi^+)^s Y\right) = 0$$

by Corollary 5.3. On the other hand,

$$(\Phi^+)^{s+1} X = 0, \ Y \cong (\Phi^-)^{s+1} (\Phi^+)^{s+1} Y$$

and

$$\operatorname{Hom}_Q \left((\Phi^-)^{s+1} (\Phi^+)^{s+1} Y, X\right) = \operatorname{Hom}_Q \left((\Phi^+)^{s+1} Y, (\Phi^+)^{s+1} X\right) = 0.$$

For the preinjective case, we use duality.   □

---

[1] Some authors prefer the term postprojective instead of preprojective, see [13].

From now on, we assume that Q is affine. Note that if X is indecomposable and $x := \dim X = m\delta$, then $r_i(x) = x$ for all $i \in Q_0$. Therefore X is regular. Define the *defect* of X by

$$\mathrm{def}\,(X) := \langle \delta, x \rangle_Q = -\langle x, c(\delta) \rangle_Q = -\langle x, \delta \rangle_Q.$$

We write $x \leq y$ if $y - x \in \mathbb{N}^{Q_0}$.

LEMMA 7.5. *If $x < \delta$ and if X is indecomposable of dimension x, then X is a brick, x is a root and $\mathrm{Ext}^1_Q(X, X) = 0$.*

PROOF. If X is not a brick, then there is a brick $Y \subset X$ such that $\mathrm{Ext}^1(Y, Y) \neq 0$, see Lemma 4.4 of Chapter 7. But then $q(y) \leq 0$, which is impossible as $y < \delta$. Hence X is a brick. Since $q(x) > 0$, we have $\mathrm{Ext}^1(X, X) = 0$ and $q(x) = 1$. □

LEMMA 7.6. *There exists an indecomposable representation of dimension δ.*

PROOF. Pick an orbit O_Z in $\mathrm{Rep}\,(\delta)$ of maximal dimension. Then by Corollary 5.5 of Chapter 7, $Z = X^1 \oplus \cdots \oplus X^p$, where every X^i is indecomposable and $\mathrm{Ext}^1(X^i, X^j) = 0$ if $i \neq j$. If $p > 1$, then $q(z) > 0$ which is impossible. □

LEMMA 7.7. *If X is regular, then there exists $m > 0$ such that $c^m(x) = x$.*

PROOF. From the description of affine root systems (Lemma 6.1), we know that the W-orbits on $\mathbb{Z}^{Q_0}/\mathbb{Z}\delta$ are finite in number. Therefore one can find m such that $c^m(x) = x + l\delta$ for some $l \in \mathbb{Z}$. If $l \neq 0$, then $c^{md}(x) = x + ld\delta$. Hence there exists $d \in \mathbb{Z}$ such that $c^d(x) < 0$. This contradicts the regularity of X. □

THEOREM 7.8. *Let X be an indecomposable representation of Q.*

(1) *If X is preprojective, then $\mathrm{def}\,(X) < 0$;*
(2) *If X is regular, then $\mathrm{def}\,(X) = 0$;*
(3) *If X is preinjective, then $\mathrm{def}\,(X) > 0$.*

PROOF. Let X be preprojective, Z be an indecomposable representation of dimension δ constructed in Lemma 7.6. Then $\mathrm{Ext}^1_Q(X, Z) = 0$ by Lemma 7.4. On the other hand, $X = (\Phi^-)^s Ae_i$. Therefore

$$\mathrm{Hom}_Q(X, Z) = \mathrm{Hom}_Q\left((\Phi^-)^s Ae_i, Z\right) = \mathrm{Hom}_Q\left(Ae_i, (\Phi^+)^s Z\right) \neq 0,$$

and hence $\langle x, \delta \rangle = \dim \mathrm{Hom}_Q(X, Z) > 0$. The statement for preinjective X follows by duality.

Finally, let X be regular. Assume $\mathrm{def}\,(X) \neq 0$, say $\mathrm{def}\,(X) > 0$. Since x is regular $c^p(x) = x$ for some p. Then $y = x + c(x) + \cdots + c^{p-1}(x)$ is c-invariant, therefore $x + c(x) + \cdots + c^{p-1}(x) = m\delta$ by Lemma 3.4. But, for every integer l, we have

$$\langle \delta, c^l(x) \rangle = \langle c^{-l}(\delta), x \rangle = \langle \delta, x \rangle > 0,$$

hence $\langle \delta, m\delta \rangle > 0$. But $\langle \delta, \delta \rangle = q(\delta) = 0$. Contradiction. □

8. Regular representations

8.1. The abelian category of regular representations. In this section we still assume that Q is an affine quiver. We say that a representation of Q is *regular* if it is a direct sum of indecomposable regular representations.

PROPOSITION 8.1. *Let X and Y be regular representations of Q and $\varphi \in \mathrm{Hom}_Q(X, Y)$. Then $\mathrm{Ker}\,\varphi$ and $\mathrm{Coker}\,\varphi$ are regular. In particular, the full subcategory of regular representations of an affine quiver Q is abelian.*

PROOF. Consider the exact sequence
$$0 \to \mathrm{Ker}\,\varphi \to X \xrightarrow{\varphi} Y \to \mathrm{Coker}\,\varphi \to 0.$$
Then
$$\dim \mathrm{Ker}\,\varphi - \dim X + \dim Y - \dim \mathrm{Coker}\,\varphi = 0$$
and therefore, by Theorem 7.8,
$$\mathrm{def}(\mathrm{Ker}\,\varphi) = \mathrm{def}(\mathrm{Coker}\,\varphi).$$
Assume that $\mathrm{Ker}\,\varphi$ is not regular. Lemma 7.4 implies that, for any preinjective W,
$$\mathrm{Hom}_Q(W, \mathrm{Ker}\,\varphi) \hookrightarrow \mathrm{Hom}_Q(W, X) = 0.$$
Therefore every non-regular indecomposable direct summand of $\mathrm{Ker}\,\varphi$ is preprojective. Hence
$$\mathrm{def}(\mathrm{Ker}\,\varphi) = \mathrm{def}(\mathrm{Coker}\,\varphi) < 0.$$
Therefore $\mathrm{Coker}\,\varphi$ has an indecomposable preprojective direct summand U. But then we have a surjective morphism $Y \to \mathrm{Coker}\,\varphi \to U$. However, Lemma 7.4 implies $\mathrm{Hom}_Q(Y, U) = 0$. This is a contradiction.

The proof that $\mathrm{Coker}\,\varphi$ is regular is similar and can be done using duality. □

LEMMA 8.2. *Let*
$$0 \to X \to Z \to Y \to 0$$
be an exact sequence of $k(Q)$-modules. If X and Y are regular, then Z is also regular.

PROOF. Suppose that Z has a preprojective direct summand Z_i. Then we have an exact sequence
$$\mathrm{Hom}_Q(Y, Z_i) = 0 \to \mathrm{Hom}_Q(Z, Z_i) \to \mathrm{Hom}_Q(X, Z_i) = 0,$$
which is a contradiction.

Similarly, one can show that Z cannot have any preinjective direct summand. □

A regular representation X is called *regular simple* if X has no proper non-trivial regular subrepresentations.

EXERCISE 8.3. Show that a regular simple representation X is a brick, hence $q(x) \leq 1$. In particular, $\dim X$ is a root.

EXERCISE 8.4. Prove that any regular simple representation of the Kronecker quiver is isomorphic to $X_{(a:b)}$ for some $(a : b) \in \mathbb{P}^1$.

EXERCISE 8.5. Check that the following analogue of Jordan-Hölder theorem holds:

let Y be a regular representation of Q. There exists a filtration

$$0 = \mathcal{F}^0(Y) \subset \mathcal{F}^1(Y) \subset \cdots \subset \mathcal{F}^k(Y) = Y,$$

such that $Y_i = \mathcal{F}^i(Y)/\mathcal{F}^{i-1}(Y)$ is simple for all $i > 0$. Furthermore, two such filtrations give the same set of regular simple subquotients counted with multiplicities. Therefore for any regular X and any regular simple S the *regular multiplicity* $[X : S]_r$ is well defined.

8.2. A one-parameter family of indecomposable representations of dimension δ. Next we will need the following lemma, the proof is similar to the argument in the proof of Gabriel's theorem (Theorem 7.2 in Chapter 7)

LEMMA 8.6. *Let α be a real positive root such that either $\langle \alpha, \delta \rangle \neq 0$ or $\alpha < \delta$, then $\mathrm{Rep}(\alpha)$ contains an open orbit whose elements are bricks.*

PROOF. Let X be a representation of Q of dimension α whose orbit in $\mathrm{Rep}(\alpha)$ is of maximal dimension. Write the decomposition $X = X^1 \oplus \ldots \oplus X^p$ of X into indecomposables. By maximality of the orbit dimension $\mathrm{Ext}^1(X^j, X^i) = 0$ whenever $j \neq i$. Therefore we have

$$1 = q(\dim X) = \sum_j q(\dim X^j) + \sum_{j \neq i} \dim \mathrm{Hom}(X^j, X^i).$$

Hence every term in the sum is 0 except one, which may be chosen as $q(\dim X^1)$ since at least one of $\dim X^j$ is not a multiple of δ; all the other $\dim X^j$ are proportional to δ. This also implies $\langle \dim X, \delta \rangle_Q = 0$. By our assumptions on α we have $p = 1$ and, thus, X is indecomposable. Furthermore, we claim that X is a brick. Indeed, otherwise X would contain a brick Z with non-trivial self-extension. Then $\dim Z$ is an imaginary root and Z is regular. But this is impossible since X is preprojective or preinjective by Theorem 7.8. Since $\mathrm{Ext}^1(X, X) = 0$, we see that the orbit of X in $\mathrm{Rep}(\alpha)$ is open. $\square$

Consider an affine quiver Q and let $s \in Q_0$ be such that $a_s = 1$. Let $P := Ae_s$ be the projective cover of L_s and $p := \dim P$. Corollary 5.6 ensures that P is preprojective, and hence by Corollary 7.3 we know that p is a real root. Furthermore, if Z is a representation of dimension δ, then

$$\mathrm{Ext}^1_Q(P, Z) = 0, \quad \dim \mathrm{Hom}_Q(P, Z) = [Z : L_0] = 1$$

and we obtain that $\langle p, \delta \rangle_Q = 1$.

Note that $r := p + \delta$ is a real root satisfying the conditions of Lemma 8.6. There exists an indecomposable representation R of dimension r. Since P is the projective cover of L_s and the Jordan-Hölder multiplicity of L_s in R equals 2, we have

$$\mathrm{Hom}_Q (P, R) = k^2.$$

LEMMA 8.7. Let $\theta \in \mathrm{Hom}_Q (P, R)$. If $\theta \neq 0$, then θ is injective.

PROOF. Recall that both P and R are preprojective of defect -1. Lemma 7.4 implies that $\mathrm{Ker}\, \theta$ and $\mathrm{Im}\, \theta$ are direct sums of preprojective representations, each summand has negative defect. Since

$$-1 = \mathrm{def}\, (P) = \mathrm{def}\, (\mathrm{Ker}\, \theta) + \mathrm{def}\, (\mathrm{Im}\, \theta),$$

there is only one such summand. □

EXERCISE 8.8.

(1) Let $\eta \in \mathrm{Hom}_Q (R, P)$. Prove that if $\eta \neq 0$ then η is surjective.
(2) Show that $\mathrm{Hom}_Q (R, P) = \mathrm{Ext}^1_Q (R, P) = 0$.

LEMMA 8.9. Let $\theta \in \mathrm{Hom}_Q (P, R)$, $\theta \neq 0$. Then $Z_\theta = \mathrm{Coker}\, \theta$ is regular indecomposable.

PROOF. Consider the exact sequence

$$0 \to P \xrightarrow{\theta} R \to Z_\theta \to 0.$$

The corresponding long exact sequence,

$$0 = \mathrm{Hom}_Q (R, P) \to \mathrm{Hom}_Q (R, R) \to \mathrm{Hom}_Q (R, Z_\theta) \to \mathrm{Ext}^1_Q (R, P) = 0,$$

provides an isomorphism $\mathrm{Hom}_Q (R, Z_\theta) \simeq \mathrm{Hom}_Q (R, R)$.

Thus, we have $\mathrm{Hom}_Q (R, Z_\theta) = k$. Furthermore, the embedding $\mathrm{End}_Q(Z_\theta) \hookrightarrow \mathrm{Hom}_Q (R, Z_\theta)$ implies $\mathrm{End}_Q(Z_\theta) = k$. Therefore Z_θ is indecomposable. Furthermore, $\mathrm{def}\, (Z_\theta) = 0$, hence Z_θ is regular. □

LEMMA 8.10. If Z_θ is isomorphic to Z_τ, then $\theta = C\tau$ for some $C \in k^*$.

PROOF. Above we have proved that $\mathrm{Hom}_Q(R, Z_\theta) = k$. Therefore the isomorphism between Z_θ and Z_τ implies $\mathrm{Im}\, \theta = \mathrm{Im}\, \tau$. Since P is a brick θ and τ are proportional. □

8.3. Properties of regular simple representations.

LEMMA 8.11. Let X be a regular indecomposable representation of Q of dimension x such that $x_s \neq 0$. Then there exists $\theta \in \mathrm{Hom}_Q (P, R)$ such that $\mathrm{Hom}_Q (Z_\theta, X) \neq 0$ (the notations are those of the previous subsection).

PROOF. First note that

$$\dim \operatorname{Hom}_Q (P, X) = x_s = \dim \operatorname{Hom}_Q (R, X)$$

since $\langle p, x \rangle = \langle r, x \rangle = x_s$.

Let $u : \operatorname{Hom}_Q(P, R) \to \operatorname{Hom}_k(\operatorname{Hom}_Q(R, X), \operatorname{Hom}_Q(P, X))$ be the composition map. This is a k-linear map, and therefore $\operatorname{Det}(u)$ is a homogeneous polynomial of degree x_s on $\operatorname{Hom}_Q(P, R) = k^2$. Take any $\theta \neq 0$ such that $\operatorname{Det}(u)(\theta) = 0$. There exists a non-zero $\varphi \in \operatorname{Hom}_Q(R, X)$ such that $\varphi \circ \theta = 0$. Then φ induces a non-zero morphism $Z_\theta \to X$. □

COROLLARY 8.12. *Let X be regular simple of dimension x. Then $x \leq \delta$.*

PROOF. By Exercise 8.3, x is a root. If $x_s \neq 0$, then $\operatorname{Hom}_Q (Z_\theta, X) \neq 0$ for some θ. Consider a non-zero homomorphism $\varphi : Z_\theta \to X$. Since the image of φ is regular and X is regular simple, φ is surjective. Therefore X is a quotient of Z_θ, hence $x \leq \delta$. Finally, if $x_s = 0$, then $x < \delta$ (see Lemma 6.1). □

EXERCISE 8.13. Let Q be the quiver of Exercise 6.4. Check that the regular simple representations have dimensions $\varepsilon_i + \varepsilon_j + \varepsilon_0$ for all $0 < i < j \leq 4$ and $\delta = \varepsilon_1 + \varepsilon_2 + \varepsilon_3 + \varepsilon_4 + 2\varepsilon_0$. Show that an indecomposable representation ρ of dimension δ is regular simple if and only if $\operatorname{Im} \rho_\gamma \neq \operatorname{Im} \rho_{\gamma'}$ for all distinct arrows $\gamma, \gamma' \in Q_1$.

LEMMA 8.14. *Let X and Y be two regular simple representations of Q. Then*

$$\operatorname{Hom}_Q (X, Y) = \begin{cases} k, & \text{if } X \simeq Y \\ 0 & \text{otherwise} \end{cases}$$

and

$$\operatorname{Ext}^1_Q (X, Y) = \begin{cases} k, & \text{if } \Phi^+ X \simeq Y \\ 0 & \text{otherwise} \end{cases}.$$

PROOF. For any homomorphism $\varphi : X \to Y$, both $\operatorname{Ker} \varphi$ and $\operatorname{Im} \varphi$ are regular. Therefore either φ is an isomorphism or $\varphi = 0$. Moreover, $\operatorname{Hom}_Q (X, X) = k$ because X is a brick (see Exercise 8.3).

To prove the second assertion, use Lemma 5.7:

$$\langle x, y \rangle_Q = - \langle y, c(x) \rangle_Q.$$

If Y is not isomorphic to X or $\Phi^+ X$, then

$$\langle x, y \rangle_Q = \dim \operatorname{Hom}_Q (X, Y) - \dim \operatorname{Ext}^1_Q (X, Y) \leq 0$$

and

$$\langle y, c(x) \rangle_Q = \dim \operatorname{Hom}_Q (Y, \Phi^+ X) - \dim \operatorname{Ext}^1_Q (Y, \Phi^+ X) \leq 0.$$

Therefore we must have $\langle x, y \rangle_Q = 0$. Hence $\operatorname{Ext}^1_Q (X, Y) = 0$.

If $Y \simeq X$, then

$$\langle x, x \rangle_Q = 1 = - \langle x, c(x) \rangle_Q$$

if x is a real root, and

$$\langle x, x \rangle_Q = 0 = - \langle x, c(x) \rangle_Q$$

if $x = c(x)$ is imaginary. In both cases we obtain $\dim \mathrm{Ext}^1 (Y, \Phi^+ X) = 1$. The case $Y \simeq \Phi^+ X$ is similar. $\square$

PROPOSITION 8.15. *Let X be a regular simple representation of Q. Then*

(1) $\Phi^+ X$ *is regular simple.*
(2) *If $x < \delta$, then $(\Phi^+)^s X \cong X$ for some $s > 1$.*
(3) *If $x = \delta$, then $\Phi^+ X \cong X$.*

PROOF. Set $Y := \Phi^+ X$. Then

$$\langle x, y \rangle_Q = - \langle c(x), c(x) \rangle_Q = \begin{cases} -1, & \text{if } x \text{ is real} \\ 0, & \text{if } x \text{ is imaginary.} \end{cases}$$

If x is real, then $\mathrm{Ext}_Q^1(X, Y) \neq 0$. Consider a Jordan-Hölder filtration of Y as in Exercise 8.5. Then $\mathrm{Ext}_Q^1(X, Y_i) \neq 0$ for at least one i. But, by Lemma 8.14, this implies that $Y_i \simeq \Phi^+ X$. Hence Y is regular simple.

Since the set of real roots $\alpha < \delta$ is finite, there exists an integer s such that $c^s(x) = x$. Then $\langle x, x \rangle_Q = \langle x, c^s(x) \rangle_Q = 1$. Therefore $\mathrm{Hom}_Q(X, (\Phi^+)^s X) \neq 0$. This proves the statement (1) in the case of a real x, and the statement (2).

Let $x = \delta$. If Y is not simple then it contains a proper regular simple submodule Z and we have

$$\mathrm{Hom}_Q(Z, Y) = \mathrm{Hom}_Q(Z, \Phi^+ X) \simeq \mathrm{Hom}_Q(\Phi^- Z, X) = 0.$$

A contradiction. Hence (1) holds for imaginary x. Furthermore, $\mathrm{Ext}_Q^1(X, Y) \simeq \mathrm{Hom}_Q(X, Y) = k$ by Lemma 8.14. Thus, (3) is proven. $\square$

The minimal number p such that $(\Phi^+)^p X \cong X$ is called the *period* of X. Regular simple representations can be divided into orbits with respect to the action of Φ^+.

COROLLARY 8.16. *Let X be a regular indecomposable representation. Then all the regular composition factors of X belong to the same Φ^+-orbit.*

PROOF. Let r denote the regular length of X, i.e. the length of a Jordan-Hölder filtration introduced in Exercise 8.5. Assume that the statement is false and pick up an X, with minimal regular length r, for which it fails. Let S be a regular simple submodule of X. Consider the exact sequence

$$0 \to S \to X \to Y \to 0$$

and let $Y = \oplus_{i=1}^k Y_i$ be a direct sum of indecomposables. Then all the regular simple constituents of Y_i belong to the same Φ^+-orbit. The indecomposability of X

implies that $\operatorname{Ext}^1_Q(Y_i, S) \neq 0$ for all $i = 1, \ldots k$. Lemma 8.14 implies that S lies in the Φ^+-orbit of the regular simple constituents of Y_i. Therefore all regular simple constituents of X are in the same orbit. A contradiction. $\qquad\square$

EXERCISE 8.17. Let $\{S_1, \ldots S_p\}$ be an orbit of Φ^+ in the set of regular simple representations and let $\alpha_i = \dim S_i$. Assume that $S_i = (\Phi^+)^{i-1}S_1$. If $p = 1$, then $\alpha_1 = \delta$, if $p = 2$, then $(\alpha_1, \alpha_2) = -2$. If $p > 2$ then $(\alpha_i, \alpha_j) = 0$ if $i \neq j \pm 1$ and $(i, j) \neq (p, 1), (1, p)$ and

$$(\alpha_i, \alpha_{i+1}) = -1, \ i = 1, \ldots, p-1, \ (\alpha_1, \alpha_p) = -1.$$

Furthermore, $\alpha_1, \ldots, \alpha_p$ are linearly independent and hence form a set of simple roots for a root system of type $\tilde{A}_{p-1}$.

8.4. Indecomposable regular representations and tubes. As follows from Corollary 8.16, simple regular representations which are constituents of some indecomposable regular representation belong to the same orbit of Φ^+. Thus, each orbit of Φ^+ in the set of simple regular representations defines a family of indecomposables called a *tube*.

EXERCISE 8.18. Let Q be the quiver of Exercise 6.4. Then there are 3 orbits of regular simple representations of period 2 and infinitely many orbits of period 1.

THEOREM 8.19. *Let T be a tube and S be the set of isomorphism classes of regular simple representations in T. Then for every $r > 0$ and $S \in S$ there exists a unique regular representation $X^{(r)}(S) \in T$ of regular length r, up to isomorphism, such that*

(1) *$X^{(r)}(S)$ has a unique regular simple quotient which is isomorphic to S,*
(2) *every non-zero regular submodule of $X^{(r)}(S)$ has a unique regular simple quotient.*

Moreover, every indecomposable representation in T is isomorphic to $X^{(r)}(S)$ for some $r > 0$ and $S \in S$.

PROOF. We first prove the existence of $X^{(r)}(S)$ by induction on r, with the base case $X^{(1)}(S) = S$. To construct $X^{(r)}(S)$ for an arbitrary r, use $X^{(r-1)}(\Phi^+S)$ and the surjection $p : X^{(r-1)}(\Phi^+S) \to \Phi^+S$ which defines the surjection:

$$\bar{p} : \operatorname{Ext}^1_Q(S, X^{(r-1)}(\Phi^+S)) \to \operatorname{Ext}^1_Q(S, \Phi^+S) = k.$$

Let $\varphi \in \operatorname{Ext}^1_Q(S, X^{(r-1)}(\Phi^+S))$ be such that $\bar{p}(\varphi) \neq 0$: it induces a non-split exact sequence

$$0 \to X^{(r-1)}(\Phi^+S) \to Z \xrightarrow{\pi} S \to 0.$$

By the induction assumption, the module $X^{(r-1)}(\Phi^+S)$ has a unique maximal proper regular submodule N. Since $\bar{p}(\varphi) \neq 0$, the sequence

(8.2) $$0 \to \Phi^+S \to Z/N \to S \to 0$$

does not split.

We set $X^{(r)}(S) := Z$ and claim that Z satisfies the requirements of the theorem. It suffices to check that Z has a unique proper maximal regular submodule. Indeed, let M be a proper maximal regular submodule of Z. If $M \neq X^{(r-1)}(\Phi^+S)$ then $M \cap X^{(r-1)}(\Phi^+S) = N$ and M/N is a submodule of Z/N which does not coincide with $\Phi^+(S)$. Since the sequence (8.2) does not split we get a contradiction.

Now let us show that $X^{(r)}(S)$ is unique up to isomorphism again by induction on r. We will need the following Lemma before we can go on with the proof.

LEMMA 8.20. *For any $r > 0$ and $S, S' \in \mathcal{S}$ we have*

$$\operatorname{Ext}^1_Q(S, X^{(r)}(S')) = \begin{cases} 0, & \text{if } S' \neq \Phi^+S \\ k, & \text{if } S' = \Phi^+S \end{cases}.$$

PROOF. We prove the Lemma by induction on r. If $r = 1$, the statement is Lemma 8.14. Let S'' be the unique regular simple submodule of $X^{(r)}(S')$. We consider the exact sequence

$$0 \to S'' \to X^{(r)}(S') \to X^{(r-1)}(S') \to 0$$

and the corresponding long exact sequence for $\operatorname{Ext}_Q(S, \cdot)$. Note that $\operatorname{Hom}_Q(S, S'') \to \operatorname{Hom}_Q(S, X^{(r)}(S'))$ is an isomorphism since S'' is the unique regular simple submodule of $X^{(r)}(S')$. Furthermore, $\operatorname{Ext}^1_Q(S, S'') = 0$ if $S'' \neq \Phi^+S$. So in this case we obtain an isomorphism

$$\operatorname{Ext}^1_Q(S, X^{(r)}(S')) \xrightarrow{\sim} \operatorname{Ext}^1_Q(S, X^{(r-1)}(S'))$$

and the statement follows by induction. On the other hand, if $S'' = \Phi^+S$, the unique simple submodule of $X^{(r-1)}(S')$ is isomorphic to S and therefore we have

$$0 \to \operatorname{Hom}_Q(S, X^{(r-1)}(S')) \xrightarrow{\sim} \operatorname{Ext}^1_Q(S, S'') \xrightarrow{0} \operatorname{Ext}^1_Q(S, X^{(r)}(S'))$$
$$\xrightarrow{\sim} \operatorname{Ext}^1_Q(S, X^{(r-1)}(S')) \to 0.$$

$\square$

Now by induction assumption $X^{(r)}(S)$ has a unique maximal regular submodule isomorphic to $X^{(r-1)}(\Phi^+(S))$. By Lemma 8.20, $\operatorname{Ext}^1_Q(S, X^{(r-1)}(\Phi^+)) = k$, which implies the uniqueness of $X^{(r)}(S)$.

Finally let us prove the last assertion. Let $X \in T$ be an indecomposable representation of regular length r and let S be some regular simple quotient of X. We will prove that X is isomorphic to $X^{(r)}(S)$ by induction on r. Consider the exact sequence

(8.3) $$0 \to Y \to X \to S \to 0$$

and let $Y_1, ..., Y_m$ denote the indecomposable summands of Y. Assume that $m \geq 2$. The indecomposability of X implies that $\operatorname{Ext}^1_Q(S, Y_i) \neq 0$ for all $i = 1, \ldots, m$. So by induction assumption and Lemma 8.20, each summand Y_i is isomorphic to $X^{(r_i)}(\Phi^+S)$ for some r_i. Without loss of generality we may assume that r_1 is maximal. Then for every $i \geq 2$ we have a surjective homomorphism $f_i : Y_1 \to Y_i$ which induces an isomorphism $\tilde{f}_i : \mathrm{Ext}^1_Q(S, Y_1) \to \mathrm{Ext}^1_Q(S, Y_i)$. Assume that the sequence (8.3) corresponds to the element $\sum_{i=1}^m \psi_i$ for some $\psi_i \in \mathrm{Ext}^1_Q(S, Y_i)$. Then $\psi_i = c_i \tilde{f}_i(\psi_1)$. Let

$$Z_2 := \{c_2 m - f_2(m) \mid m \in Y_1\}.$$

Then Z_2 splits as a direct summand in X and we obtain a contradiction.

Thus Y is indecomposable, isomorphic to $X^{(r-1)}(S)$ and hence X is isomorphic to $X^{(r)}(S)$. □

LEMMA 8.21. *Every tube contains exactly one indecomposable representation isomorphic to Z_θ.*

PROOF. Pick up a simple regular X in a tube T. Let p be the period of X, i.e. $(\Phi^+)^p X \cong X$. If $x = \dim X$, then

$$x + c(x) + \cdots + c^{p-1}(x) = m\delta$$

There exists i such that $y := c^i(x)$ has the coordinate $y_s \neq 0$. Let $Y := (\Phi^+)^i X$. By Lemma 8.11 there exists $\theta \in \mathrm{Hom}_Q(P, R)$ and a non-zero homomorphism $\varphi : Z_\theta \to Y$. Hence Z_θ is in the tube T. Moreover Z_θ is isomorphic to $X^{(p)}(Y)$. This implies $m = 1$ and the choice of i is unique.

To prove the uniqueness of Z_θ, recall preprojective representations P and R introduced in Section 8.2. Note that for any regular simple Z of dimension z we have

$$\langle p, z \rangle_Q = \langle r, z \rangle_Q = z_s.$$

Since R is preprojective, $\mathrm{Ext}^1_Q(R, Z) = 0$. Therefore if Z is not isomorphic to Y we have $\mathrm{Hom}_Q(R, Z) = 0$. Since any Z_τ is a quotient of R we also have $\mathrm{Hom}_Q(Z_\tau, Z) = 0$. Therefore any indecomposable representation in T isomorphic to Z_τ must be isomorphic to $X^{(p)}(Y)$. Hence the uniqueness part of the statement. □

COROLLARY 8.22. *Let T be a tube and $S_1, \ldots, S_p$ be all regular simple representations of T up to isomorphism. Then $\dim S_1 + \cdots + \dim S_p = \delta$.*

LEMMA 8.23. *Let X be regular indecomposable, then $\dim X$ is a root.*

PROOF. The result follows from Theorem 8.19 and Exercise 8.17. Indeed, for any S and r we have

$$\dim X^{(r)}(S) = m\delta + \alpha_1 + \cdots + \alpha_q,$$

where $\alpha_i := \dim(\Phi^+)^{i-1} X$, $r = mp + q$, $0 \leq q < p$. Then $\dim X^{(r)}$ is a root by Exercise 8.17. □

9. Indecomposable representations of affine quivers

THEOREM 9.1. *Let Q be a quiver whose underlying graph is affine.*

(1) *The dimension of every indecomposable representation of Q is a root.*

(2) If α is a real root, then there exists exactly one indecomposable representation of dimension α (up to isomorphism).

(3) If $\alpha = m\delta$ is an imaginary root, then there are infinitely many indecomposable representations of dimension α. Every tube contains finitely many indecomposable representations of dimension α.

PROOF. (1) Let α be the dimension of an indecomposable representation X. If $\langle \alpha, \delta \rangle \neq 0$, then X is preprojective or preinjective, and α is a real root by Corollary 7.3. If $\langle \alpha, \delta \rangle = 0$, then X is regular and α is a root by Lemma 8.23.

(2) Let α be a real root. If $\langle \alpha, \delta \rangle \neq 0$ or if $\alpha < \delta$ then the statement follows from Lemma 8.6. If $\langle \alpha, \delta \rangle = 0$, then there is a unique integer m such that $\alpha = m\delta + \beta$ with $0 < \beta < \delta$; let Y be a regular indecomposable representation of dimension β, let S be the regular simple quotient of Y, denote by p its period and by l the regular length of Y. By Theorem 8.19, we every indecomposable representation of dimension α is isomorphic to $X = X^{(mp+l)}(S)$. Hence the statement.

(3) The first assertion follows from Lemma 8.10 and the second one is a simple consequence of Theorem 8.19. $\square$

EXAMPLE 9.2. Let Q be a quiver with underlying graph $\Gamma = \tilde{A}_n$ for $n > 1$. Then we have $\delta = \varepsilon_0 + \cdots + \varepsilon_n$. The roots are in bijection with counterclockwise paths on Γ disregarding the orientation of the arrows. Indeed, write a path as a sequence of vertices $i_1, \ldots, i_s$ such that $(i_j, i_{j+1}) \in \Gamma_1$ and i_{j+1} is next to i_j in the counterclockwise direction. Then the corresponding root α equals $\varepsilon_{i_1} + \cdots + \varepsilon_{i_s}$. It is easy to see that α is imaginary if and only if $s = m(n+1)$, in which case i_1 is next to i_s in the counterclockwise direction.

Let α be a real root. We define the unique indecomposable representation of dimension α as follows. Let $V = k^s$ with fixed basis $v_1, \ldots, v_s$. For every $i \in Q_0$ define V_i to be the span of all v_j such that $i_j = i$. For every arrow γ, define $\rho_\gamma : V_{s(\gamma)} \to V_{t(\gamma)}$ by setting for each j such that $i_j = s(\gamma)$

$$\rho_\gamma(v_j) = \begin{cases} v_{j+1}, & \text{if } i_{j+1} = t(\gamma) \\ v_{j-1}, & \text{if } i_{j-1} = t(\gamma) \\ 0 \text{ otherwise} \end{cases} .$$

In particular, $\rho_\gamma(v_1) = 0$ if and only if i_1 is the target of the first arrow of the path and $\rho_\gamma(v_s) = 0$ if and only if i_s is the target of the last arrow of the path.

EXERCISE 9.3. Show that ρ is indecomposable.

Now let $\alpha = m\delta$ be an imaginary root. We define a family of indecomposable representation $\rho^{t,\beta}$ where $\beta \in Q_1$ and $t \in k$. We identify V_i with k^m for each $i \in Q_0$ and set

$$\rho_\gamma^{t,\beta} = \begin{cases} I_m, & \text{if } \gamma \neq \beta \\ J_m(t), & \text{if } \gamma = \beta \end{cases} ,$$

where I_m is the identity matrix and $J_m(t)$ is the Jordan block with eigenvalue t.

EXERCISE 9.4.

(1) Show that $\rho^{t,\beta}$ are indecomposable.
(2) If $t \neq 0$, then $\rho^{t,\beta}$ is isomorphic to $\rho^{t^{-1},\beta'}$ if β and β' have the same direction (clockwise or counterclockwise) and is isomorphic to $\rho^{t,\beta'}$ if β and β' have opposite directions.
(3) The representations $\rho^{t,\beta}$ for $t \neq 0$ and fixed β together with $\rho^{0,\beta}$ for all $\beta \in Q_1$ provide a complete list of all indecomposable representations of Q of dimension $m\delta$ up to isomorphism.

CHAPTER 9

Applications of quivers

In which we define quivers with relations and use them to study certain abelian categories. We do not resist the temptation to use quivers with relations in order to study Harish-Chandra modules over the Lie algebra $\mathfrak{sl}_2(\mathbb{C})$.

1. From abelian categories to algebras

In this section we work with abelian categories, see Chapter 5, Section 8 for definitions. If X, Y are objects of an abelian category $\mathcal{C}$, then by $\mathrm{Hom}_{\mathcal{C}}(X, Y)$ we denote the abelian group of morphisms from X to Y.

Consider an abelian category $\mathcal{C}$ which satisfies the following properties:

(1) the category is k-linear, i.e. for any pair of objects X, Y, the abelian group $\mathrm{Hom}_{\mathcal{C}}(X, Y)$ is a finite-dimensional vector space over the ground field k;
(2) there are finitely many isomorphism classes of simple objects $L_1, \ldots, L_n$;
(3) every object in $\mathcal{C}$ has finite length;
(4) the category $\mathcal{C}$ has enough projective objects, i.e. for every object X there exists a projective object P and a surjective morphism $\psi : P \to X$.

EXERCISE 1.1. Check that the category of finite-dimensional representations of a finite group G over k and the category of finite-dimensional representations of a quiver Q without loops or oriented cycles satisfy the conditions above. More generally, let A be a finite-dimensional k-algebra, then the category of finite-dimensional A-modules satisfies the conditions above.

EXERCISE 1.2. Let $\mathcal{C}$ satisfy (1)–(4). If X is an indecomposable object of $\mathcal{C}$, then any $\varphi \in \mathrm{End}_{\mathcal{C}}(X)$ is either nilpotent or invertible. Therefore the Krull–Schmidt theorem (Theorem 4.19, Chapter 5) holds for any object of $\mathcal{C}$.

EXERCISE 1.3. Prove that any object of $\mathcal{C}$ satisfying our assumptions has a projective resolution and use it to define the extension groups $\mathrm{Ext}^i_{\mathcal{C}}(X, Y)$ between two objects $X, Y \in \mathcal{C}$. Then prove that $\mathrm{Ext}^i_{\mathcal{C}}(X, Y)$ is a finite-dimensional vector space for all $i \geq 0$ and all objects $X, Y \in \mathcal{C}$.

Let A be a finite-dimensional k-algebra. An A-module structure on an object X of $\mathcal{C}$ is a k-algebra homomorphism $A \to \mathrm{End}_{\mathcal{C}}(X)$. If X has an A-module structure, then the functor $Y \mapsto \mathrm{Hom}_{\mathcal{C}}(X, Y)$ goes from $\mathcal{C}$ to the category of right A-modules

© Springer Nature Switzerland AG 2018
C. Gruson and V. Serganova, *A Journey Through Representation Theory*,
Universitext, https://doi.org/10.1007/978-3-319-98271-7_9

mod$_A$. This functor admits a left-adjoint functor, which we denote $M \mapsto M \boxtimes_A X$: this means that there exists a canonical isomorphism

$$\mathrm{Hom}_{\mathcal{C}}(M \boxtimes_A X, Y) \simeq \mathrm{Hom}_{\mathrm{mod}_A}(M, \mathrm{Hom}_{\mathcal{C}}(X, Y)),$$

in particular, $M \boxtimes_A X$ is well-defined up to a unique isomorphism.

Let us define the functor $\cdot \boxtimes_A X$. If $A = k$, or more generally if M is a free right A-module of rank n, we identify M with A^n and we may take $M \boxtimes_A X = X^{\oplus n}$. In the general case, we may consider M as the cokernel of an A-linear map $d : A^m \to A^n$, then we take for $M \boxtimes_A X$ the cokernel of the block matrix $d_X : X^{\oplus m} \to X^{\oplus n}$ in $\mathcal{C}$. There is a natural choice for d, namely the map $M \otimes_k A \otimes_k A \to M \otimes_k A$, sending $m \otimes a \otimes b$ to $ma \otimes b - m \otimes ab$ (this map is A-linear if we decide that A acts on the right factor of the tensor product).

A *projective generator* of $\mathcal{C}$ is a projective object P in $\mathcal{C}$ such that $\mathrm{Hom}_{\mathcal{C}}(P, S) \neq 0$ for every simple object S of $\mathcal{C}$.

THEOREM 1.4. *Let P be a projective generator of $\mathcal{C}$, set $A = \mathrm{End}_{\mathcal{C}}(P)$. The pair of mutually adjoint functors $\Phi = \mathrm{Hom}_{\mathcal{C}}(P, .) : \mathcal{C} \to \mathrm{mod}_A$ and $\Psi = . \boxtimes_A P : \mathrm{mod}_A \to \mathcal{C}$ define an equivalence of categories.*

PROOF. Since Φ and Ψ are mutually adjoint functors, we have canonical morphisms $\alpha_M : M \to \Phi\Psi M$ for any right-A-module M and $\beta_X : \Psi\Phi X \to X$ for any object X in $\mathcal{C}$, and we want to show that they are isomorphisms.

First, we show that $\mathrm{Coker}\,\beta_X = 0$. Indeed, if it is not true, then $\mathrm{Coker}\,\beta_X$ has a simple quotient L. Since the map β_L is not zero (because it is the image of Id_L by the isomorphism $\mathrm{Hom}_A(\Phi L, \Phi L) \simeq \mathrm{Hom}_{\mathcal{C}}(\Psi\Phi L, L)$, and $\Phi L \neq 0$ since P is a projective generator of $\mathcal{C}$), and since $\Psi\Phi$ is right-exact, the following diagram induces a contradiction:

$$
\begin{array}{ccc}
\Psi\Phi X & \xrightarrow{\beta_X} & X \\
\downarrow & & \downarrow \\
\Psi\Phi L & \xrightarrow{\beta_L} & L
\end{array}
$$

Next, we prove that $\mathrm{Ker}\,\beta_X = 0$. Indeed, by the surjectivity of β_X and the Snake lemma, we know that $\mathrm{Ker}\,\beta_X$ is a right-exact functor in the variable X. By construction, $\mathrm{Ker}\,\beta_P = 0$. Since every object in $\mathcal{C}$ is a quotient of direct sum of copies of P (once more by the surjectivity of β_X), we conclude that $\mathrm{Ker}\,\beta_X = 0$.

Finally, we show that α_M is an isomorphism. By construction, α_A is an isomorphism so α_M is an isomorphism when M is free; the statement follows since every A-module of finite length is the cokernel of a map between two free A-modules and $\Phi\Psi$ is right-exact. $\square$

Note that the above Theorem allows one to prove many facts about $\mathcal{C}$ using the results of Section 7 in Chapter 5. For example, we have the following

COROLLARY 1.5. *For any simple object L_i there exists a unique indecomposable projective object P_i (up to isomorphism), whose unique simple quotient is isomorphic to L_i. Every indecomposable projective object of $\mathcal{C}$ is isomorphic to one of $P_1, \ldots, P_n$.*

2. From categories to quivers

2.1. Quivers with relations. Let us assume that $\mathcal{C}$ satisfies conditions (1)–(4) of Section 9.1. Assume also that $\mathrm{End}_{\mathcal{C}}(L_i) = k$ for every simple object $L_1, \ldots, L_n$ of $\mathcal{C}$. We choose the direct sum $P_1 \oplus \cdots \oplus P_n$ as a projective generator P (see Corollary 1.5). The *extension quiver* (Ext-quiver for short) of $\mathcal{C}$ is the quiver with vertices $\{1, \ldots, n\}$ such that for every ordered pair of vertices (i, j), the number of arrows with source i and target j equals $\dim \mathrm{Ext}^1_{\mathcal{C}}(L_i, L_j)$.

LEMMA 2.1. *Let $P := P_1 \oplus \cdots \oplus P_n$ and $A := \mathrm{End}_{\mathcal{C}}(P)$. Then A^{op} is isomorphic to a quotient $k(Q)/\mathcal{R}$, for some two-sided-ideal $\mathcal{R}$ contained in $\mathrm{rad}^2 k(Q)$.*

PROOF. For every indecomposable P_i consider the radical filtration

$$P_i \supset \nabla^1(P_i) \supset \nabla^2(P_i) \supset \ldots.$$

Recall that $\nabla^s(P_i)$ is the minimal submodule in $\nabla^{s-1}(P_i)$ such that the quotient $\nabla^{s-1}(P_i)/\nabla^s(P_i)$ is semisimple. By Theorem 1.4, we have the following identity for the radical filtration of P (see Corollary 7.11 Chapter 5):

(9.1) $$(\mathrm{rad}^s A)P = \nabla^s(P).$$

Then $P_i/\nabla^1(P_i) \simeq L_i$ and for any simple object L_j, we consider the long exact sequence (Theorem 5.7 chapter 5):

$$0 \to \mathrm{Hom}_{\mathcal{C}}(L_i, L_j) \to \mathrm{Hom}_{\mathcal{C}}(P_i, L_j)$$
$$\to \mathrm{Hom}_{\mathcal{C}}(\nabla^1(P_i), L_j) \to \mathrm{Ext}^1_{\mathcal{C}}(L_i, L_j) \to \mathrm{Ext}^1_{\mathcal{C}}(P_i, L_j) = 0.$$

Since by definition of P_i the map $\mathrm{Hom}_{\mathcal{C}}(L_i, L_j) \to \mathrm{Hom}_{\mathcal{C}}(P_i, L_j)$ is an isomorphism, we also have an isomorphism

$$\mathrm{Ext}^1_{\mathcal{C}}(L_i, L_j) \simeq \mathrm{Hom}_{\mathcal{C}}(\nabla^1(P_i), L_j) \simeq \mathrm{Hom}_{\mathcal{C}}(\nabla^1(P_i)/\nabla^2(P_i), L_j).$$

Since $\nabla^1(P_i)/\nabla^2(P_i)$ is semisimple,

$$\mathrm{Hom}_{\mathcal{C}}(\nabla^1(P_i)/\nabla^2(P_i), L_j) \simeq \mathrm{Hom}_{\mathcal{C}}(L_j, \nabla^1(P_i)/\nabla^2(P_i)).$$

Combining all this together we finally get

$$\mathrm{Ext}^1_{\mathcal{C}}(L_i, L_j) \simeq \mathrm{Hom}_{\mathcal{C}}(L_j, \nabla^1(P_i)/\nabla^2(P_i)) \simeq \mathrm{Hom}_{\mathcal{C}}(P_j, \nabla^1(P_i)/\nabla^2(P_i)).$$

Now we fix a basis in $\mathrm{Hom}_{\mathcal{C}}(P_j, \nabla^1(P_i)/\nabla^2(P_i))$ enumerated by arrows of the opposite quiver Q^{op} with source j and target i. By a slight abuse of notation we can consider every arrow $\gamma = (j \to i)$ as an element of $\mathrm{Hom}_{\mathcal{C}}(P_j, \nabla^1(P_i)/\nabla^2(P_i))$.

By projectivity of P_j, the natural map

$$\theta : \mathrm{Hom}_{\mathcal{C}}(P_j, \nabla^1(P_i)) \to \mathrm{Hom}_{\mathcal{C}}(P_j, \nabla^1(P_i)/\nabla^2(P_i))$$

is surjective.

We construct a homomorphism $F : k(Q^{op}) \to A$ by setting $F(e_i)$ to be the projector $P \to P_i$ with kernel $\oplus_{j \neq i} P_j$ and $F(\gamma)$ to be a element in the preimage $\theta^{-1}(\gamma)$. This homomorphism depends on the choice of bases in $\mathrm{Hom}_{\mathcal{C}}(P_j, \nabla^1(P_i)/\nabla^2(P_i))$ and on the choice of a lift of γ by θ.

It follows immediately from the construction of F and (9.1) that the induced homomorphism $k(Q^{op}) \to A/\mathrm{rad}^2 A$ is surjective. It is easy to show by induction on s that $k(Q^{op}) \to A/\mathrm{rad}^s A$ is surjective for all s. Since $\mathrm{rad}^s A = 0$ for sufficiently large s, F is surjective. $\qquad\square$

COROLLARY 2.2. *The category $\mathcal{C}$ with Ext-quiver Q is equivalent to the category of finite-dimensional left $B_{\mathcal{C}}$-modules, where $B_{\mathcal{C}} = k(Q)/R$ for some ideal R of $k(Q)$.*

PROOF. Follows from Lemma 2.1 and Theorem 1.4, since $B_{\mathcal{C}}$ is isomorphic to A^{op}. $\qquad\square$

The representations of $k(Q)/R$ are called representations of the quiver Q with relations R. In general, classifying indecomposable representations of $k(Q)/R$ may be a very difficult problem. There are however two boundary cases, when the problem can be reduced to the ordinary representation theory of quivers. The first case is $R = 0$, where one can use the results of two previous chapters. The second case is when R is maximal possible, i.e. the product of any two arrows is zero. In this case the following trick of *duplicating the quiver* may be useful:

Let Q be a quiver with set of vertices $Q_0 = \{1, \dots, n\}$ and assume that the ideal $R \subset k(Q)$ is generated by all paths of length 2. Consider the quiver $\bar{Q}$, with set of vertices $\{1^-, \dots, n^-, 1^+, \dots n^+\}$ and with set of arrows $\bar{Q}_1$ defined by

$$\bar{Q}_1 := \{(i^- \to j^+) \,|\, (i \to j) \in Q_1\}.$$

For example, if Q is a loop $\overset{\gamma}{\underset{\bullet^1}{\curvearrowright}}$, then $\bar{Q}$ is $\bullet^{1^-} \to \bullet^{1^+}$.

Let (V, ρ) be a representation of $k(Q)/R$, then for every $i \in Q_0$ we have

$$\sum_{\gamma \in Q_+(i)} \mathrm{Im}\,\gamma \subset \bigcap_{\gamma \in Q_-(i)} \mathrm{Ker}\,\gamma.$$

If we assume that (V, ρ) is indecomposable but not irreducible, then the above embedding becomes the equality

$$\sum_{\gamma \in Q_+(i)} \mathrm{Im}\,\gamma = \bigcap_{\gamma \in Q_-(i)} \mathrm{Ker}\,\gamma.$$

We construct a representation $(\bar{V}, \bar{\rho})$ of $\bar{Q}$ in the following way. We set

$$\bar{V}_{i^+} := \bigcap_{\gamma \in Q_-(i)} \mathrm{Ker}\,\gamma, \qquad \bar{V}_{i^-} := V_i/\bar{V}_{i^+}.$$

For any $\gamma = (i \to j) \in Q_1$, the map $\rho_\gamma : V_i \to V_j$ induces in the natural way the map $\bar{\rho}_{\bar{\gamma}} : V_{i^-} \to V_{j^+}$, where $\bar{\gamma} = (i^- \to j^+) \in \bar{Q}_1$.

Conversely, let $(\bar{V}, \bar{\rho})$ be a representation of $\bar{Q}$. We can construct a representation (V, ρ) of $k(Q)/R$ by setting

$$V_i := \bar{V}_{i^-} \oplus \bar{V}_{i^+}, \quad \rho_\gamma := \bar{\rho}_{\bar{\gamma}} \text{ for all } \bar{\gamma} = (i^- \to j^+), \gamma = (i \to j).$$

The following statement is straightforward.

LEMMA 2.3. *The map* $(V, \rho) \mapsto (\bar{V}, \bar{\rho})$ *defines a bijection between isomorphism classes of non-irreducible indecomposable representations of* $k(Q)/R$ *and isomorphism classes of non-irreducible indecomposable representations of* $\bar{Q}$.

EXAMPLE 2.4. Let us revisit Example 7.17 from Chapter 5. Let $\mathcal{C}$ be the category of finite-dimensional representations of S_3 over $\mathbb{F}_3$. There are exactly two simple objects (up to isomorphism) in $\mathcal{C}$, the trivial representation and the sign representation. The corresponding indecomposable projectives are $P_+ := \mathrm{Ind}_{S_2}^{S_3} \mathrm{triv}$ and $P_- := \mathrm{Ind}_{S_2}^{S_3} \mathrm{sgn}$. The Ext-quiver Q is

$$\bullet \underset{\beta}{\overset{\alpha}{\rightleftarrows}} \bullet.$$

It is not difficult to check, using the radical filtration of $P_\pm$ calculated in Example 7.17 in Chapter 5, that the ideal R is generated by $\alpha\beta\alpha$ and $\beta\alpha\beta$. The classification of all indecomposable representations of S_3 (up to isomorphism) is therefore equivalent to the classification of indecomposable left $B_{\mathcal{C}}$-modules. If we set $P_1 := \Phi P_+$ and $P_2 := \Phi P_-$, then $P_1 = Ae_1$ and $P_2 = Ae_2$, where e_1, e_2 are the primitive idempotents of $B_{\mathcal{C}}$ corresponding to the vertices of Q. Let us note also that in this case $B_{\mathcal{C}}$ is isomorphic to $B_{\mathcal{C}}^{op}$. If P is an indecomposable projective module, then its dual P^* is an indecomposable injective $B_{\mathcal{C}}^{op}$-module. By direct inspection one can see that P_1 and P_2 are self-dual, hence they are injective.

Suppose that M is an indecomposable $B_{\mathcal{C}}$-module. Assume that $\beta\alpha M \neq 0$. There exists a non-zero vector $m \in M$ such that $m, \alpha m, \beta\alpha m$ are linearly independent. This implies that M contains a submodule isomorphic to P_1. Since P_1 is injective and M is indecomposable we obtain that M is isomorphic to P_1. Similarly, if $\alpha\beta M \neq 0$ we obtain that M is isomorphic to P_2.

Next, we assume that $\beta\alpha M = \alpha\beta M = 0$. Then we can use Lemma 2.3. Indeed, $\bar{Q}$ is the disjoint union of two A_2 quivers:

$$\bullet \to \bullet \quad \bullet \to \bullet.$$

Hence $k(Q)/(\alpha\beta, \beta\alpha)$ has four indecomposable representations: two of dimension $(1,1)$ and two, which are irreducible, of dimension $(0,1)$ and $(1,0)$, respectively. In particular, we obtain that in this case every indecomposable representation is a quotient of an indecomposable projective representation.

If we drop the assumption that the number of simples of $\mathcal{C}$ is finite, we still have that $\mathcal{C}$ is equivalent to the category of representations of the (infinite) Ext-quiver with relations. This can be proven by passing to the truncated categories which have just finitely many simple objects and for which Corollary 2.2 holds. Instead of giving a proof here we illustrate this point by the following example.

EXAMPLE 2.5. Let Λ be the Grassmann algebra with two generators, i.e.

$$\Lambda = k < x, y > / \left(x^2, y^2, xy + yx \right).$$

Consider the $\mathbb{Z}$-grading $\Lambda = \Lambda_0 \oplus \Lambda_1 \oplus \Lambda_2$, where

$$\Lambda_0 = k \quad \Lambda_1 = kx + ky, \quad \Lambda_2 = kxy.$$

Let $\mathcal{C}$ be the category of graded Λ-modules. The objects of $\mathcal{C}$ are Λ-modules $M = \bigoplus_{i \in \mathbb{Z}} M_i$, such that $\Lambda_i M_j \subset M_{i+j}$, and we assume that morphisms preserve the grading. We leave it to the reader to check that all simple modules (up to isomorphism) are one-dimensional and hence are determined uniquely by the degree. Indecomposable projective modules are free of rank 1 and are parametrized by integers. An indecomposable projective module P_i is isomorphic to Λ with shifted grading: $\deg(1) = i$. The Ext-quiver Q has infinitely many vertices:

$$\cdots \bullet \underset{\beta_i}{\overset{\alpha_i}{\rightrightarrows}} \bullet \underset{\beta_{i+1}}{\overset{\alpha_{i+1}}{\rightrightarrows}} \bullet \underset{\beta_{i+2}}{\overset{\alpha_{i+2}}{\rightrightarrows}} \bullet \cdots,$$

and it is not hard to check that the defining relations are

$$\alpha_{i+1}\alpha_i = \beta_{i+1}\beta_i = 0, \quad \alpha_{i+1}\beta_i + \beta_{i+1}\alpha_i = 0$$

for all $i \in \mathbb{Z}$.

Let us classify the indecomposable representations of this quiver with relations. First observe that, as in the previous example, any indecomposable projective is injective. As in the previous example we first assume that $\alpha_{i+1}\beta_i M \neq 0$ for some $i \in \mathbb{Z}$. Then if M is indecomposable, M is isomorphic P_i by an argument similar to the one in the previous example.

Next we consider the case when $\alpha_{i+1}\beta_i M = 0$ for all i. Then we should apply again Lemma 2.3. Note that $\bar{Q}$ is a disjoint union of infinitely many Kronecker quivers. Thus, if M is an indecomposable object which is neither projective nor simple, then M is isomorphic to the corresponding representation of the Kronecker quiver.

REMARK 2.6. The reader who is familiar with algebraic geometry could read [3] in order to understand the link between the "derived category" of coherent sheaves over the projective line and the "derived category" of $\mathcal{C}$.

3. Finitely represented, tame and wild algebras

Let C be a finite-dimensional k-algebra. We say that C is *finitely represented* if C has finitely many indecomposable representations. We call C *tame* if for each $d \subset \mathbb{Z}_{>0}$, there exists a finite set $M_1, \ldots, M_r$ of $C - k[x]$ bimodules (free of rank d over $k[x]$) such that every indecomposable representation of C of dimension d is isomorphic to $M_i \otimes_{k[x]} k[x]/(x-\lambda)$ for some $i \le r$, $\lambda \in k$. Finally, C is *wild* if there exists a $C - k<x,y>$ bimodule M such that the functor $X \mapsto M \otimes_{k<x,y>} X$ preserves indecomposability and is faithful. We formulate here without proof the following fundamental results. The first one is called Drozd's trichotomy.

THEOREM 3.1. [9] *Every finite-dimensional algebra over an algebraically closed field k is either finitely represented or tame or wild.*

THEOREM 3.2. [9] *Let Alg_n be the algebraic variety of all associative algebra structures on a given n-dimensional k-vector space. Then the subset of finitely represented algebras is Zariski open in Alg_n.*

REMARK 3.3. If C is a category satisfying the assumptions (1)–(4) of section 1, then by Theorem 1.4, it is equivalent to the category of representations of some finite-dimensional algebra. Therefore, the notions of finitely-represented, tame and wild are well-defined for such categories.

Let us consider the case where $C = k(Q)$ is the path algebra of a quiver Q. Gabriel's theorem (Theorem 7.2, Chapter 7) implies that C is finitely represented if and only if all connected components of Q are Coxeter–Dynkin graphs. We have seen in Chapter 8 that the path algebra of an affine quiver is not finitely represented. On the other hand, it is tame, since indecomposable representation in each dimension can be parametrized by at most one continuous parameter. The following result claims that all other quivers are wild.

THEOREM 3.4. [13] *Let Q be a connected quiver without oriented cycles. Then $k(Q)$ is finitely represented if and only if Q is Dynkin, $k(Q)$ is tame if and only if Q is affine.*

In general, the classification of tame algebras seems to be an impossible problem. If the Ext-quiver has just a few vertices such a classification can be found in K. Erdmann's book [10]. However, in applications it is often possible to determine whether the category in question is tame or wild.

EXERCISE 3.5. If an algebra C is finitely represented then any quotient algebra C/I is finitely represented. If C is tame, then any quotient algebra C/I is finitely represented or tame.

EXERCISE 3.6. Let C be a category satisfying the assumptions (1)–(4) and $\operatorname{End}_C(S) \simeq k$ for any simple object S. Let Q be its Ext-quiver.

(1) If $\mathcal{C}$ is tame, then $\bar{Q}$ is a disjoint union of Dynkin and affine quivers.
(2) If Q is a disjoint union of Dynkin and affine quivers, then $\mathcal{C}$ is tame.
(3) If Q is a disjoint union of Dynkin quivers, then $\mathcal{C}$ is finitely represented.

EXERCISE 3.7. Generalize Example 2.5 to the Grassman algebra

$$\Lambda = k < x_1, \ldots, x_n > / \left(x_i^2, x_i x_j + x_j x_i \right)$$

with n generators. Describe the Ext-quiver for the category $\mathcal{C}$ of $\mathbb{Z}$-graded Λ-modules and show that for $n > 2$ the category $\mathcal{C}$ is wild.

4. Frobenius algebras

Let A be a finite-dimensional algebra over k. We use the notation $_A$ mod (resp. mod$_A$) for the categories of finite-dimensional left (resp. right) A-modules. We denote by $_A A$ (resp. A_A) the algebra A considered as a left (resp. right) module over itself. We have an exact contravariant functor

$$D : _A \text{mod} \to \text{mod}_A$$

defined by $D(M) = M^*$. Obviously, D sends projective objects to injective objects and vice versa.

A finite-dimensional A algebra over k is called a *Frobenius algebra* if $D(A_A)$ is isomorphic to $_A A$.

THEOREM 4.1. *The following conditions on A are equivalent*

(1) *A is a Frobenius algebra;*
(2) *There exists a non-degenerate bilinear form $\langle \cdot, \cdot \rangle$ on A such that $\langle ab, c \rangle = \langle a, bc \rangle$;*
(3) *There exists $\lambda \in A^*$ such that $\operatorname{Ker} \lambda$ does not contain non-trivial left or right ideals.*

PROOF. A form $\langle \cdot, \cdot \rangle$ gives an isomorphism $\mu \colon A \to A^*$ by the formula $x \to \langle \cdot, x \rangle$. The condition $\langle ab, c \rangle = \langle a, bc \rangle$ is equivalent to μ being a homomorphism of modules. The linear functional λ can be constructed by $\lambda(x) = \langle 1, x \rangle$. Conversely, given λ, one can define $\langle x, y \rangle = \lambda(xy)$. The assumption that $\operatorname{Ker} \lambda$ does not contain non-trivial one-sided ideals is equivalent to the condition that the left and right kernels of $\langle \cdot, \cdot \rangle$ are zero. $\square$

LEMMA 4.2. *Let A be a Frobenius algebra. A left A-module X is projective if and only if it is injective.*

PROOF. A projective module X is a direct summand of a free module, but a free module is injective since $D(A_A)$ is isomorphic to $_A A$. Hence, X is injective. By duality, an injective module is projective. $\square$

EXAMPLE 4.3. A group algebra $k(G)$ is Frobenius. Take

$$\lambda \left(\sum_{g \in G} a_g g \right) = a_1.$$

The corresponding bilinear form is symmetric.

EXAMPLE 4.4. The Grassmann algebra $\Lambda = k < x_1, \ldots, x_n > / \left(x_i^2, x_i x_j + x_j x_i \right)$ is Frobenius. Put

$$\lambda \left(\sum_{i_1 < \cdots < i_k} c_{i_1 \ldots i_k} x_{i_1} \ldots x_{i_k} \right) = c_{12 \ldots n}.$$

In a sense Frobenius algebras generalize group algebras. For example, if $T \in \mathrm{Hom}_k (X, Y)$ for two $k(G)$-modules X and Y then

$$\bar{T} := \sum_{g \in G} g T g^{-1} \in \mathrm{Hom}_G (X, Y).$$

The idea of taking average over the group is very important in representation theory. It can be generalized for Frobenius algebras.

Choose a basis $e_1, \ldots, e_n$ in a Frobenius algebra A. Let $f_1, \ldots, f_n$ be the dual basis, i.e.

(9.2) $$\langle f_i, e_j \rangle = \delta_{ij}.$$

Every $a \in A$ can be written

(9.3) $$a = \sum \langle f_i, a \rangle e_i = \sum \langle a, e_i \rangle f_i.$$

and

(9.4) $$\sum a e_i \otimes f_i = \sum \langle f_j, a e_i \rangle e_j \otimes f_i = \sum \langle f_j a, e_i \rangle e_j \otimes f_i = \sum e_j \otimes f_j a.$$

LEMMA 4.5. Let X and Y be A-modules, $T \in \mathrm{Hom}_k (X, Y)$. Define $\bar{T}$ by the formula

$$\bar{T} = \sum e_i T f_i.$$

Then $\bar{T} \in \mathrm{Hom}_A (X, Y)$.

PROOF. This is a direct calculation using (9.3) and (9.4). $\square$

EXAMPLE 4.6. If $A = k(G)$, the dual bases can be chosen as $\{g\}_{g \in G}$ and $\{g^{-1}\}_{g \in G}$. Hence $\bar{T} = \sum g T g^{-1}$.

In a Frobenius algebra one can use the following criterion of projectivity:

THEOREM 4.7. An A-module X is injective (hence projective) if there exists $T \in \mathrm{End}_k (X)$ such that $\bar{T} = \mathrm{Id}_X$.

PROOF. First, let us assume the existence of T. We have to show that X is injective, i.e. for any embedding $\varepsilon\colon X \to Y$ there exists $\pi \in \mathrm{Hom}_A(Y, X)$ such that $\pi \circ \varepsilon = Id$. Consider a k-linear map $p \in \mathrm{Hom}_k(Y, X)$ such that $p \circ \varepsilon = Id$. Put $\pi = \sum e_i T p f_i$. Then for any $x \in X$ we have

$$\pi(\varepsilon(x)) = \sum e_i T p f_i(\varepsilon(x)) = \sum e_i T(p\varepsilon(f_i x)) = \sum e_i T(f_i x) = \bar{T}x = \mathrm{Id}_X,$$

using $f_i \varepsilon = \varepsilon f_i$. By Lemma 4.5, $\pi \in \mathrm{Hom}_A(X, Y)$.

Now assume that X is injective. We define a map $\delta\colon X \to A \otimes_k X$ by the formula

$$\delta(x) := \sum e_i \otimes f_i x.$$

Then $\delta \in \mathrm{Hom}_A(X, A \otimes_k X)$ by (9.4). Furthermore, since $f_i x$ span Ax, δ is injective. Since X is an injective A-module, there exists $\tau\colon A \otimes_k X \to X$ such that $\tau \circ \delta = \mathrm{Id}_X$. We define $S \in \mathrm{Hom}_k(A \otimes_k X, A \otimes_k X)$ by the formula

$$S(a \otimes x) = \langle 1, a \rangle 1 \otimes x.$$

Then

$$\bar{S}(a \otimes x) = \sum e_i S(f_i a \otimes x) = \sum \langle 1, f_i a \rangle e_i \otimes x = \sum \langle f_i, a \rangle e_i \otimes x = a \otimes x$$

due to (9.3). Set $T := \tau \circ S \circ \delta$. Then $\bar{T} = \mathrm{Id}_X$. $\qquad\square$

5. Application to group algebras

Let G be a finite group. The goal of this section is to determine when the group algebra $k(G)$ is finitely represented. If $\mathrm{char}\, k = 0$, or $\mathrm{char}\, k$ does not divide $|G|$, then $k(G)$ is finitely represented because it is semisimple. Hence in this section we assume that $\mathrm{char}\, k = p > 0$ and that p divides $|G|$.

5.1. Relative projective and injective modules in a group algebra. Let H be a subgroup of a group G. A $k(G)$-module X is H-*injective* if any exact sequence of $k(G)$-modules

$$0 \to X \to Y \to Z \to 0,$$

which splits over $k(H)$, splits over $k(G)$.

In the similar way one defines H-*projective* module.

Let $\{g_1, \ldots, g_r\}$ be a set of representatives in the set of left cosets G/H. For any $k(G)$-modules X, Y, and $T \in \mathrm{Hom}_H(X, Y)$ put

$$\bar{T} = \sum g_i T g_i^{-1}.$$

EXERCISE 5.1. Show that $\bar{T}$ does not depend on the choice of representatives and $\bar{T} \in \mathrm{Hom}_G(X, Y)$.

THEOREM 5.2. *The following conditions on* $k(G)$-*module* X *are equivalent*

(1) X *is* H-*injective*;

(2) X *is a direct summand in* $\mathrm{Ind}_H^G X$;

(3) X is H-projective;

(4) There exists $T \in \operatorname{End}_H (X)$ such that $\bar{T} = \operatorname{Id}$.

PROOF. This theorem is very similar to Theorem 4.7. To prove $1 \Rightarrow 2$ we check that $\delta \colon X \to \operatorname{Ind}_H^G X$, defined by the formula

$$\delta (x) = \sum g_i \otimes g_i^{-1} x,$$

defines an embedding of X. By injectivity X is a direct summand of $\operatorname{Ind}_H^G X$.

To prove $3 \Rightarrow 2$ we use the projection $\operatorname{Ind}_H^G X \to X$ defined by $g \otimes x \mapsto gx$.

Now let us prove $2 \Rightarrow 4$. We define $S \colon \operatorname{Ind}_H^G X \to \operatorname{Ind}_H^G X$ by

$$S \left(\sum g_i \otimes x_i \right) = 1 \otimes x_1,$$

here we assume that $g_1 = 1$. It is easily checked that $S \in \operatorname{End}_H \left(\operatorname{Ind}_H^G X \right)$ and $\bar{S} = Id$. Then we obtain $T = \tau \circ S \circ \delta$, where $\tau \colon \operatorname{Ind}_H^G X \to X$ is the projection such that $\tau \circ \delta = \operatorname{Id}$.

It remains to prove $4 \Rightarrow 1$ and $4 \Rightarrow 3$. Both proofs are similar to the first part of the proof of Theorem 4.7 and we leave them to the reader. $\qquad \square$

Let p be prime. Recall that if $|G| = p^s r$ with $(p,r) = 1$, then there exists a subgroup P of order p^s. It is called a Sylow p-subgroup. All Sylow p-subgroups are conjugate in G. In what follows we will use the following statement:

COROLLARY 5.3. Let char $k = p$ and P be a Sylow p-subgroup. Then every $k(G)$-module X is P-projective.

PROOF. We have to check condition (4) from Theorem 5.2. But $r = [G : P]$ is invertible in k. So we can put $T = \frac{1}{r} \operatorname{Id}$. $\qquad \square$

5.2. Finitely represented group algebras. We assume that char $k = p$ and $|G| = p^s r$ with $(p,r) = 1$ and $s > 0$. In this subsection we will prove the following:

THEOREM 5.4. The group algebra $k(G)$ is finitely represented over a field of characteristic p if and only if any Sylow p-subgroup of G is cyclic.

LEMMA 5.5. Let H be a cyclic p-group, i.e. $|H| = p^s$. Then there are exactly p^s isomorphism classes of indecomposable representations of H over k, exactly one for each dimension. More precisely each indecomposable L_m of dimension $m \leq p^s$ is isomorphic to $k(H) / (g - 1)^m$, where g is a generator of H.

PROOF. We use

$$k(H) \cong k[g]/(g^{p^s} - 1) \cong k[\alpha] / \alpha^{p^s},$$

where $\alpha = g - 1$ Let M be an m-dimensional indecomposable H-module. Then the matrix of α in a suitable basis of M is the nilpotent Jordan block of size m and $m \leq p^s$. $\qquad \square$

LEMMA 5.6. *If a Sylow p-subgroup P of G is cyclic, then $k(G)$ is finitely represented. Moreover, the number of indecomposable $k(G)$-modules is not greater than $|G|$.*

PROOF. Let $\{Y_i \mid i = 1, \ldots, p^s\}$ be the set of all indecomposable P-modules (up to isomorphism), $\dim Y_i = i$. Let Θ_i denote the set of isomorphism classes of indecomposable G-modules such that the restriction of this module to P has indecomposable components of dimension i or higher. Let m_i denote the number of G-indecomposable components from Θ_i which appear in the induced module $\operatorname{Ind}_P^G Y_i$. By comparing dimensions we know that $m_i \leq r$.

By Corollary 5.3 every indecomposable $k(G)$-module X is P-projective. We have a canonical exact sequence of G-modules:

$$0 \to N \to \operatorname{Ind}_P^G X \to X \to 0,$$

which splits over P and hence over G. Therefore X is isomorphic to a direct summand of $\operatorname{Ind}_P^G X$ and hence of $\operatorname{Ind}_P^G Y_i$ for some i.

This implies that the number of all indecomposable $k(G)$-modules (up to isomorphism) does not exceed $\sum_{i=1}^{p^s} m_i \leq |G|$.   □

LEMMA 5.7. *If P is a non-cyclic p-group, then P contains a normal subgroup N such that $P/N \cong \mathbb{Z}_p \times \mathbb{Z}_p$.*

PROOF. This statement can be obtained easily by induction on $|P|$. If P is abelian, the statement follows from the classification of finite abelian groups. If P is not abelian, then P has a non-trivial center Z and the quotient P/Z is not cyclic. By induction assumption we can choose a normal subgroup $N' \subset P/Z$ such that the quotient $(P/Z)/N'$ is isomorphic to $\mathbb{Z}_p \times \mathbb{Z}_p$. We set $N = \pi^{-1}(N')$, where $\pi : P \to P/Z$ is the canonical projection.   □

LEMMA 5.8. *The group $Z = \mathbb{Z}_p \times \mathbb{Z}_p$ has an indecomposable representation of dimension n for each $n \in \mathbb{Z}_{>0}$.*

PROOF. Note that $k(Z)$ has a unique simple module (up to isomorphism). Hence the unique indecomposable projective Z-module is isomorphic to $k(Z)$. Let g and h be generators of Z, $\alpha = g - 1$, $\beta = h - 1$. Then $k(Z) \simeq k[\alpha, \beta]/(\alpha^p, \beta^p)$. Therefore the Ext-quiver Q of the category of finite-dimensional Z-modules has one vertex and two loops. The quiver $\bar{Q}$ is the Kronecker quiver. It has at least one indecomposable representation in each dimension.   □

LEMMA 5.9. *If a Sylow p-subgroup P of G is not cyclic, then G has an indecomposable representation of arbitrary high dimension.*

PROOF. By Lemma 5.7 and Lemma 5.8, P has an indecomposable representation Y of dimension n for any positive integer n. Decompose $\operatorname{Ind}_P^G Y$ into direct sum of indecomposable $k(G)$-modules. At least one indecomposable component X contains Y as an indecomposable $k(P)$-component. Hence $\dim X \geq n$.   □

Lemma 5.6 and Lemma 5.9 imply Theorem 5.4.

6. On certain categories of $\mathfrak{sl}_2$-modules

In this section we show how quivers can be used to study the category of Harish-Chandra modules for the Lie algebra $\mathfrak{sl}_2$. Let us point out that representation theory of Lie algebras is a very deep and popular topic with multitude of applications. Here we just touch it, hoping that it will work as a motivation for further study of the subject.

6.1. Lie algebras.
A *Lie algebra* $\mathfrak{g}$ is a vector space equipped with a bilinear operation (the *Lie bracket*), $[\cdot, \cdot] : \mathfrak{g} \otimes \mathfrak{g} \to k$, satisfying the following conditions for all $x, y, z \in \mathfrak{g}$:

(1) $[x, y] = -[y, x]$;
(2) $[[x, y], z] + [[y, z], x] + [[z, x], y] = 0$.

EXERCISE 6.1. Let A be an associative algebra. Show that the bracket $[a, b] := ab - ba$ defines a Lie algebra structure on A.

For instance:

EXAMPLE 6.2. The Lie algebra $\mathfrak{sl}_n(k)$ is the algebra of traceless $n \times n$ matrices, the Lie bracket is given by the commutator $[x, y] = xy - yx$, and its dimension is equal to $n^2 - 1$.

The importance of Lie algebras is related to the fact that they can be seen as the tangent spaces to continuous groups at the identity element. Let us illustrate this with an example. Let $k = \mathbb{R}$ or $\mathbb{C}$. Consider the group $G = SL_n(\mathbb{R})$ or $SL_n(\mathbb{C})$ consisting of $n \times n$ matrices with determinant 1. Let $X(t) : \mathbb{R} \to G$ be a smooth function such that $X(0) = I_n$. Let $x := X'(0) = \frac{d}{dt} X(t)|_{t=0}$. By differentiating the relation $\det X(t) = 1$ we obtain

$$\frac{d}{dt} \det X(t)|_{t=0} = \operatorname{Tr} X'(0) = 0.$$

Therefore $x \in \mathfrak{sl}_n(k)$. Conversely, if $x \in \mathfrak{sl}_n(k)$, then $X(t) = e^{tx}$ is a path in G (called a *1-parameter subgroup*) such that $X(0) = 1$ and $X'(0) = x$.

EXERCISE 6.3. Consider two paths $X(t), Y(s) : \mathbb{R} \to G$ such that $X(0) = Y(0) = 1$, $X'(0) = x$, $Y'(0) = y$, then show that

(1) $\frac{d}{dt} X(t) Y(t)|_{t=0} = x + y$,
(2) $\frac{\partial^2}{\partial t \partial s} X(t) Y(s) X^{-1}(t) Y^{-1}(s)|_{t=0, s=0} = [x, y]$.

Let $\rho : G \to GL(V)$ be a finite-dimensional representation of G. We can differentiate ρ to define a linear map $\mathfrak{g} \otimes V \to V$, by the formula

$$xv := \frac{d}{dt} X(t) v|_{t=0}.$$

It follows from Exercise 6.3 that this linear map satisfies the relation

$$[x, y]v = x(yv) - y(xv).$$

This leads to the notion of $\mathfrak{g}$-module: a vector space M is a $\mathfrak{g}$-module if there is an action $\mathfrak{g} \otimes M \to M$ such that $[x, y]m = xym - yxm$ for all $x, y \in \mathfrak{g}$ and $m \in M$.

EXERCISE 6.4. The *adjoint* module. Consider the action ad of $\mathfrak{g}$ on itself given by $\mathrm{ad}(x)y := [x, y]$. Check that ad defines a $\mathfrak{g}$-module structure on $\mathfrak{g}$. Let $G = SL_n$, $X(t) : \mathbb{R} \to G$, $X(0) = 1$, $X'(0) = x$. Check that for all $y \in \mathfrak{g}$

$$\mathrm{ad}(x)y = \frac{d}{dt}X(t)yX^{-1}(t)|_{t=0}.$$

EXERCISE 6.5. Let M and N be $\mathfrak{g}$-modules. Define a $\mathfrak{g}$-action on M^* and on $M \otimes N$ by the formulae

$$x\varphi(m) := -\varphi(xm), \quad x(m \otimes n) := xm \otimes n + m \otimes xn, \quad x \in \mathfrak{g}, \; m \in M, \varphi \in M^*, n \in N.$$

Check that M^* and $M \otimes N$ are $\mathfrak{g}$-modules.

6.2. Harish-Chandra modules: motivation.

This subsection is a sketch of the construction of Harish-Chandra modules over $\mathfrak{sl}_n$, it is not necessary to understand the details in order to read the remaining part of the chapter.

We explained above how to define the $\mathfrak{g}$-module associated to a finite-dimensional representation of the group G. But what to do if the representation is infinite-dimensional?

Let $\rho : G \to U(V)$ be a unitary representation of G. We would like to differentiate matrix coefficients. It is clear, however, that $\langle gv, w \rangle$ can not be a smooth function of $g \in G$ for all $v, w \in V$. One can prove the existence of a G-stable dense subspace $V_0 \subset V$ such that $\langle gv, w \rangle$ is a smooth function of $g \in G$ for all $v, w \in V_0$.[1]

Now for any $x \in \mathfrak{g}$ and $v \in V_0$ we can define $xv \in V_0^*$ by setting

$$\langle xv, w \rangle := \frac{d}{dt}\langle X(t)v, w \rangle|_{t=0},$$

where as usual $X(t)$ is a path in G tangent to x at $t = 0$.

In this way we define a linear map $\mathfrak{g} \otimes V_0 \to V_0^*$ where V_0^* is the algebraic dual of V_0. Harish-Chandra discovered a remarkable dense subspace M in V_0 which is $\mathfrak{g}$-invariant.

Let us consider a maximal compact subgroup K of G: if $G = SL_n(\mathbb{R})$ then we can take $K = SO_n$. Since all irreducible unitary representations of K are finite-dimensional, we can write $V = \hat{\oplus}_{\rho \in \hat{K}} W_\rho$, where W_ρ denote K-isotypic components. Let us assume that every W_ρ is finite-dimensional (this can be proven for irreducible V).

---

[1] For instance we may take for V_0 the subspace of V spanned by the integrals $\int_G f(g)gvdg$ ($v \in V$ f being a compactly supported smooth function on G); or the space of C^∞-vectors in V, see [18].

Consider the usual direct sum $M = \oplus_{\rho \in \hat{K}} W_\rho$. Then M is dense in V, it is the sum of all irreducible K-submodules in V. Furthermore, $M \subset V_0$ since V_0 is stable under the projector on every isotypic K-component (use Exercise 2.10, Chapter 3). Let us show that M is invariant under the action of $\mathfrak{g}$. Indeed, every $v \in M$ belongs to a finite-dimensional K-stable W. Consider the map $\alpha : \mathfrak{g} \otimes W \to V_0^*$ defined as the restriction of the map $\mathfrak{g} \otimes V_0 \to V_0^*$. For any $w \in W$, $g \in K$ and $x \in \mathfrak{g}$ we have $g(xw) = gxg^{-1}gw$. Therefore α is a homomorphism of K-modules, where K acts on $\mathfrak{g}$ by conjugation. So the image of α is a finite-dimensional K-submodule of V^*. The assumption about finite multiplicities of K-modules ensures that $\operatorname{Im} \alpha \subset M$.

This construction motivates the following definition. A *Harish-Chandra module* is a $\mathfrak{g}$-module M, which is a direct sum of irreducible K-modules with finite multiplicities, such that for any smooth path $X(t) : \mathbb{R} \to K$, with $X(0) = 1$ and $X'(0) = x$, and $m \in M$ we have

$$xm = \frac{d}{dt}X(t)m|_{t=0}.$$

For any topologically irreducible unitary representation V of G, we can construct the Harish-Chandra module $M \subset V$. It is not hard to show that M is simple in the algebraic sense, i.e. M does not contain any proper non-zero $\mathfrak{g}$-invariant subspace. We do not construct in this way every simple Harish-Chandra $\mathfrak{g}$-module, but anyway one can consider the classification of simple Harish-Chandra modules as a first step towards describing the unitary dual of G.

From now on, we assume that $G = SL_2(\mathbb{R})$ and $K = SO_2 \simeq S^1$. In matrix form

$$K = \{\begin{pmatrix} a & b \\ -b & a \end{pmatrix} \mid a^2 + b^2 = 1\},$$

we set $z = a + bi$. Let $\mathfrak{k}$ be the Lie algebra of K. It is easy to see that $\mathfrak{k} = \mathbb{R}u$, where $u = \begin{pmatrix} 0 & 1 \\ -1 & 0 \end{pmatrix}$. In this way every element of K can be written in a unique way

$$g_\theta := \begin{pmatrix} \cos\theta & \sin\theta \\ -\sin\theta & \cos\theta \end{pmatrix} = e^{u\theta},$$

where $0 \le \theta < 2\pi$. All irreducible unitary representations of K are one-dimensional: they are defined by the maps $g_\theta \mapsto e^{in\theta}$ for $n \in \mathbb{Z}$. Therefore an $\mathfrak{sl}_2(\mathbb{R})$-module M is a Harish-Chandra module if

$$M = \bigoplus_{n \in \mathbb{Z}} M_n, \quad M_n = \{m \in M | um = inm\}, \quad \dim M_n < \infty.$$

If M is a simple $\mathfrak{sl}_2(\mathbb{R})$-module, then the condition $\dim M_n < \infty$ holds automatically.

Since M is a vector space over $\mathbb{C}$, it makes sense to complexify the Lie algebra and consider the $\mathfrak{sl}_2(\mathbb{C})$-action on M. Consider the basis $\{e, h, f\}$ in $\mathfrak{sl}_2(\mathbb{C})$ given by

$$e = \begin{pmatrix} 0 & 1 \\ 0 & 0 \end{pmatrix}, \quad h = \begin{pmatrix} 1 & 0 \\ 0 & -1 \end{pmatrix}, \quad f = \begin{pmatrix} 0 & 0 \\ 1 & 0 \end{pmatrix}.$$

This basis satisfies the relations

(9.5) $[h, e] = 2e, \quad [h, f] = -2f, \quad [e, f] = h,$

which we have seen already in Section 3.4 of Chapter 3. Furthermore, the conjugation by the matrix $\begin{pmatrix} 1 & 1 \\ i & -i \end{pmatrix}$ maps $\mathfrak{k}$ to the diagonal subalgebra of $\mathfrak{sl}_2$.

6.3. Weight modules over $\mathfrak{sl}_2(\mathbb{C})$.

DEFINITION 6.6. Let $\mathfrak{g} = \mathfrak{sl}_2(\mathbb{C})$. A $\overset{\bullet}{\mathfrak{g}}$-module M is called a *weight module* if the action of h in M is diagonalizable and all the corresponding eigenspaces are finite-dimensional. By M_λ we denote the h-eigenspace with eigenvalue $\lambda \in \mathbb{C}$. By supp M we denote the set of all $\lambda \in \mathbb{C}$ such that $M_\lambda \neq 0$.

As we have shown in Section 6.2, a complexified Harish-Chandra module is a weight module with integral eigenvalues of h.

EXERCISE 6.7. Show that submodules and quotients of weight modules are weight modules.

LEMMA 6.8. *Let M be a weight module. Then $eM_\lambda \subset M_{\lambda+2}$ and $fM(\lambda) \subset M_{\lambda-2}$.*

PROOF. This is a simple consequence of (9.5). If $hm = \lambda m$, then

$$hem = ([h, e] + eh)m = 2em + ehm = (\lambda + 2)em,$$

$$hfm = ([h, f] + fh)m = -2fm + fhm = (\lambda - 2)fm.$$

$\square$

COROLLARY 6.9. *If M is an indecomposable weight module, then supp $M \subset \nu + 2\mathbb{Z}$ for some $\nu \in \mathbb{C}$.*

Now we are going to construct an important family $\mathcal{F}(\lambda, \mu)$ of weight modules for all $\lambda, \mu \in \mathbb{C}$. We identify $\mathcal{F}(\lambda, \mu)$ with $t^\lambda \mathbb{C}[t, t^{-1}]$ and define the action of f, h, e by setting

$$f \mapsto \frac{\partial}{\partial t}, \quad h \mapsto 2t\frac{\partial}{\partial t} + \mu, \quad e \mapsto -t^2\frac{\partial}{\partial t} - \mu t,$$

with the convention that $\frac{\partial}{\partial t}t^\nu := \nu t^{\nu-1}$.

EXERCISE 6.10. Prove that:
  (1) $\mathcal{F}(\lambda, \mu)$ is a weight module and supp $\mathcal{F}(\lambda, \mu) = 2\lambda + \mu + 2\mathbb{Z}$;
  (2) All h-eigenspaces in $\mathcal{F}(\lambda, \mu)$ are one-dimensional;
  (3) $\mathcal{F}(\lambda, \mu)$ is simple if and only if $\lambda, \lambda + \mu \notin \mathbb{Z}$.

Now let $\mathcal{U}$ be the associative algebra over $\mathbb{C}$ with generators e, h, f and relations (9.5). Then every $\mathfrak{g}$-module is automatically a $\mathcal{U}$-module.

EXERCISE 6.11. Define the adjoint action of $\mathfrak{g}$ on $\mathcal{U}$ by setting

$$\mathrm{ad}_x\, y := xy - yx \quad \text{for all} \quad x \in \mathfrak{g},\, y \in \mathcal{U}.$$

Check that $\mathrm{ad}_x(yz) = (\mathrm{ad}_x\, y)z + y\,\mathrm{ad}_x\, z$ and that $\mathcal{U}$ is a $\mathfrak{g}$-module with respect to this action.

Consider the adjoint action of h on $\mathcal{U}$ and set

$$\mathcal{U}_n = \{y \in \mathcal{U} \mid \mathrm{ad}_h\, y = ny\}.$$

Exercise 6.11 implies that $\mathcal{U}_n \neq 0$ if and only if $n \in 2\mathbb{Z}$ and

$$\mathcal{U} = \bigoplus_{n \in \mathbb{Z}} \mathcal{U}_n$$

defines a grading of the associative algebra $\mathcal{U}$. In particular, it is clear that $\mathcal{U}_0$ is a commutative subalgebra of $\mathcal{U}$ generated by h and ef.

EXERCISE 6.12. If M is a weight module, then $\mathcal{U}_n M_\lambda \subset M_{\lambda+n}$.

Now we describe the center $\mathcal{Z}$ of $\mathcal{U}$. Note first that $\mathcal{Z}$ is a subalgebra of $\mathcal{U}_0$ since $\mathcal{U}_0$ is the centralizer of h in $\mathcal{U}$. Let

$$\Omega := \frac{h^2}{2} + h + 2fe = \frac{h^2}{2} - h + 2ef.$$

This Ω is called the Casimir element. By a straightforward computation we see that $\Omega \in \mathcal{Z}$. (It suffices to check that it commutes with e and f).

EXERCISE 6.13. Check that Ω acts on $\mathcal{F}(\lambda, \mu)$ by the scalar operator $\frac{\mu^2}{2} - \mu$.

In fact, any element of $\mathcal{Z}$ acts by a scalar operator on any simple $\mathfrak{g}$-module. This is a consequence of the following useful analogue of Schur lemma in the infinite-dimensional case.

LEMMA 6.14. *Let A be a countable dimensional algebra over $\mathbb{C}$ and M be a simple A-module. Then $\mathrm{End}_A(M) = \mathbb{C}$.*

PROOF. Note that M is countable dimensional and any $\varphi \in \mathrm{End}_A(M)$ is completely determined by its value on a generator of M. Hence $\mathrm{End}_A(M)$ is a countable dimensional $\mathbb{C}$-algebra. Since $\mathrm{End}_A(M)$ is a division ring, it suffices to show that every $\varphi \in \mathrm{End}_A(M) \setminus \mathbb{C}$ is algebraic over $\mathbb{C}$. If φ is transcendental, then $\mathrm{End}_A(M)$ contains a subfield field $\mathbb{C}(z)$ of rational functions. This field is not of countable dimensional over $\mathbb{C}$ since the set $\{\frac{1}{\varphi-a} \mid a \in \mathbb{C}\}$ is linearly independent.

LEMMA 6.15. (1) *The commutative algebra $\mathcal{U}_0$ is isomorphic to the polynomial algebra $\mathbb{C}[h, \Omega]$.*
    (2) *Let $n = 2m$. Then $\mathcal{U}_n$ is a free $\mathcal{U}_0$-module of rank 1 with generator e^m for $m > 0$ and f^m for $m < 0$.*
    (3) *The center $\mathcal{Z}$ of $\mathcal{U}$ coincides with $\mathbb{C}[\Omega]$.*

PROOF. The relations (9.5) easily imply that the monomials $e^a f^b h^c$ for all $a, b, c \in \mathbb{N}$ span $\mathcal{U}$. The monomial $e^a f^b h^c$ is an eigenvector of ad_h with eigenvalue $2(a - b)$. Hence $\mathcal{U}_0$ is generated by h and ef or equivalently by h and Ω.

To show that h and Ω are algebraically independent we use $\mathcal{F}(\lambda, \mu)$. Suppose that there exists a polynomial relation $p(h, \Omega) = 0$. Then $p(h, \Omega)$ acts by zero on $\mathcal{F}(\lambda, \mu)$ for all $\lambda, \mu \in \mathbb{C}$. This implies

$$p(2\lambda + \mu + 2n, \frac{\mu^2}{2} - \mu) = 0$$

for all $n \in \mathbb{Z}$, $\lambda, \mu \in \mathbb{C}$. This implies $p \equiv 0$. This completes the proof of (1).

Let us show (2). Assume that $m > 0$ (the case $m < 0$ is similar). Then by the same argument involving monomials, we easily obtain that e^m is a generator of $\mathcal{U}_n$ seen as a $\mathcal{U}_0$-module. Assume that $p(h, \Omega)e^m = 0$. Then $p(h, \Omega)e^m$ acts by zero on $\mathcal{F}(\lambda, \mu)$ for all $\lambda, \mu \in \mathbb{C}$, which is only possible in the case $p \equiv 0$.

Finally, let us prove (3). Let $z \in \mathcal{Z}$. By Lemma 6.14 z acts by a scalar operator on all simple $\mathcal{F}(\lambda, \mu)$. Let $z = p(h, \Omega)$ for some polynomial p. Fix $\lambda, \mu \in \mathbb{C}$ such that that $\mathcal{F}(\lambda, \mu)$ is simple. Then $p(2\lambda + \mu + 2n, \frac{\mu^2}{2} - \mu)$ is a constant function of $n \in \mathbb{Z}$. Since this holds for generic λ and μ, we obtain $p = p(\Omega)$.   □

6.4. The category of weight $\mathfrak{sl}_2$-modules with semisimple action of Ω.

Denote by $\mathcal{C}$ the category of finite length $\mathcal{U}$-modules which are semisimple over $\mathcal{U}_0$. Any object of $\mathcal{C}$ is a weight $\mathfrak{sl}_2$-module with semisimple action of Ω.

Let us fix $\theta \in \mathbb{C}/2\mathbb{Z}$ and $\chi \in \mathbb{C}$. Denote by $\mathcal{C}_{\theta,\chi}$ the subcategory of $\mathcal{C}$ of all modules such that

- If $\lambda \in \mathrm{supp}\, M$, then $\lambda \in \theta$.
- Ω acts on M by the scalar operator $\chi \, \mathrm{Id}$.

LEMMA 6.16. *For every $M \in \mathcal{C}$ there exists a unique decomposition*

$$M = \oplus_{(\theta,\chi)\in\mathbb{C}/2\mathbb{Z}\times\mathbb{C}}M(\theta, \chi),$$

with $M(\theta, \chi) \in \mathcal{C}_{\theta,\chi}$, and almost every $M(\theta, \chi) = 0$.

PROOF. For every $\theta \in \mathbb{C}/2\mathbb{Z}$ and $\chi \in \mathbb{C}$ set

$$M(\theta, \chi) := \{m \in M \,|\, \Omega m = \chi m,\, hm = \lambda m \text{ for some } \lambda \in \theta\}.$$

By Lemma 6.8, $M(\theta, \chi)$ is $\mathfrak{g}$-stable, and clearly M is the direct sum of all $M(\theta, \chi)$. Since M has finite length, $M(\theta, \chi) \neq 0$ for finitely many (θ, χ).   □

The subcategories $\mathcal{C}_{\theta,\chi}$ are called, in the literature, the *blocks* of $\mathcal{C}$. We are going to show that they satisfy the conditions of Sections 1 and 2 and hence they can be described by quivers with relations. In fact, we only have to check that every block has finitely many isomorphism classes of simple objects and that $\mathcal{C}$ has enough projectives. The condition that all objects have finite length and Lemma 6.14 imply all other properties.

LEMMA 6.17. *Let $\mathbb{C}_{\nu,\chi}$ denote the one-dimensional $\mathcal{U}_0$-module on which h acts by ν and Ω by χ, and $P(\nu,\chi) := \mathcal{U} \otimes_{\mathcal{U}_0} \mathbb{C}_{\nu,\chi}$. Then:*

(1) *$P(\nu,\chi)$ is projective in $\mathcal{C}$;*
(2) *$\operatorname{supp} P(\nu,\chi) = \nu + 2\mathbb{Z}$ and all h-eigenspaces are one-dimensional;*
(3) *$P(\nu,\chi)$ has a unique proper maximal submodule.*

PROOF. The first assertion follows immediately from the Frobenius reciprocity (see Theorem 5.3 Chapter 2). Indeed, for every $M \in \mathcal{C}$ we have

$$\operatorname{Hom}_{\mathcal{U}}(P(\nu,\chi), M) \simeq \operatorname{Hom}_{\mathcal{U}_0}(\mathbb{C}_{\nu,\chi}, M).$$

Since every M is a semisimple $\mathcal{U}_0$-module, the functor $\operatorname{Hom}_{\mathcal{U}_0}(\mathbb{C}_{\nu,\chi}, \cdot)$ is exact on $\mathcal{C}$ and therefore $\operatorname{Hom}_{\mathcal{U}}(P(\nu,\chi), \cdot)$ is also exact.

The second assertion is a consequence of Lemma 6.15 (1). Indeed, we have

$$P(\nu,\chi) = 1 \otimes \mathbb{C}_{\nu,\chi} \oplus \bigoplus_{m>0} e^m \otimes \mathbb{C}_{\nu,\chi} \oplus \bigoplus_{m>0} f^m \otimes \mathbb{C}_{\nu,\chi},$$

and $e^m \otimes \mathbb{C}_{\nu,\chi}$ is the h-eigenspace with eigenvalue $\nu + 2m$ while $f^m \otimes \mathbb{C}_{\nu,\chi}$ is the h-eigenspace with eigenvalue $\nu - 2m$.

To prove (3), let F be the sum of all submodules M of $P(\nu,\chi)$ such that $\nu \notin \operatorname{supp} M$. It is obvious that F is the unique maximal proper submodule of $P(\nu,\chi)$. $\square$

COROLLARY 6.18. *The category $\mathcal{C}$ has enough projective modules.*

PROOF. Every $M \in \mathcal{C}$ is finitely generated, since it has finite length. Therefore there exists a finite-dimensional $\mathcal{U}_0$-module M' which generates M. Set $P = \mathcal{U} \otimes_{\mathcal{U}_0} M'$. By Frobenius reciprocity the embedding $M' \hookrightarrow M$ induces the surjection $P \to M$. Since M' is a direct sum of finitely many $\mathbb{C}_{\nu,\chi}$, P is a direct sum of finitely many $P(\nu,\chi)$. $\square$

LEMMA 6.19. *Let L be a simple module in $\mathcal{C}_{\theta,\chi}$. Then all h-eigenspaces in L are one dimensional. Furthermore, if L' is another simple module in $\mathcal{C}_{\theta,\chi}$ and the intersection $\operatorname{supp} L \cap \operatorname{supp} L'$ is not empty, then L' and L are isomorphic.*

PROOF. Let $\nu \in \operatorname{supp} L$. Then by Lemma 6.17, L is isomorphic to the unique simple quotient of $P(\nu,\chi)$. Hence the statement. $\square$

LEMMA 6.20. *Let M be a simple weight module. Then we have the following four possibilities for $\operatorname{supp} M$:*

(1) *$\operatorname{supp} M = \{m, m-2, \ldots, 2-m, -m\}$ for some $m \in \mathbb{N}$. In this case M is finite-dimensional;*
(2) *$\operatorname{supp} M = \nu + 2\mathbb{N}$ for some $\nu \in \mathbb{C}$;*
(3) *$\operatorname{supp} M = \nu - 2\mathbb{N}$ for some $\nu \in \mathbb{C}$;*
(4) *$\operatorname{supp} M = \nu + 2\mathbb{Z}$ for some $\nu \in \mathbb{C}$.*

PROOF. Consider the decomposition $M = \bigoplus_{\lambda \in \operatorname{supp} M} M_\lambda$, by Lemma 6.19 every M_λ is 1-dimensional. There are four possibilities

(1) Both e and f have a non-trivial kernel in M;
(2) f acts injectively on M and $eM_\nu = 0$ for some $\nu \in \mathbb{C}$;
(3) e acts injectively on M and $fM_\nu = 0$ for some $\nu \in \mathbb{C}$;
(4) Both e and f act injectively on M.

Cases (2), (3) and (4) immediately imply the corresponding cases in the lemma. Now let us consider the first case. Take a non-zero $v \in M_\nu$ such that $ev = 0$. Since f does not act injectively on M, there exists $m \in \mathbb{N}$ such that $v, fv, f^2v, \ldots f^m v$ are linearly independent and $f^{m+1}v = 0$. Then we have

$$ef^{m+1}v = \sum_{j=0}^{m} f^{m-j}hf^j v = \left(\sum_{j=0}^{m}(-2j + \lambda) \right) f^m v = ((m+1)\lambda - m(m+1)) f^m v = 0.$$

Hence $\lambda = m$ and the proof of the lemma is complete. $\qquad\square$

Now we can classify simple modules in all blocks $\mathcal{C}_{\theta,\chi}$.

PROPOSITION 6.21. *Consider the category $\mathcal{C}_{\theta,\chi}$. Then we have the following three cases:*

(1) *Assume that $2\chi \neq \nu(\nu+2)$ for any $\nu \in \theta$. Then $\mathcal{C}_{\theta,\chi}$ has one simple module M (up to isomorphism) with supp $M = \theta$. Furthermore M is projective.*
(2) *Assume that $2\chi = \nu(\nu + 2)$ for some $\nu \in \theta$ and $\theta \neq 2\mathbb{Z}$ or $1 + 2\mathbb{Z}$ or $\chi = -\frac{1}{2}$, $\theta = 1 + 2\mathbb{Z}$, $\nu = -1$. Then $\mathcal{C}_{\theta,\chi}$ has two simple modules (up to isomorphism):*
    $L^-(\nu)$ *with supp $L^-(\nu) = \nu - 2\mathbb{N}$,*
    $L^+(\nu + 2)$ *with supp $L^+(\nu) = \nu + 2 + 2\mathbb{N}$.*
(3) *Finally if $2\chi = m(m + 2)$ for some $m \in \mathbb{N}$ such that $m \in \theta$. Then $\mathcal{C}_{\theta,\chi}$ has three simple modules (up to isomorphism):*
    $L^-(-m - 2)$ *with supp $L^-(-m - 2) = -m - 2 - 2\mathbb{N}$,*
    $L^+(m + 2)$ *with supp $L^+(m + 2) = m + 2 + 2\mathbb{N}$,*
    $V(m)$ *with supp $V(m) = \{m, m - 2, \ldots, 2 - m, -m\}$.*

PROOF. We use Lemma 6.20. If v is an h-eigenvector with eigenvalue ν such that $ev = 0$, then

$$\Omega v = \frac{h^2 + 2h}{2}v = \chi v$$

which implies the relation $2\chi = \nu(\nu + 2)$. Similarly, if $fv = 0$ we have

$$\Omega v = \frac{h^2 - 2h}{2}v = \chi v$$

which implies the relation $2\chi = \nu(\nu - 2)$. Now the Proposition follows from Lemma 6.20. $\qquad\square$

REMARK 6.22. As a consequence, we obtained a classification of all simple finite-dimensional $\mathfrak{sl}_2(\mathbb{C})$-modules. Observe that $V(m)$ is a submodule of $\mathcal{F}(0, -m)$. The corresponding representation of the group $SL_2(\mathbb{C})$ can be realized as the m-th symmetric power of the natural 2-dimensional representation.

Furthermore, if $\theta = 2\mathbb{Z}$ or $1 + 2\mathbb{Z}$, then Proposition 6.21 provides the classification of simple Harish-Chandra modules. It makes sense to compare this classification with the list of irreducible unitary representations of $SL_2(\mathbb{R})$ given in Section 3, Chapter 4. The reader can check that the discrete series representations $\mathcal{H}_m^{\pm}$ correspond to the Harish-Chandra modules $L^{\pm}(m)$ and that the principal (resp. complementary) series correspond to the simple projective modules $P(0, \chi)$ (resp. $P(1, \chi)$), for specific values of the parameter χ.

6.5. Ext-quivers of blocks. Now we will compute the quiver with relations corresponding to each block $\mathcal{C}_{\theta,\chi}$. Note that the blocks satisfying (1) of Proposition 6.21 are semisimple and have only one simple object. Therefore this case is trivial.

PROPOSITION 6.23. *(a) If (θ, χ) satisfies the condition (2) of Proposition 6.21, then the category $\mathcal{C}_{\theta,\chi}$ is equivalent to the category of representations of the quiver*

$$\bullet \underset{\beta}{\overset{\alpha}{\rightleftarrows}} \bullet,$$

submitted to the relations $\alpha\beta = \beta\alpha = 0$.
(b) If (θ, χ) satisfies the condition (3) of Proposition 6.21, then the category $\mathcal{C}_{\theta,\chi}$ is equivalent to the category of representations of the quiver

$$\bullet^1 \underset{\beta}{\overset{\alpha}{\rightleftarrows}} \bullet^2 \underset{\beta}{\overset{\alpha}{\rightleftarrows}} \bullet^3.$$

submitted to the relations $\alpha\beta = \beta\alpha = 0$.

PROOF. (a) Assume (θ, χ) satisfies the condition (2) of Proposition 6.21. Then the Ext-quiver Q has two vertices corresponding to the two simple modules $L^-(\nu)$ and $L^+(\nu+2)$, the indecomposable projective modules are $P^- := P(\nu, \chi)$ and $P^+ := P(\nu+2, \chi)$. Furthermore, the radical filtration of P^- (resp. P^+) has two layers and is determined from the exact sequence

$$0 \to L^+(\nu+2) \to P^- \to L^-(\nu) \to 0 \quad (\text{resp.}\, 0 \to L^-(\nu) \to P^+ \to L^+(\nu+2) \to 0).$$

The statement follows.

(b) Now let us consider the most interesting case: when $\mathcal{C}_{\theta,\chi}$ has three simple modules $L^-(-n-2), V(n), L^+(n+2)$. Therefore the block has three indecomposable projective modules P^-, P^f, P^+ which cover $L^-(-m-2), V(m), L^+(m+2)$, respectively. In this case the statement also follows from the calculation of the radical filtration of the indecomposable projective modules.

The following exercise completes the proof:

EXERCISE 6.24. Show that the radical filtration of P^+ (resp. P^-) has three layers:

$$P^+/\nabla(P^+) \simeq L^+(m+2), \quad \nabla(P^+)/\nabla^2(P^+) \simeq V(m), \quad \nabla^2(P^+) \simeq L^-(-m-2),$$

$$P^-/\nabla(P^-) \simeq L^-(-m-2), \quad \nabla(P^-)/\nabla^2(P^-) \simeq V(m), \quad \nabla^2(P^-) \simeq L^+(m+2).$$

The radical filtration of P^f has two layers

$$P^f/\nabla(P^f) \simeq V(m), \ \nabla(P^f) \simeq L^+(m+2) \oplus L^-(-2-m).$$

$\square$

PROPOSITION 6.25. *All the blocks of the category $\mathcal{C}$ are finitely represented.*

PROOF. All we have to show is that the quivers with relations appearing in Proposition 6.23 are finitely represented. In case (a), we have a finitely represented algebra since $\bar{Q}$ is the disjoint union of two quivers of type A_2. There are 4 indecomposable representations $P^+, P^-, L^+(\nu+2)$ and $L^-(\nu)$ (up to isomorphism).

In case (b) we use the same trick as in Example 2.5. Assume that $V = V_0 \oplus V_1 \oplus V_2$ is an indecomposable representation. If $\alpha^2 v \neq 0$ for some $v \in V_0$, then V contains a submodule spanned by $v, \alpha v, \alpha^2 v$ which is both injective and projective. Therefore $V \simeq P_1$. Similarly, if there is $w \in V_2$ such that $\beta^2 w \neq 0$, then $V \simeq P_3$. If we exclude these two cases, then we may assume the relation $\alpha^2 = \beta^2 = 0$. Hence the classification of indecomposable modules can be reduced to the same problem for $\bar{Q}$, see Lemma 2.3. In this case $\bar{Q}$ is the disjoint union of two Dynkin quivers:

$$\bullet \rightarrow \bullet \leftarrow \bullet, \quad \bullet \leftarrow \bullet \rightarrow \bullet.$$

The reader can check that there are 9 indecomposable modules. $\square$

6.6. The category of all weight modules. Now we drop the assumption that Ω is diagonalizable. Let $\tilde{\mathcal{C}}$ denote the category of all weight $\mathfrak{sl}_2$-modules of finite length. Obviously $\mathcal{C}$ is a subcategory of $\tilde{\mathcal{C}}$, and both categories have the same simple modules by Lemma 6.14.

Our first observation is that for any $M \in \tilde{\mathcal{C}}$ there exists a polynomial $p(x) \in \mathbb{C}[x]$ such that $p(\Omega)M = 0$. Therefore the category $\tilde{\mathcal{C}}$ decomposes into the following direct sum of blocks

$$\tilde{\mathcal{C}} = \bigoplus \tilde{\mathcal{C}}_{\theta,\chi},$$

where $\tilde{\mathcal{C}}_{\theta,\chi}$ is the subcategory consisting of modules M such that $\operatorname{supp} M \subset \theta$ and $(\Omega - \chi)^n M = 0$ for sufficiently large n.

Our next observation concerns projective modules. It is not difficult to see that $\tilde{\mathcal{C}}_{\theta,\chi}$ does not have enough projectives. To overcome this obstacle we will consider the categories $\mathcal{C}_{\theta,\chi}^{(n)}$ for all $n > 0$, consisiting of modules M which satisfy the additional condition $(\Omega - \chi)^n M = 0$. Then we have a relation

(9.6) $$\tilde{\mathcal{C}}_{\theta,\chi} = \varinjlim \mathcal{C}_{\theta,\chi}^{(n)},$$

i.e. every object of $\tilde{\mathcal{C}}_{\theta,\chi}$ is an object of $\mathcal{C}_{\theta,\chi}^{(n)}$ for sufficiently large n. Thus, the problem of classifying indecomposable weight modules is reduced to the same problem for $\mathcal{C}_{\theta,\chi}^{(n)}$. To construct projective modules in $\mathcal{C}_{\theta,\chi}^{(n)}$ we use the induction

$$\mathcal{U} \otimes_{\mathcal{U}_0} \left(\mathcal{U}_0/(h - \nu, (\Omega - \chi)^n) \right).$$

EXERCISE 6.26. Use the Frobenius reciprocity to show that

(1) If $\nu \in \theta$, then $\mathcal{U} \otimes_{\mathcal{U}_0} (\mathcal{U}_0/(h - \nu, (\Omega - \chi)^n))$ is an indecomposable projective module in $\tilde{\mathcal{C}}_{\theta,\chi}^{(n)}$.

(2) If $z = \Omega - \chi$, then

$$\operatorname{End}_{\tilde{\mathcal{C}}} \left(\mathcal{U} \otimes_{\mathcal{U}_0} (\mathcal{U}_0/(h - \nu, (\Omega - \chi)^n)) \right) \simeq \mathbb{C}[z]/(z^n).$$

LEMMA 6.27. (a) If (θ, χ) satisfies condition (1) of Proposition 6.21, then the category $\mathcal{C}_{\theta,\chi}^{(n)}$ is equivalent to the category of representations of the quiver $\overset{\alpha}{\underset{}{\circlearrowright}}\ \bullet$ with the relation $\alpha^n = 0$.

(b) If (θ, χ) satisfies condition (2) of Proposition 6.21, then the category $\mathcal{C}_{\theta,\chi}^{(n)}$ is equivalent to the category of representations of the quiver

$$\bullet \underset{\beta}{\overset{\alpha}{\rightleftarrows}} \bullet,$$

with relations $(\alpha\beta)^n = (\beta\alpha)^n = 0$.

(c) If (θ, χ) satisfies condition (3) of Proposition 6.21, then the category $\mathcal{C}_{\theta,\chi}^{(n)}$ is equivalent to the category of representations of the quiver

$$\bullet^1 \underset{\beta}{\overset{\alpha}{\rightleftarrows}} \bullet^2 \underset{\delta}{\overset{\gamma}{\rightleftarrows}} \bullet^3.$$

with relations

$$(\beta\alpha)^n = (\gamma\delta)^n = 0, \quad \alpha\beta = \delta\gamma, \quad (\alpha\beta)^n = 0.$$

PROOF. In case (a), there is only one indecomposable projective module and its endomorphism algebra is isomorphic $\mathbb{C}[z]/(z^n)$ by Exercise 6.26 (2).

In case (b), there are two non-isomorphic indecomposable projective modules, call them P_1 and P_2, the corresponding simple modules will be denoted by L_1 and L_2. Note that $z := \Omega - \chi$ is nilpotent on P_1 and P_2 and P_1/zP_1, P_2/zP_2 are indecomposable projectives in $\mathcal{C}_{\theta,\chi}$. Then, using induction on n, one can compute the radical filtrations of P_1 and P_2:

$$\nabla^i(P_1)/\nabla^{i+1}(P_1) = \begin{cases} L_1 & \text{if } i \text{ is even} \\ L_2 & \text{if } i \text{ is odd} \end{cases},$$

$$\nabla^i(P_2)/\nabla^{i+1}(P_2) = \begin{cases} L_2 & \text{if } i \text{ is even} \\ L_1 & \text{if } i \text{ is odd} \end{cases},$$

for $i < 2n$, and $\nabla^{2n}(P_1) = \nabla^{2n}(P_2) = 0$. The statement follows.

In case (c) there are 3 non-isomorphic indecomposable projective modules: P_1, P_2, P_3, with simple quotients L_1, L_2, L_3, respectively. We assume that L_2 is the finite-dimensional module isomorphic to $V(m)$. In order to calculate the radical filtration of projective modules we consider first the filtration:

$$P_i \supset zP_i \supset \ldots \supset z^{n-1}P_i \supset z^n P_i = 0.$$

Then every quotient $z^j P_i/z^{j+1}P_i$ is isomorphic to the projective cover of L_i in $\mathcal{C}_{\theta,\chi}$. After application of Exercise 6.24 and induction on n, we obtain the radical filtration of P_2:

$$\nabla^i(P_2)/\nabla^{i+1}(P_2) = \begin{cases} L_2 \text{ if } i \text{ is even} \\ L_1 \oplus L_3 \text{ if } i \text{ is odd} \end{cases},$$

for $i < 2n$ and $\nabla^{2n}(P_2) = 0$. The radical filtrations of P_1 and P_3 can be obtained similarly:

$$\nabla^i(P_j)/\nabla^{i+1}(P_j) = \begin{cases} L_1 \oplus L_3 \text{ if } i \text{ is even and } i \neq 0, 2n \\ L_2 \text{ if } i \text{ is odd} \\ L_j \text{ if } i = 0 \\ L_{j'} \text{ if } i = 2n \end{cases},$$

where $j = 1, 3$ and $j' = 3, 1$, respectively, and $i \leq 2n$, $\nabla^{2n+1}(P_j) = 0$. That implies (c). □

A representation of a quiver Q is called nilpotent of every path acts as a nilpotent endomorphism. Lemma 6.27 implies the following:

PROPOSITION 6.28. *Each block $\tilde{\mathcal{C}}_{\theta,\chi}$ is equivalent to the category of nilpotent representation of one of the following quivers:*

(1) $\bullet$;

(2) $\bullet \rightleftarrows \bullet$;

(3) $\bullet \overset{\alpha}{\underset{\beta}{\rightleftarrows}} \bullet \overset{\gamma}{\underset{\delta}{\rightleftarrows}} \bullet$ *with relation* $\alpha\beta = \delta\gamma$.

Let us finish with the following proposition.

PROPOSITION 6.29. *For every $\theta \in \mathbb{C}/2\mathbb{Z}$, $\chi \in \mathbb{C}$ and $n > 0$, the block $\mathcal{C}_{\theta,\chi}^{(n)}$ is finitely represented and $\tilde{\mathcal{C}}_{\theta,\chi}$ is tame.*

PROOF. Note that the second assertion is an immediate consequence of the first one by (9.6). Moreover, the first assertion in case (1) is obvious: there is exactly one indecomposable representation in every dimension, given by the nilpotent Jordan block.

In case (2), the block is equivalent to the category of $\mathbb{Z}/2\mathbb{Z}$-graded modules over the $\mathbb{Z}/2\mathbb{Z}$-graded algebra $\mathbb{C}[\xi]/(\xi^{2n})$, where the degree of ξ is equal to 1. For every dimension vector of the shape (p, p), $(p + 1, p)$ and $(p, p + 1)$ for $p \leq n$, there is exactly one indecomposable representation.

In case (3), we give a description of all indecomposable nilpotent representations of the quiver (3) of Proposition 6.28 and leave the proof to the reader. This is a particular case of so called Gelfand-Ponomarev quiver (see [17]).

We call a quiver Γ admissible if the following conditions hold:

- The set Γ_0 of vertices is a finite subset of $\mathbb{N} \times \mathbb{N}$ such that if $(a, b) \in \Gamma_0$, then $b = a$ or $a \pm 1$.
- If $(a, a), (a + 1, a + 1) \in \Gamma_0$ then $(a, a + 1), (a + 1, a) \in \Gamma_0$.
- At least one of $(0, 0)$, $(1, 0)$ and $(0, 1)$ belongs to Γ_0.
- The arrows of Γ are of the form $(a, b) \to (c, d)$ where $c = a$, $d = b + 1$ or $c = a + 1$, $d = b$.
- Γ is connected.
- Each arrow has a mark α, β, γ or δ according to the following rule:

$$(a + 1, a) \xrightarrow{\alpha} (a + 1, a + 1),$$
$$(a, a) \xrightarrow{\beta} (a + 1, a),$$
$$(a, a) \xrightarrow{\gamma} (a, a + 1),$$
$$(a, a + 1) \xrightarrow{\delta} (a + 1, a + 1).$$

Here is an example of an admissible quiver:

$$
\begin{array}{ccc}
(1, 2) & \xrightarrow{\delta} & (2, 2) \\
\gamma \uparrow & & \alpha \uparrow \\
(0, 1) \xrightarrow{\delta} (1, 1) & \xrightarrow{\beta} & (2, 1) \\
\gamma \uparrow \quad\quad \alpha \uparrow & & \\
(0, 0) \xrightarrow{\beta} (1, 0) & &
\end{array}
$$

To every admissible quiver Γ, we associate a representation (V^Γ, ρ^γ) as follows: we set V^Γ to be the formal span of e_s, for all $s \in \Gamma_0$, assuming $e_s \in V_1^\Gamma$ for $s = (a + 1, a)$, $e_s \in V_2^\Gamma$ for $s = (a, a)$ and $e_s \in V_3^\Gamma$ for $s = (a, a + 1)$. For every $u \in \{\alpha, \beta, \gamma, \delta\} = Q_1$, we set $\rho^\Gamma e_s = e_t$ if $(s \xrightarrow{u} t)$ is an arrow of Γ, and zero otherwise.

It is not difficult to check that (V^Γ, ρ^Γ) is indeed an indecomposable nilpotent representation of the quiver (3). We claim that (V^Γ, ρ^Γ) for all admissible Γ are pairwise non-isomorphic and form a complete list of all indecomposable representations of (3), up to isomorphism. $\square$

With these results, we hope we have shown that representation theory of quivers is a powerful tool in relation with various classical problems and we believe that it is time for us to end this book.

Bibliography

1. E. Artin, *Galois Theory*. Notre-Dame Mathematical Lectures (1942)
2. H. Bass, *Algebraic K-Theory* (Benjamin, New York, 1968)
3. A.A. Beilinson, The derived category of coherent sheaves on $\mathbb{P}^n$. Selected translations. Selecta Math. Soviet. **3**(3) (1983/1984)
4. I.N. Bernstein, I.M. Gel'fand, V.A. Ponomarev, Coxeter functors, and Gabriel's theorem. Uspekhi Mat. Nauk 28 **2**(170) (1973) (Russian)
5. N. Bourbaki, *Lie Groups and Lie Algebras, Chapters 4–6* (Springer, Berlin, 2002)
6. W. Crawley-Boevey, Lectures on representations of quivers, PDF available on his web-page (1992)
7. C.W. Curtis, I. Reiner, Representation theory of finite groups and associative algebras. Reprint of the 1962 original (AMS Chelsea Publishing, Providence, 2006)
8. J. Dixmier, *Enveloping Algebras*. North-Holland Mathematical Library, vol. 14. Translated from French (North-Holland Publishing Co., Amsterdam, 1977)
9. Ju.A. Drozd, Tame and wild matrix problems, *Representation Theory II (Proceedings of the Second International Conference, Carleton University, Ottawa, 1979)*. Lecture Notes in Mathematics, vol. 832 (Springer, Berlin, 1980)
10. K. Erdmann, *Blocks of Tame Representation Type and Related Algebras*. Lecture Notes in Mathematics, vol. 1428 (Springer, Berlin, 1990)
11. P. Etingof, O. Schiffmann, *Lectures on Quantum Groups*, 2nd edn. (Lectures in Mathematical Physics (International Press, Somerville, 2002)
12. W. Fulton, J. Harris, Representation theory, a first course, GTM 129, Springer (1991)
13. P. Gabriel, A.V. Roiter, *Representations of Finite-Dimensional Algebras*. With a chapter by B. Keller. Encyclopaedia Mathematical Science, Algebra VIII, vol. 73 (Springer, Berlin, 1992)
14. L. Geissinger, *Hopf algebras of symmetric functions and class functions, LNM*, vol. 579 (Springer, Berlin, 1977), pp. 168–181
15. I. Gel'fand, I. Graev, N. Vilenkin, *Generalized Functions, Volume 5: Integral Geometry and Representation Theory*, Translated from the Russian by E. Saletan (Academic Press, New York, 1966)
16. I. Gel'fand, I. Graev, Construction of irreducible representations of simple algebraic groups over a finite field. Dokl. Akad. Nauk SSSR **147** (1962) (Russian)

© Springer Nature Switzerland AG 2018

C. Gruson and V. Serganova, *A Journey Through Representation Theory*,
Universitext, https://doi.org/10.1007/978-3-319-98271-7

17. I.M. Gel'fand, V.A. Ponomarev, Indecomposable representations of the Lorentz group. Uspekh Mat. Nauk. **28**, 1–60 (1968)
18. R. Godement, *Analysis IV, Universitext* (Springer, Berlin, 2015)
19. A. Grothendieck, Sur quelques points d'algèbre homologique. Tohoku Math. J. **2**(9), 119–221 (1957) (French)
20. S. Helgason, *Groups and Geometric Analysis: Integral Geometry, Invariant Differential Operators and Spherical Functions* (Academic Press, New York, 1984); (AMS, 1994)
21. J.E. Humphreys, *Linear Algebraic Groups*. GTM, vol. 21 (Springer, Berlin, 1975)
22. J.C. Jantzen, *Lectures on Quantum Groups*. Graduate Studies in Mathematics, vol. 6 (American Mathematical Society, Providence, 1996)
23. T.-Y. Lam, *Lectures on Modules and Rings*. Graduate Texts in Mathematics, vol. 189 (Springer, Berlin, 1999)
24. G. Lusztig, *Introduction to Quantum Groups* (Bikhäuser, Basel, 1993)
25. I.G. Macdonald, *Symmetric Functions and Hall Polynomials*, 2nd edn. (Oxford University Press, Oxford, 1995)
26. S. Mac Lane, *Homology*, Die Grundlehren der mathematischen Wissenschaften, vol. 114 (Academic Press, New York; Springer, Berlin, 1963)
27. D. Mumford, *Tata Lectures on Theta III*. With collaboration of M. Nori and P. Norman. Reprint of the 1991 original. Modern Birkhäuser Classics (Birkhäuser Inc., Boston, 2007)
28. J. Rotman, *An Introduction to Homological Algebra*. Pure and Applied Mathematics, vol. 85 (Academic Press, Boston, 1979)
29. W. Rudin, *Real and Complex Analysis*, 3rd edn. (McGraw-Hill Book Co., New York, 1987)
30. J.P. Serre, *Linear Representations of Finite Groups* (Springer, Berlin, 1977)
31. T.A. Springer, A.V. Zelevinsky, Characters of $GL(n, \mathbb{F}_q)$ and Hopf algebras. J. Lond. Math. Soc. **2**(30), 27–43 (1984)
32. T.A. Springer, *Invariant Theory*. Lecture Notes in Mathematics, vol. 585 (Springer, Berlin, 1977)
33. T.A. Springer, *Linear Algebraic Groups*, 2nd edn. Progress in Mathematics, vol. 9 (Birkhäuser, Boston, 1998)
34. R. Steinberg, The representations of $GL(3, q)$, $GL(4, q)$, $PGL(3, q)$ and $PGL(4, q)$. Can. J. Math. **3**, 225–235 (1951)
35. D. Vogan, *Unitary Representations of Reductive Lie Groups* (Princeton University Press, Princeton, 1987)
36. D. Vogan, *Representations of Real Reductive Lie Groups*. P.M. 15 (Birkhäuser, Basel, 1981)
37. C. Weibel, *An Introduction to Homological Algebra*. Cambridge Studies in Advanced Mathematics, vol. 38 (Cambridge University Press, Cambridge, 1994)
38. A.V. Zelevinsky, *Representations of Finite Classical Groups: A Hopf Algebra Approach*. LNM, vol. 869 (Springer, Berlin, 1981)

Index

© Springer Nature Switzerland AG 2018
C. Gruson and V. Serganova, *A Journey Through Representation Theory*,
Universitext, https://doi.org/10.1007/978-3-319-98271-7

Schur functor, 113
Schur's lemma, 5
Schur-Weyl duality, 111
self-adjoint operator, 52
self-adjointness of a Hopf algebra, 118
semisimple module (over a ring), 81
semisimple ring, 84
semistandard tableau, 129
simple module, 26
simple reflection, 172
snake lemma, 92
source, 149
Springer-Zelevinsky theorem, 145
Stone–von Neumann theorem, 70

T
tail, 149
tame algebra, 199
target, 149
Tits form (of a graph), 163
Tits form (of a quiver), 159
topological group, 47

torsion element, 26
torsion free module, 26
trivial representation, 2

U
unitary dual, 55
unitary representation, 49

W
Wedderburn-Artin theorem, 85
weight module, 208
weight of a semistandard tableau, 129
Weyl group, 172
wild algebra, 199

Y
Young diagram, 106
Young symmetrizer, 106
Young tableau, 106
Young tableau (shape of), 106

Z
Zelevinsky's decomposition theorem, 121

Printed in the United States
By Bookmasters